SSADM Version 4

WILEY SERIES IN SOFTWARE ENGINEERING PRACTICE

Series Editors:

Patrick A.V. Hall, *The Open University, UK*
Martyn A. Ould, *Praxis Systems plc, UK*
William E. Riddle, *Software Design & Analysis, Inc., USA*

Aims and Scope

The focus of this series is the software creation and evolution processes and related organizational and automated systems necessary to support them. The aim is to produce books dealing with all aspects of software engineering, particularly the practical exploitation of the best methods and tools for the development process.

The series covers the following topics:

- process models and software lifecycle
- project management, quality assurance, configuration management, process and product standards
- the external business environment and legal constraints
- computer aided software engineering (CASE) and integrated project support environments (IPSES)
- requirements analysis, specification and validation
- architectural design techniques, software components and re-use
- system design methods and verification
- system implementation, build and test
- maintenance and enhancement

For full list of titles in this series, see back pages.

SSADM Version 4

The Advanced Practitioner's Guide

John S. Hares
Information Technology Associates, UK

JOHN WILEY & SONS
Chichester · New York · Brisbane · Toronto · Singapore

Copyright © 1994 by John Wiley & Sons Ltd,
Baffins Lane, Chichester,
West Sussex PO19 1UD, England
Telephone (+44) (243) 779777

All rights reserved.

No part of this book may be reproduced by any means,
or transmitted, or translated into a machine language
without the written permission of the publisher.

Designations used by companies to distinguish their products
are often claimed as trademarks. In all instances where John
Wiley & Sons Ltd is aware of a claim, the product names appear
in initial capital or all capital letters. Readers, however, should
contact the appropriate companies for more complete information regarding trademarks and registration.

Other Wiley Editorial Offices

John Wiley & Sons, Inc., 605 Third Avenue,
New York, NY 10158-0012, USA

Jacaranda Wiley Ltd, 33 Park Road, Milton,
Queensland 4064, Australia

John Wiley & Sons (Canada) Ltd, 22 Worcester Road,
Rexdale, Ontario M9W 1L1, Canada

John Wiley & Sons (SEA) Pte Ltd, 37 Jalan Pemimpin #05-04,
Block B, Union Industrial Building, Singapore 2057

Library of Congress Cataloging-in-Publication Data:

Hares, John S.
    SSADM version 4 : the advanced practitioner's guide / John S. Hares.
      p. cm. — (Wiley series in software engineering practice)
    Includes index.
    ISBN 0 471 93564 6
    1. Electronic data processing—Structured techniques. 2. System analysis. 3. System design. I. Title. II. Series.
QA76.9.S84H38   1993
004.2'1—dc20   93-10575
                                                     CIP

A catalogue record for this book is available from the British Library.

ISBN 0 471 93564 6

Typeset in 10/12pt Palatino from author's disks by Text Processing Deptartment,
John Wiley & Sons Ltd, Chichester
Printed in Great Britain by Biddles Ltd, Guildford and King's Lynn

The book has been assessed and passed by the User group Technical Committee and the SSADM Design Authority Board as being compliant with core SSADM version 4. Since that time additions and some enhancements have been made to the chapter on object orientation and how to make SSADM version 4 object oriented. The additions relate to modelling polymorphism, private methods and genericity. Enhancements have been made to the section on messaging strategies. None of these additions or enhancements in any way affect core SSADM version 4.

CONTENTS

1	**SSADM—ITS PAST, CURRENT AND FUTURE POSITION IN THE STRUCTURED DESIGN AND DEVELOPMENT METHODS INDUSTRY**	**1**
1.1	The Basis and Purpose of this Book	1
1.2	The Data Processing Environments and their Concepts, Technologies and Techniques	3
1.3	The Scope of Version 4	9
1.4	SSADM Structure	11
1.5	The SSADM Design Techniques	16
2	**STRENGTHS AND WEAKNESSES**	**21**
2.1	The Basis of Evaluation	21
2.3	Design Concepts for SSADM	22
	2.2.1 Generic Concepts	22
	2.2.2 Specific Concepts	32
2.3	SSADM Strengths	35
	2.3.1 The General Strengths	35
	2.3.2 The Systematic Strengths	41
2.4	SSADM Weaknesses	69
	2.4.1 The General Weaknesses	69
	2.4.2 The Systematic Weaknesses	70
3	**"PEARLS OF PRACTICAL WISDOM"**	**107**
3.1	Conceptual Tricks	109
	3.1.1 Testing for Logical Design Efficiency	109

	3.1.2	Producing Summary Access Path Maps	118
	3.1.3	Interpreting the Summary Access Path Maps	120
	3.1.4	Producing a Balanced Database Design	129
3.2	Systematic Tricks		132
	3.2.1	Ensuring Process Decomposition is Consistent and Sensible	132
	3.2.2	Matching DFD Processes, the I/O Structures, the Dialogue Structures and the Menu Structures	137
	3.2.3	Taking Relational Data Analysis to Sixth Normal Form	142
	3.2.4	Combining Entities	146
	3.2.5	Additional Points of Detail	147
3.3	Tricks Specific to Version 4 of SSADM		159
	3.3.1	Client-server Architecture	160
	3.3.2	Transaction Access Path Analysis	161
	3.3.3	The One-to-One Lassoing of Entities	161
	3.3.4	Dialogue Design	167
	3.3.5	The Enquiry Process Models	168
	3.3.6	The Need to Support Full Jackson Standards	169
	3.3.7	Structure Clashes	171
	3.3.8	The Function Component Implementation Map (FCIM)	172

4 ADDITIONAL DATA PROCESSING ENVIRONMENTS FOR SSADM—DISTRIBUTED AND REALTIME 175

4.1	Distributed Systems		175
	4.1.1	Distributed Database Concepts	177
	4.1.2	Types of Distributed System	178
	4.1.3	SSADM and Distributed Systems	186
	4.1.4	Distributed Database Technology	186
	4.1.5	Enhancements to SSADM	207
	4.1.6	Redefining the Logical Design	231
4.2	Realtime Processing		232
	4.2.1	Realtime Concepts	233
	4.2.2	The Impact of Realtime Concepts	235
	4.2.3	Integrating the Yourdon Method: Logical Design Techniques	237

5 ADDITIONAL DATA PROCESSING ENVIRONMENTS FOR SSADM—EXPERT SYSTEMS 249

5.1	The Common Link with Object Orientation		249
5.2	Expert System Concept		252
5.3	Knowledge Acquisition and Construction		255
5.4	Expert System Components		256
5.5	The Knowledgebase		259
	5.5.1	Rules	259
	5.5.2	Rule Types	260
	5.5.3	Ruleset Access Strategies	261
	5.5.4	Ruleset Access Mechanisms	262
	5.5.5	Rule Storage and Access	266
	5.5.6	Rule I/O	268

5.6	Semantic Nets	270
	5.6.1 Property Inheritance	270
	5.6.2 Conflict Resolution	270
5.7	Retraction	271
5.8	Uncertainty	271
	5.8.1 Uncertain Logic	271
	5.8.2 Uncertain Data	277
5.9	Dictionary and Database Synchronisation	277
5.10	Worked Examples	277
	5.10.1 New Query Languages (with GENERIS)	277
	5.10.2 GOLDWORKS	280
5.11	Enhancements to SSADM	287
	5.11.1 How to Identify Expert System Applications	288
	5.11.2 Provide a Set of Structural Standards	291
	5.11.3 Provide a Set of Knowledge Engineering Standards	291

6 ADDITIONAL DATA PROCESSING ENVIRONMENTS FOR SSADM—OBJECT ORIENTATION 305

6.1	The Problem	305
6.2	The Scope of Object Orientation	306
6.3	The Concepts of Object Orientation	307
6.4	Does Object Orientation Require a New Way of Thinking?	310
6.5	An Object—What Is It?	311
6.6	The Object Oriented Facilities	313
	6.6.1 Class, Composition/Aggregation, Property Inheritance and Relationship Semantics → Information Abstraction	313
	6.6.2 Concrete and Instance Objects	328
	6.6.3 Property Instantiation	331
	6.6.4 Encapsulation	331
	6.6.5 Messages and Responses	336
	6.6.6 Methods	337
	6.6.7 Polymorphism	340
6.7	The Class Model	343
	6.7.1 The Business Classes Object	343
	6.7.2 The System Classes	349
6.8	Enhancements to SSADM	351
	6.8.1 The Class Model	351
	6.8.2 Information Normalisation (Formerly Relational Data Analysis)	362
	6.8.3 Dataflow Diagrams	368
	6.8.4 The I/O Structures	368
	6.8.5 Modelling Object Oriented Application Programs	368
	6.8.6 Human/Computer Interface Design	376
	6.8.7 Message Strategies	379
	6.8.8 Modelling Pre and Post Conditions	382
	6.8.9 Private Methods	386
	6.8.10 Genericity	387
	6.8.11 The Deliverables Integration	387
	6.8.12 Objectbase Design	388
	6.8.13 Object Oriented Program Design	389

INDEX 391

1

SSADM—Its Past, Current and Future Position in the Structured Design and Development Methods Industry

1.1 The Basis and Purpose of this Book

This book is an assessment of version 4 of the SSADM structured method and how it can be applied more effectively and more widely by experienced practitioners. The book:

- provides an independent and practical assessment of the strengths and weaknesses of SSADM;

- provides "pearls of practical wisdom" advice regarding "tricks of the practitioner's trade" that can be applied to the existing SSADM design techniques, procedures and standards;

- provides new and additional techniques where, in the author's opinion, SSADM requires enhancement;

- shows how SSADM can be used without change, only enhancement by addition, for new and additional data processing environments for which it is not currently targeted, namely the distributed, realtime, expert systems, object oriented and conversational environments;

- compares SSADM, where appropriate, to other competing structured methods;

- reviews its current and likely future position in the structured methods market place.

This book is therefore targeted towards:

- experienced practitioners of SSADM;
- those practitioners who wish to enhance the method so that it can be used to design application systems requiring distributed, realtime, expert system, object oriented or conversational processing.

The assessment is not yet another rewrite of SSADM. The reader will not learn the method from this book. It is assumed that the reader has not only a good theoretical understanding of the method in whatever version, but up to a year of effective practical experience in applying the procedures, techniques and standards. No explanation of SSADM technical terms is provided. It is also assumed the reader has a reasonable understanding of database technology.

The book is targeted to the procedures and, in particular, the techniques for the logical and physical design of computer application systems. It does not address other associated aspects, such as project management, software package selection, contract negotiation and integration and quality management.

The book is the author's opinion of SSADM, its strengths and weaknesses, where and how it can be improved and how it can be extended to support additional data processing environments. Some may find the results controversial.

There are two ways in which a product, such as a structured method, can be enhanced—by addition or by modification. If a structured method suffers from sins of omission and not from sins of commission, then enhancement by addition should be applied. The benefit of this is that nothing about the method needs be unlearned, and the procedures, techniques and standards of the method are preserved. There is therefore no risk to the method in incorporating the enhancements. Only when there are sins of commission should there be enhancement by modification.

Version 3 and version 4 of SSADM suffer overwhelmingly from sins of omission and not commission. What is there is well thought out and thorough. Where this book makes suggestions for the enhancement of the method, the enhancements are almost all those of addition. The reader of this book does not require to unlearn any aspect of SSADM.

All of the suggestions concerning enhancements to the existing techniques, additions of new techniques, "pearls of practical wisdom" in applying these techniques and in the application of the method in data processing environments for which it is not targeted are based on real world experiences. *None of this book is theory.* Each suggestion is supported by one or more examples. Almost all of the examples are themselves based on real world

The Data Processing Environments

situations encountered by the author in applying SSADM. The examples have been "pasteurised" to protect client confidentiality. The pasteurisation has been achieved via a process of making the examples *generic to the industry being described*, simplification of and modification to the real world functionality, the addition of extra but realistic functionality to the real world examples and, at times, alterations to the naming conventions used.

While this assessment of SSADM is based on the latest version of the method, the suggestions for the application of and enhancements to the method are appropriate to any future version. This is because the suggestions are generic to designing and developing any computer application and are therefore not specific to a particular version of the method.

The source for SSADM is the manuals issued by the National Computing Centre (NCC) under the Crown Copyright on behalf of the Central Computing and Telecommunications Agency (CCTA). The CCTA is a United Kingdom government body. It has many functions, one of which is to act as a consultant on matters relating to information technology in general and, *inter alia*, on structured methods. The CCTA is the design authority for SSADM and, as such, produced the manuals for the first three versions of the method. The CCTA continues to act as the design authority for the method with editorial control over the authorship of the manuals. The publication of manuals has been delegated by the CCTA to the NCC, which issued the first version 4 manuals in October 1990.

1.2 THE DATA PROCESSING ENVIRONMENTS AND THEIR CONCEPTS, TECHNOLOGIES AND TECHNIQUES

Six data processing environments can be identified. Two of the environments—centralised and realtime processing—have been long established and are well recognised and identified by widely used and understood terms. A other four—distributed, expert system, neural and object oriented—have only recently been recognised, but will become established during the 1990s. The environments are diagrammatically represented in figure 1.1.

Purists would say that there is a considerable degree of overlap in the scope of and technology used by the seven data processing environments and that any classification on these lines is therefore artificial. The purists are in part right, but any classification, particularly high level, is open to adverse comment. It will be seen that there is, in fact, no overlap between the environments in either scope or technology, except for the one environment that is generic. All the other environments are unique.

Centralised processing occurs where all data processing is to be executed on a single processor, with remote dumb terminals linked to the processor. The terminals are dumb in that they contain no data processing facilities. The terminals may be spread over a large geographical area, even at multiple sites.

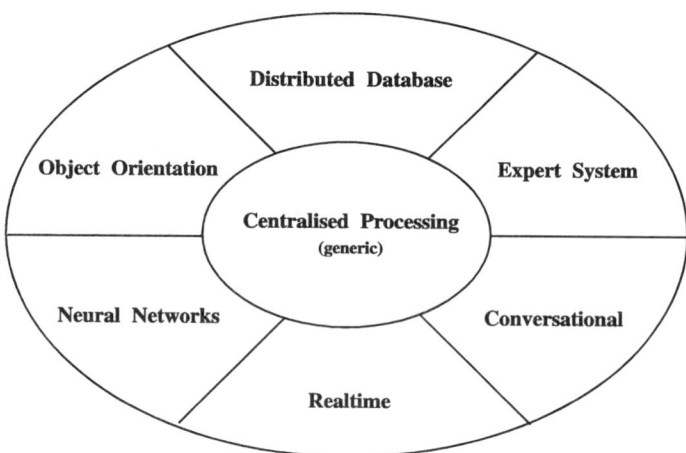

Figure 1.1 The data processing environments

The centralised environment is generic, with its technologies being used by the other data processing environments. The facilities such as the file handler, screen painter, query language and the report writer are all used by these other environments. All these environments contain batch and online processing. They function within a centralised environment. *They are, in fact, nothing more than centralised processing "plus a little bit".* This is because the vendors of the technology for these other environments have followed a policy of *using the existing technology of centralised processing without change and adding additional software facilities on top.*

Distributed processing occurs where multiple processors support a set of multi-site application systems, each with a local database. The processors are widely distributed geographically, one to each site, and are linked dynamically, such that the applications and their databases function as a synchronised whole.

Much confusion exists as to the difference between realtime and online processing. The Yourdon definition of realtime best identifies the difference. Realtime is defined as "immediate output of current input". This means that as an input message is received it is immediately processed in a processor and the output from that process is immediately displayed.

For example, a radar aerial monitoring the movement of aircraft through the sky records the electrical signals received and passes the information to the processor, which then immediately displays the information on a visual display unit. The process may access a standard database to retrieve user data to ascertain, for example, the type of aircraft which matches the signal received. This part is online processing and not realtime. Much realtime data has a limited life-span, even as little as a few seconds. The radar system can discard the signal received after a few revolutions of the aerial if the task of the system is merely to monitor the flight movement of aircraft.

With online processing the output is not the result of immediate input. Modifying Yourdon's quotation "Online processing is immediate processing of previous input". For example, the data for an online request to display customer's orders may well have been stored in the database for some considerable time. Online processing produces immediate output, but not on the basis of immediate input.

Conversational processing is the least clear cut environment. It is an extension of online processing, where a user is requesting immediate response to database access requests. The difference is that the user is having a dialogue conversation with the processor at a terminal and hence requires to monitor previous iterations of the conversation. For example, a user may wish to terminate a conversation but record the state of the conversation, typically as represented by the data currently displayed at the terminal at the point of the termination, for subsequent recall. When the conversation is renewed at a later point in time, the user wishes to recall the state of the conversation, which may have been forgotten, by retrieving the relevant information from the last screen. The application program/system software therefore requires to recognise explicitly that the user requires continuous running iterations of online interactions with the application system, each iteration being represented by one or more screens. An early form of conversational processing was that used in IMS/DC teleprocessing with the scratch pad facility for recording information from one conversation to another. The most widely used form of conversational processing today is windows technology, with each window acting as a scratch pad. This is different from online processing, where the application program switches itself off between output and input screens and the data from the previous screen is lost.

There are no design techniques for conversational technology, in particular for the windows technology. The design techniques for the human/computer interface is currently limited to online processing with dumb terminals. The subject is not taken further in this book.

Expert systems have the ability to store and access knowledge as well as data. This means that they can represent expertise regarding a particular application domain, for example how to service a car engine or diagnose an illness. Expert systems can therefore be relevant to different types of computer applications from those which have traditionally been developed. They are particularly suited to those applications that require expertise to solve problems. Traditionally data processing has been for those applications where dumb data is presented to users, for them as *homo sapiens*, to interpret. Expert systems, by contrast, can offer advice as well as merely present data. They can therefore "assist" and "advise" in such expertise based tasks as diagnosis, planning, design and interpretation.

The object oriented environment is a "modernised" version of centralised processing. Until recently all application systems were designed and developed on the basis that data and logic were kept separate—the data in the database and the logic in the application programs and that data is based on keys and logic on events. It is now realised that this separation is illogical (after all both data and logic are information) and inefficient (data has to be moved to the logic), with a raft of other inadequacies and inconsistencies. Object oriented technology has been developed in the database file handlers and application programming languages to bring both types of information together and to add additional information modelling facilities, such as class and aggregation.

Given that object orientation is another form of centralised processing why has it been treated as another data processing environment? The reason is that, of all the other environments, it is the one that is more than centralised processing plus a little bit. There is some modification. There are, as described in chapter 6, some differences in the technologies and design techniques for those parts that overlap centralised processing.

The neural network environment is not addressed in this book as the author does not have experience of it. It is the ability of computers to learn from the information they are processing. The environment is not considered further.

The techniques in turn have been designed on the basis of the technologies. There is a cascading relationship. The concepts drive the technologies and the technologies drive the techniques. *These relationships of the concepts to the technologies and the technologies to the techniques are rarely recognised—yet they are the basis of this book.*

The reason for the relationship dependency of the design techniques on the implementation technologies is easy to explain. Throughout the history of computing the technologies have preceded the techniques. The reason is simple. There is no purpose in developing a structured method containing a set of logical and physical design techniques if there is no physical technology to implement the logical design specification. There has been technology to support batch and online processing for the last three decades or more but structured methods with design techniques have only gained wide acceptance in the last 15 years. The time gap between technology and techniques is the same for the other environments. Object oriented technology has been available for the last 5 or more years, but the techniques are only now appearing.

The relationship of the technology to the concepts has not been addressed before and is more difficult to explain. The way this is done in this book is to explain the concepts and the supporting technologies and point out where the concepts are not supported and what technology is required. Suffice it to say at this stage that it is obvious that the concepts require to be supported by the technology.

The Data Processing Environments

The various concepts for each of the environments are described in the subsequent chapters, as are the technology facilities required. Suffice it to say that if the technology has not been developed to support the concepts then the technology is unstable. The technology for centralised and realtime processing is fully developed. Chapter 5 shows that the technology to support the concepts underlying expert systems is also fully, and beautifully, provided. The author believes that there is thus little improvement to be found in expert systems. The technology will remain stable. This is in contrast to the technology for distributed databases. Neither of the concepts are fully supported, for example there is no synchronisation of recovery for hardware across locations. Such technology is therefore unstable, because improvements are required.

What is true of the technology is true of the structured methods containing the design techniques. Consider the Yourdon method. It is specifically targeted for realtime processing and explicitly recognises the concepts on which this data processing environment is based. The method was developed in its initial form by Ed Yourdon. The method has been further upgraded by others, such as Ward and Mellor. The author finds the method a delight to apply, with the techniques directly reflecting the concepts underlying realtime systems, and hence relating to each other most attractively and producing a set of excellent integrated deliverables. The concepts for each of the data processing environments, the related technologies and the appropriate techniques are illustrated in figures 1.2–1.4.

SSADM currently provides design techniques solely for the centralised processing environment. Supporting Interface Guide manuals will be issued after this book is published on distributed processing and in the fullness of

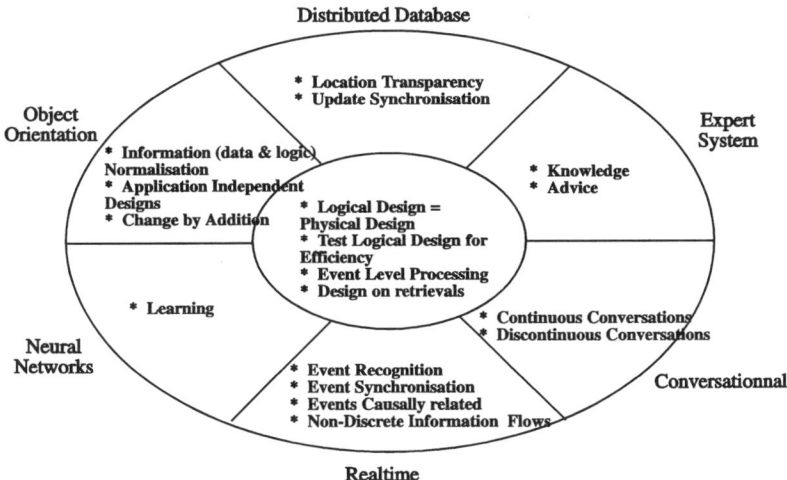

Figure 1.2 The data processing environment concepts

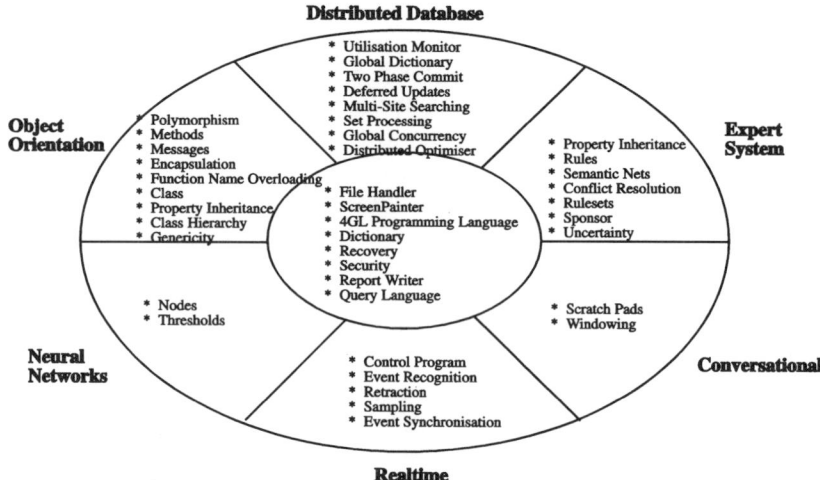

Figure 1.3 The data processing environment technologies

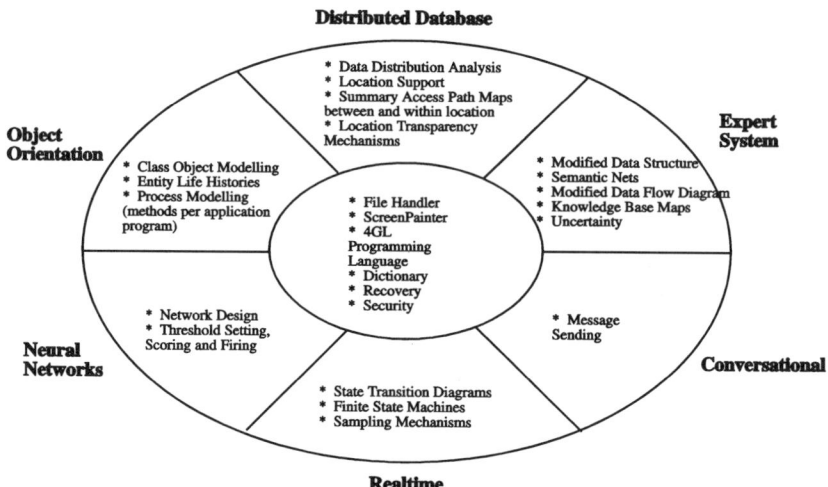

Figure 1.4 The data processing environment techniques

time the other data processing environments will be supported. The CCTA had announced that version 5 of SSADM will support object oriented and realtime processing before major funding of new versions of the method were stopped. There is also a major study underway on behalf of the CCTA to develop a method which will support expert system processing. The study is called GEMINI (General Expert systems Method INItiative). There is little doubt that SSADM will become major force in a number of data processing environments in the next few years.

The Scope of Version 4 9

The CCTA is wise to design the SSADM manuals on the basis of a set of core manuals and set of supporting guideline manuals. The core component manuals contain the techniques for the design of centralised data processing. *It will be seen that the SSADM core techniques, like the centralised data processing environment it supports, are generic to all the data processing environments. SSADM can therefore be used as the core component for these other environments.* The techniques specific to these other environments can be merged with/added to the SSADM core techniques.

1.3 THE SCOPE OF VERSION 4

SSADM is a structured method for producing logical and physical design specifications for computer systems applications. It is still not a method for conducting information systems and technology strategy studies, but, in comparison to version 3, is much better equipped for developing and implementing applications. In the context of the full life cycle of computer applications—business strategy studies, information strategy/information technology strategy studies, feasibility studies, logical design, physical design, development/construction and implementation—SSADM is certainly "headless".

This is not an oversight on the part of the CCTA. A conscious decision has been taken not to include information systems and information technology strategy techniques in the method, partly because the strategy scope is much broader than that of SSADM and partly for the reason that strategy studies are by their nature iterative, whereas logical and physical design is more a linear sequence of techniques.

The author finds this a strange argument. One needs to have an information strategy within a business enterprise or government department. Indeed the Kobler Unit at the Department of Computing, Imperial College, London stated in a report *Does Information Technology slow you down?* (1988) that one of the few factors that pointed to the success of computing was the possession of corporate information systems and technology strategies that ensure the computer applications support the business objectives of the corporation. There was a high correlation between failed investment in information technology and the failure to have an information strategy. The author is pleased to note the various management guideline reports on information strategy produced by the CCTA, such as *SSADM in an IS Strategy Environment*, which the author found very informative, and the various studies being undertaken by the CCTA, for example on corporate information architectures, that point to a recognition of the importance of information strategies.

Version 4 of the method is a major upgrade, particularly as regards the specification of logic. There is a wholesale adoption of the Michael Jackson approach for representing data structures that are input and output of a business requirement/event process and the appending of logic operations onto the data structures. It is therefore much easier to use the logical specification of the processing of a business requirement/event as identified in the version 4 Process Models than in the version 3 Process Outlines for the production of application programs. This will be discussed in chapter 2. Although not recognised by the SSADM manuals, which state that the method does not cover the development phase of the system life cycle, the method is much more suitable for systems construction and testing from the logical design specification.

Version 4 has also adopted the policy of defining the technology to be used for implementing the logical design specification in terms of the facilities that are generic to all computer systems in the form of a universal physical model. The universal model is the basis, preferably the direct basis, of the first-cut physical design. The universal model is an excellent idea, as it reflects the fact that there are certain facilities that are generic, are universal, to computing. One such facility is that it is possible to cluster detail records in the same block on disk as the master record, such that hierarchical data structures can be made more efficient as regards disk I/O than network data structures.

The production of guidelines and techniques for the detailed physical design is being left to the vendors of the target software environment. Thus, for example, it is expected that Oracle Inc. will produce specific guidelines for implementing an ORACLE design from the universal physical model. These design guidelines will be issued as supporting manuals to the core method.

Given these two developments, and notwithstanding the improvements that can still be made, it cannot be said that SSADM version 4 is "tailless". The manuals are unduly modest.

Although not explicitly recognised in the manuals SSADM is yet further focused. The method is targeted to batch and online processing in a centralised data processing environment and assumes that the application will be bespoke developed and not "purchased" off the shelf as a software package. The method is certainly optimised for, even if not explicitly stated, bespoke design rather than implementation as a package solution.

The limited range of SSADM even within the scope of logical and physical design for centralised processing is explicitly recognised by the CCTA, and is deliberate. The manuals discuss the other activities that require to be done, activities such as risk assessment, capacity planning, data take-on and testing. The interface of these activities with SSADM is achieved by defining the inputs that are required for and the outputs required from the activities.

An accusation that could be made against version 3 was that it was also a "riderless" method, because it contained no advice on project management procedures, techniques and standards. Project management is much better catered for in version 4. The new version does not contain its own "home grown" project management method, but rather has been specifically designed to interface in a "clean and properly managed" way with external project management methods. The "philosophy" is the same as for interfacing with the "activities" defined above— a set of product inputs and outputs. This is described in detail in chapter 3. Although it does not include its own project management method, and is not intended to, SSADM cannot now be said to be "riderless".

SSADM can be divided into two parts—the "core" and the satellite Interface Guides. The core is defined and described in the version 4 manuals. The core manuals are sets of procedures for the sequenced application of a set of techniques for the production of an application system logical design specification. There are less definitive techniques for the production of the physical design, particularly for the design of application program designs.

The method is, however, an open one. Anyone can develop an interface guide on any different but related techniques for the other data processing environments and activities appropriate to applications design and implementation. The SSADM Design Authority Board will recommend any guide that "reinforces" the method.

1.4 SSADM Structure

SSADM is divided into four components:

- *a dictionary*, that defines "what" should be produced. This is done in the form of the contents breakdown of and quality criteria for the product deliverables, along with a glossary of terms;
- *a structural model*, that shows "when" to undertake the stages and steps and apply the techniques to produce the product deliverables;
- *a set of procedural chapters*, that define "how" to apply the logical and physical design techniques.

The above three are the core components. The other non-SSADM component is:

- *project procedures*, for the management and control of a design project.

The dictionary performs two roles—as a repository of the specification of the deliverable products to be produced and of the definition of the terms used in SSADM. It therefore contains a Product Breakdown Structure (the PBS), which shows the relationships between the products, along with Product Descriptions explaining the purpose and content of the products. There is also a Glossary of terms. The PBS identifies three types of products: those for the management planning and control of a project, the deliverables from the application of the design techniques and supporting non-SSADM deliverables, such as human factors, education, security and risk assessment, and those concerned with the quality of the design deliverables and exceptions to the project plans.

The dictionary should not be thought of as a CASE tool encyclopedia. It does not store the products/deliverables when they are produced as the design process is undertaken. It is a paper encyclopedia of definitions, not an electronic encyclopedia of products. Clearly the PBS and Glossary make it easier for the CASE tool vendors to support SSADM.

The structural model in version 4 is new. It is represented in figures 1.5(a) and 1.5(b). Each SSADM activity "box", whether it is a Module, a Stage or a Step, is discrete and distinct, comprising all the tasks and techniques required to produce a product/deliverable(s) for the project management. Each activity has its own defined inputs and outputs, each with predefined quality criteria. Between the bottom boxes and the top boxes is the information highway. The information highway is the communication route

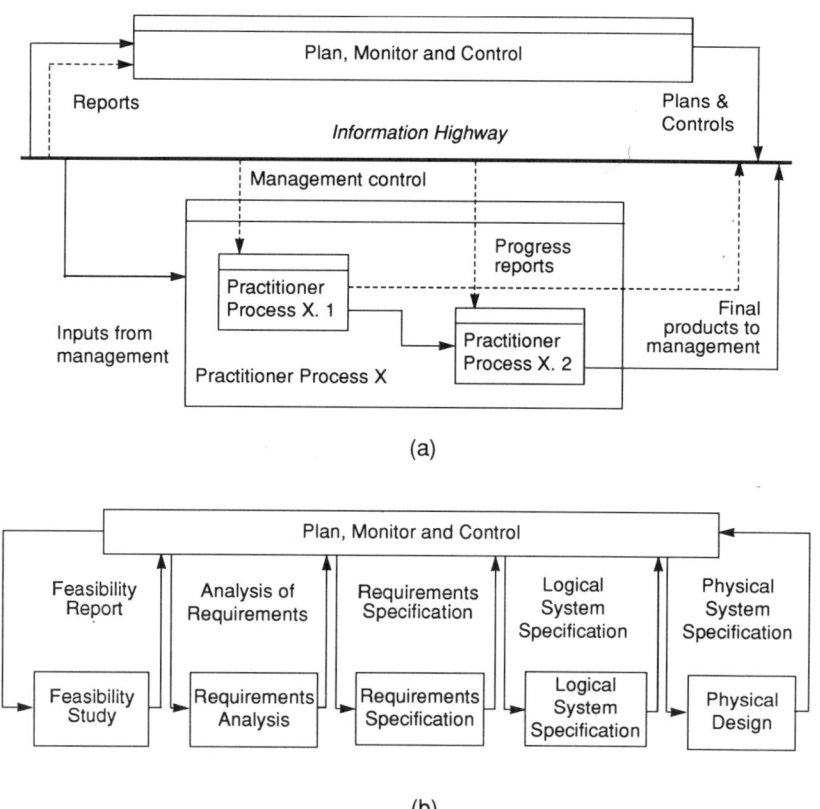

Figure 1.5 SSADM version 4 structural model

SSADM Structure

for the product deliverables (the flow lines going up from the application designers to the project management) and the control flows (the flow lines going down in reverse), and is the division between the management and technical skills within a project. *SSADM is below the line; processes to which SSADM relates are above the line.* Such planning, monitoring and control processes are known as Project Procedures.

This book concentrates on the design techniques below the information highway line, and only refers to the Project Procedures if they support the application of the techniques.

The modular structure is in three parts—the activities concerned with the application of the method (in figure 1.5(b) they are represented as the Modules); the planning, monitoring and control tasks of project management; and the two-way "flow lines" of information between the two. The flow lines represent deliverables, which are unambiguously specified as Product Descriptions in the Dictionary. (Incidentally the input flow lines to the Modules are the deliverables that are submitted to the Project Board for acceptance.)

SSADM is composed of five modules, which occur in the sequence shown in figure 1.5(b). Unlike version 3 where stages 4 and 5 occur in parallel the modules are strictly sequential. It has been argued by some that this "statement is misleading. In some cases they can be overlapped: for example, Physical Design starts after Requirements Specification but before the completion of Logical Systems Specification". The author disagrees with this. There are several diagrams in the volume one manual showing the sequential nature of the modules and figure 5 on page F-CON-17 shows that the techniques of Physical Design (physical data design and physical process specification) are restricted to the module. The only "logical" techniques that can span into the Physical Design module are function definition and requirements definition. If the statement is genuinely misleading then the course notes of some of the training material that has been presented need to be modified.

What is true is that a technique in one module can be used again in another. Logical data modelling is used in the Feasibility, the Requirements Analysis and the Requirements Specification modules. Other techniques, such as dataflow modelling and dialogue design, are also used in several modules. This may give the impression of overlap. It is also true that in stage 4 within-module Logical System Specification physical design can be produced, but this is in the context of technical options rather than a production database and application program environment.

Within a module the stages can occur in parallel. Stages 4 and 5, Technical Systems Options and Logical Design, occur in parallel in the Logical Systems Specification module. Modules are the highest level of activity in SSADM. Within each module there is a more detailed activity structure of constituent stages and constituent steps for the application of the design techniques and the production of the design deliverables. Modules are new to version 4. The stages and steps are the same as in version 3 in that they are convenient ways of compartmentalising the method.

Each module is a self-contained set of techniques that produce a coherent set of product deliverables, the "should we proceed" in module 1, the "what is required" in module 2, the "how it is to be achieved" in modules 3 and 4 and the "finished product" in module 5. The techniques in a module require a common level of skill. This, in combination with the coherent deliverables, means that *modules are designed to be controlled and operated as a contractible unit of systems analysis and design, should it be necessary to adopt outside contractors.*

The first Module is the Feasibility (FS) Module. The module has only one Stage. It fulfils a similar role to the Feasibility Stage in version 3, but is less prescriptive in the application of SSADM techniques. FS is limited to identifying the current and required environments in sufficient detail only for a Problem Definition to be developed and agreed with the project board and for Business Systems Options (BSOs) and Technical Systems Options (TSOs) to be identified. Although FS uses the dataflow and data modelling and requirements definition techniques, *these are limited to high level identification of the business, not its definition and description.* The prime output from this module is the feasibility report.

The need for the FS to interface with other non-SSADM activities is explicitly recognised in Step 010. An interface with information system strategy (planning up to 5 years in the future), tactical planning (the next 12–18 months project portfolio plans) and the need for a business case are all identified.

The business case for a proposed application is emphasised rather than the issues of implementation technology. Version 3 merely describes a cost/benefit analysis. A business case may involve a financial loss, but wider and other opportunities not normally included in a cost/benefit analysis may point towards proceeding with the project. This is where the explicit recognition of "soft" issues and non-SSADM techniques in the manuals for the FS Module, such as organisation and value chain analysis, is to be applauded. There is scope for somebody developing an interface guide on these soft issues.

The Requirements Analysis (RA) Module is divided into two Stages: the Investigation of the Current Environment followed by the Business Systems Options. This may give the impression that only the current system is investigated in the first stage. Not so. The current environment includes business requirements to be satisfied by the future required systems to overcome the current problems and deficiencies. These are primarily identified in parallel in Step 120 and recorded in the Requirements Catalogue. Additional business requirements may, however, be identified in the Requirements Specification module step 330 Derive System Functions. The RA module therefore covers the definition of the current and part of the required systems before BSOs are drawn up. The module can be regarded as similar to the FS Module, but with the same design techniques used to define and describe, not merely identify, the information needs of the current and required systems. The design deliverables/products are therefore to a more detailed level than in the FS Module.

Other deliverables are also produced to define the "what" processing and data of the business requirements, in the form of Elementary Process Descriptions (Elementary Functions Descriptions in version 3) and I/O Descriptions. The BSOs are unchanged from version 3, and define the scope and functionality of alternative logical business solutions of the proposed application system, along with risk and financial assessment, technical considerations and organisational impact. The project board selects a BSO, which may be a composite, a hybrid, of the options. The chosen BSO then forms the basis of detailed logical design of the required system in the next two modules. The prime output from this module is a selected Business System Option.

The Requirements Specification (RS) Module consists of one Stage, the Definition of Requirements. The aim is to produce a detailed requirements specification, including the prototyping of selected online menu dialogues of the man/machine interface (MMI). Previously only the data and functional aspects of the requirements are considered. Non- functional requirements are also included, such as security and service levels to be provided.

It is this module that introduces new logical design techniques to SSADM, with existing techniques being enhanced, fortunately by addition rather than by modification. The new activities (a) and techniques (t) are the identification of functions (a), the definition of the user role (a), the production of I/O Structures from the I/O Description (t), the Effect Correspondence Diagrams (ECDs)(t) and the Enquiry Access Paths (EAPs)(t). The enhancements are the adding of operations to the Entity Life History technique and the taking of Relational Data Analysis to fifth normal form. The ELHs, ECDs and EAPs are the beginnings of Jackson-like data structures for the specification of logic. Prototyping of the man/machine interface (a) is also introduced. The prime output from this module is the requirements specification.

The fourth module is the Logical Systems Specification. It is composed of two stages, the Technical Systems Options (TSOs) and the Logical Design. These two stages occur in parallel. Its purpose reflects the constituent stages, that is, to select a technical hardware/software framework on which the required system will be developed and implemented and to convert the requirements specification into an implementable detailed logical design specification against which a physical design appropriate to the target hardware/software environment can be made. The TSOs are unchanged from version 3. Logical Design further details the specification process, converting the ECDs and EAPs into their respective more precise Jackson-like Process Model structures with the strangely named Logical Database

Process Design technique—*per se* the techniques has nothing to do with database. The Dialogue Design technique is also applied for the definition of the man/machine interface. There are two parts to the technique:: the conversion of the I/O Structures into Dialogue Structures for the dialogue screens and the creation of a menu screen hierarchy diagram for the menu screens.

The final module is Physical Design. It is composed of a single stage, that of Physical Design. The module is concerned with producing the design of the database (the data component) and the application programs (the processing component) for the target hardware/software environment on which the required system(s) will run. The module is not concerned with the development of the required system. The module is in two parts—preparing the design by logical to physical mapping for both the data and processing components according to universally available physical facilities in the Physical Environment Classification ("rules of thumb applicable to any DBMS or file handler" and non-procedural versus procedural processing) and then producing a product-specific design of the target hardware/software. The last remaining aspect of what some feel is Logical Design, that of the formatting of the man/machine interface inputs and outputs, as well as syntax error checking, is also produced. Any application program transaction not meeting the performance objectives requires either the transaction or the database design to be modified, until the objectives are satisfied or a design compromise is mandated.

The project procedures are outlined as non-SSADM activities, of which there are currently nine, including project management (specifically the PRINCE method), risk assessment, quality assurance, capacity planning and database take-on. This book is not concerned with project procedures, other than to state that SSADM is now much more related to outside activities with which it needs to interface. For example, as regards the project procedures a complete chapter of the manual addresses this topic, concentrating on what requires to be done and leaving the "how it is to be done" to the specialist manuals. The method recognises it does not operate in a technical vacuum, and is correct to do so.

1.5 THE SSADM DESIGN TECHNIQUES

The design techniques require the systematic application of specific skills to produce the design product deliverables. For example, the technique of relational data analysis (popularly known as third normal form analysis) requires the application of the skills of data normalisation to produce "clean" relations/entities of data. There is a much clearer distinction in version 4 between the techniques for the two universally recognised major components of a computer system—the data and the logic.

The SSADM Design Techniques

The techniques for data modelling have been enhanced by included graphical and textual semantics in the logical data model and by some "tidying up" of the previously separate data modelling techniques. The three separate techniques of Logical Data Structuring, Relational Data Analysis and Composite Logical Data Design of version 3 have been merged to become Logical Data Modelling. The enhancements are of a minor nature, in large measure because there was little room for improvement.

The area of change, massive change, has been in the area of the specification of logic, access and process logic. There has been a wholesale adoption of the Michael Jackson approach to specifying logic operations, that is, the operations being based on the structure of the data that is input to and output from an event/business requirement. Unlike version 3, where there was one technique for the specification of logic, the Process Outlines, several techniques are used in version 4 for drawing Jackson-like structures and identifying the logic operations to be appended to the structures. These structures, suitably brought together and, for the enquiry business requirements, merged, become the major component of application program specifications.

The other difference in the specification of logic is the clear separation of the logic for processing the data in the database and the logic for processing the man/machine interface. In version 3 the logic for both aspects of processing were detailed in the Process Outlines. In version 4 they are separated into different techniques, the database component in the Process Models and the man/machine interface in Dialogue Design.

There is one further subtle difference. Database processing is at the event/business requirement level and the man/machine interface processing is at the function level.

SSADM version 4 does not apply the design techniques on a techniques by technique basis, as in version 3. Some of the techniques, such as for the modelling of data, have been bundled together under a general heading.

All told, SSADM incorporates 11 logical design techniques:

- *Logical Data Modelling (LDM)*. The LDM bundle comprises three techniques:
  — *the version 3 LDST*, the building of a logical data model (LDM) based on the data retrieval business requirements. This is the front-end part of LDM;
  — *the relational data analysis* of the data attributes. The technique is often and perhaps more popularly known as third normal form (TNF) data analysis. RDA is the front-end technique to the building of a relational data model. It is based on the analysis of the raw data items of the current and required systems and progressively "rearranges" the attributes into relation/entities. RDA decomposes the data to third normal form;
  — *composite logical data modelling*, the building of the final logical data model from the merging of the above two models.

These techniques are not recognised as being separate, but are a continuum, the RDA enhancing and validating the LDM.

There is also no longer the concept of a composite logical data model. Relational data analysis is no longer applied as a separate technique from which to build a relational data model, but "intuitively" as the data attributes are identified as LDM proceeds. The results are added "incrementally" to the logical data model.

- *Enquiry Access Paths* (EAPs), which show the access path against the LDM for each data retrieval business requirement. The EAPs become the basis for the development of the Enquiry Process Models. The technique is bundled as part of LDM and is new to SSADM.[1]

- *Dataflow Modelling*. The single technique is that of dataflow diagramming (DFD), along with the by-products of the I/O Descriptions and external entity descriptions. Dataflow diagramming models the flow of data around an enterprise. Data "flows" because it changes its state due to a modification of some kind. An order is received by the Orders department; it is then updated by the Stock Control department through a stock allocation process; the order is then despatched by the Despatch department. DFDs should therefore be only concerned with data maintenance business requirements. Data retrieval business requirements do not change the state of data. The additional component is the writing of the Elementary Process Descriptions of the lowest level processes. The EPDs (formerly the Elementary Function Descriptions) contain a "what is required" description of the processes.

- *Function Definition—I/O Structuring*. A function is all the processing required for an input data flow crossing the system boundary as seen by a user role. The I/O Descriptions from the data flow diagrams are converted into I/O Structures by the I/O Structuring technique. This is a technique new to SSADM. The I/O Structures show the sequence, selection and iteration of the input and output data items groupings.

- *Entity/Event Modelling*. This is concerned with modelling the lives of the entities on the LDM and identifying the processing operations of the events. It comprises two techniques:
    — *Entity Life Histories* (ELHs). The technique identifies the sequence of data maintenance business requirements/events that update each entity in the LDM, to which in version 4 are added some of the processing operations. These are operations to insert and update an entity and maintain the integrity of the relationships to other entities. An operation

---

[1] Access path analysis has always been part of SSADM. The use of a Jackson style to represent the access path is new.

The SSADM Design Techniques

is a statement of some processing, either a procedural statement of some single action/unit of processing, a non-procedural command or a more powerful declarative rule.
— *Effect Correspondence Diagrams* (ECDs), which show all the effects of an update event, the correspondences of the effects with each other and the access path of the event against the LDM. Although not explicitly stated in the manuals, the ECDs are to the update business requirements what the EAPs are to the data retrieval business requirements. The ECDs become the basis for the development of the Update Process Models. The technique is new to SSADM.

- *Dialogue Design*. Dialogues are the online interactions of the computer system with the users as defined in the user roles. There are two techniques, one for the menu selection screens for the user roles and one for the dialogue screens for the business requirements/events. The Dialogue Design technique has a set of supporting tasks, such as the Command Structures and the Dialogue Control Tables. Dialogue structures are built up from the I/O structures constructed in Function Definition. They do not identify the menu selection component of the man/machine interface, only the part that is concerned with the processing of the function once it has been chosen from the menu. The menu selection routines front-end a dialogue and are constructed for each user role. Both structures are in the form of Jackson-like diagrams.

- *Logical Database Process Design* (LDPD). This covers the technique of process modelling for both the data maintenance and data retrieval business requirements. It covers the processing to and from the logical database, but not the man/machine interface. The access to the logical data model has already been identified in the ECDs and EAPs, which, as has already been identified, are the basis for the development of the process models. The ELHs and the ECDs are the inputs for the Update Process Models (UPMs). The inputs to the Enquiry Process Models (EPMs) are the EAPs and the I/O Structures produced from Function Definition. The I/O Structures for the update business requirements are not used as inputs to the Update Process Models because the data that is output from an update to a database is a trivial structure, being merely an error message if the update is unsuccessful or an indicator of some kind if the update is successful.

    Some of the operations have already been specified in the ELHs for the update processes. The task is therefore is to add yet further operations for the update processes and a new set for the enquiry processes. These are the access and process logic operations to support the business of the application. Process modelling is a new technique to SSADM.

- *Physical Design* is treated differently in version 4. There is the concept of a universal data and functional design that incorporates the facilities that are generic to computing. A universal physical design is produced, then converted to a product-specific design (the design facilities being defined by the product vendor) and then optimised where the transactions do not meet the performance objectives.

An annex of the steps to produce a database design based on the universal data facilities and for the different types of database management systems (DBMS) is provided, as are some examples of the transaction timing technique (Physical Design Control in version 3). A set of first-cut design rules for the classical three types of DBMS—the hierarchical type, the CODASYL or network type and the relational type—are also provided. The same two database design techniques of version 3 of SSADM—First-Cut Design and Physical Design Control—are still used in version 4.

The technique for designing application programs from the Process Models is described, but not illustrated with worked examples. This is surprising, given that examples are used for the techniques of logical process design and database design. No examples are provided, for example, of the Function Component Implementation Map, a rather important physical design deliverable. All that is provided is a set of guideline advice about how to use the facilities of fourth generation application development software to best advantage.

2

STRENGTHS AND WEAKNESSES

2.1 THE BASIS OF EVALUATION

Like all structured design methods, SSADM does not get "ten out of ten", even for the data processing environment to which it is targeted. The perfect method still does not exist. The strengths and weaknesses of SSADM are of two kinds—conceptual and systematic.

Although the design of a computer system is based on a combination of systematic skills, as represented in applying the design techniques, and flair, as represented in the balancing act of trying to reconcile often mutually conflicting aspects of design (flexibility versus performance, for example), it is also based on a set of underlying concepts. Some of the concepts are generic to all seven data processing environments and some are specific to a single environment. The concept of logical design = physical design is generic, while the concept of event level processing is specific to batch and online systems, the concept of location specific to distributed systems and the concept of event synchronisation specific to realtime systems, to mention but a few. *Unless the techniques and procedures of a structured design method recognise and reflect these underlying concepts the resultant logical and physical design deliverables will lack an inherent soundness, elegance and rigour.*

The importance of ascertaining the concepts underlying the data processing environments should not be underestimated. The concepts reflect the purpose of the environment. Unless the implementation technologies and the design techniques recognise and reflect these underlying concepts they will be unstable, because they are unable to provide the functionality required for the purpose of the IT environment. It is therefore necessary to understand the concepts underlying the data processing environments before the strengths and weaknesses of the technologies and techniques can be appreciated.

The concepts and techniques addressed in this chapter include those that are generic to all data processing environments and to the data processing environment to which SSADM is targeted. The concepts and techniques appropriate to the other environments for which SSADM can be enhanced are addressed in chapters 4–6.

SSADM is first assessed as to its support for these generic concepts and to the concepts specific to the environment for which it was developed, namely centralised batch and online processing.

The systematic strengths and weaknesses of the design techniques are then assessed. Some of the techniques provide full or substantial support for a number of the concepts. As to be expected, some of the techniques fail to recognise any concepts at all or inadequately support a concept they are explicitly trying to support. Some of the techniques also have mechanistic problems and deficiencies in their actual application and in the quality and completeness of the deliverables they produce. The techniques also require to be applied in an orderly sequence within a structural framework of phases, steps and tasks. The quality of the framework itself must therefore also be considered.

Only occasionally will an attempt be made in this chapter to advise as to how to rectify the deficiencies or take advantage of the strengths of SSADM. The vast bulk of the advice will be presented in chapters 3 to 6.

2.2 Design Concepts for SSADM

Some concepts are universally recognised by any student of structured design and development methods and are therefore not addressed—for example, the concept that the logical design is the basis of the physical design, a concept explicitly recognised and fully supported by SSADM.

2.2.1 Generic concepts

There are certain design concepts that are generic to all data processing environments. They are as follows:

2.2.1.1 *Logical = physical*

Analysts/designers need to be aware of a fundamental change in the emphasis of designing and developing computer systems. In the past, and all too often currently, systems are designed and developed only at the physical level appropriate to a hardware/software environment by analysts who believe they understand the user requirements because a consultation process has occurred. A design specification is produced and is then manually "handcrafted" into a physical design and coded.

This "seat of the pants" approach is perfectly plausible because it has worked successfully. The trouble is, it is outmoded and has produced too many failures. The adoption of structured design techniques to produce a logical design specification independent of hardware/software constraints as the basis of a subsequent physical design is a considerable improvement on the previous approach. It does not, however, go far enough.

A statement of truth regarding computer systems design can be made and that this statement, provided a modern file handler, programming language and teleprocessing monitor software are used, will, with one exception, become a law. Modern software does not suffer the design constraints found in early generation products. The statement is *the greater the variation of the physical design from the logical design, the greater the inefficiency of the developed system*. This statement can be converted into logical design = physical design.

This statement is true from the strategic down to individual lines of application program code and for all data processing environments. Examples of the validity of this statement are based on live case studies.

A port authority illustrates the strategic truth in figure 2.1(a). The situation is that a downstream container port, built after the Second World War, currently undertakes some 80% share of the business and the upstream port dating from medieval times some 15% share. The head office generates some 5% share. Because of historical factors the allocation of computer resources across these two ports considerably mismatches the volume of business, with the share being some 15% for the container port and 80% for the medieval port. The physical allocation of resources therefore mismatches the logical demand for resources.

The inefficiency that inevitably results is, *inter alia*, unnecessarily high data communication costs. The container port is constantly issuing remote data access calls to retrieve information about its business from the database at the medieval port located some 60 kilometres away. Although only some 5% of all business of the container port was with the medieval port, logically the computer system was designed physically as if some 75% of the business was with the medieval port. From being an overwhelmingly centralised business the computer system was implemented as being overwhelmingly distributed, with very heavy access to remote data that should have been local—and remote I/O on a wide area network telecommunication line can be up to 100 times more expensive in processor overheads than local I/O to a disk pack!

Early database file handler types, particularly first generation hierarchical and hybrid types, such as IMS from IBM and TOTAL from Cincom Systems, provide numerous examples of inefficiency down to the lines of application code necessary to access the data, because they are unable to support the logical design data requirements on a one-for-one basis in the physical design.

(a) Uneven allocation of business and computer resources within port

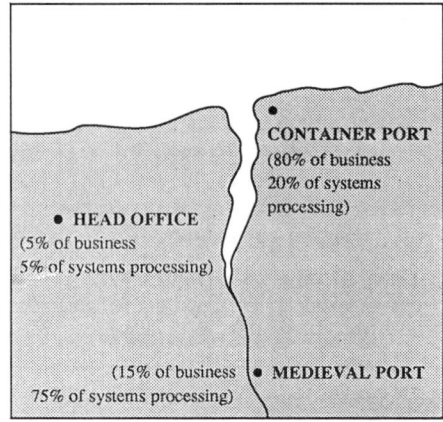

(b) Physical design not equal to logical design

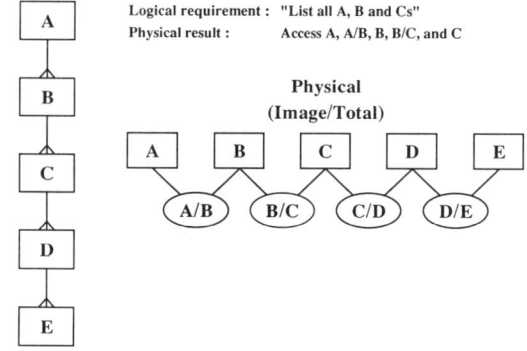

Figure 2.1 Lines of code

One example suffices. The inefficiency is illustrated in figure 2.1(b). The figure shows a logical data model that is four levels of hierarchy deep. TOTAL is only able to support a single depth of data hierarchy and therefore requires the physical design technique to squeeze the logically deep hierarchy into a shallow physical network. As one squeezes the vertical logical data structure so one elongates the horizontal physical data structure. Thus, in order to support the logical data model TOTAL requires to create link record types A/B, B/C, C/D and D/E as illustrated.

These record types are solely in the physical design as link records between record types A–E. They contain no user data, support no business requirement and have no meaning to the user. For all accesses to the logical data model that require to access more than one record type it is necessary to access one of the link records. For example, if the business requirement is "For a specified A retrieve all related Bs", it will be necessary at the physical level to access record type A, then to access record type A/B (whatever that is, thinks the user) and then access record type B. The link records not only add extra data maintenance costs whenever a record type A-E is inserted, but also add to data retrieval overheads. There is inefficiency in extra data storage, extra disk I/O to modify and retrieve the link record types, extra database access commands/lines of code in the application programs to access the link record types and higher user learning curves for data that has no meaning.

The author can hear cries of disagreement with the above law. The argument of using data redundancy to increase performance has often been put forward in discussion with many reputable database administrators (DBAs). Even DBAs of modern relational file handlers have argued that it is often worthwhile to store redundant data (derived or duplicated) in non-third normal form to increase performance. But consider this. The holding of redundant data in a table to which it doesn't rightly belong (for example, the order delivery date of the latest order received as redundant data inside the customer table) creates "storage anomalies". Figure 2.2 illustrates a number of such anomalies. Here a table is in first normal form. Strictly speaking there should be three third normal form tables—project, person and a link table showing the time that a person has spent on a project.

By producing a physical design from a table in first rather than third normal form the following inefficiencies occur:

- One cannot insert a person until that person has been allocated to a project.
- Deleting a project also wipes out all associated person data.

FNF Relation - Project/Person							
Project Code	Person No.	Name	Grade	Salary Scale	Date Joined	Alloc Time
ABC001	2146	JONES	A1	3	1.11.86	36
ABC001	3145	SMITH	A2	3	2.10.87	36
ABC001	6126	BLACK	B1	9	3.10.87	17
ABC001	1214	BROWN	A2	3	4.10.87	17
ABC001	8191	GREEN	A1	3	1.11.87	12
XYZ002	6142	JACKS	A2	3	1.11.87	5
XYZ002	3169	WHITE	B2	10	2.11.87	5
XYZ002	6145	DEAN	B3	10	14.11.87	3

Figure 2.2 Storage anomalies

- If grade A1 is switched to salary scale 4 it would be necessary to read the entire table and make multiple amendments where only one should be necessary.

The ability to support a logical design on a one-for-one basis in a physical design has effectively only been possible with the advent of second generation Codasyl file handlers, such as IDMS, and even more so with third generation relational file handlers. Such file handlers have no data structure and virtually no data access limitations. The entities and relationships between entities in a logical data model would convert directly into tables and indexes in a relational design. The extra disk I/O and data access commands to support the link record types and storage anomalies would not be incurred. Efficiency would inherently be increased.

A comment has been made to the author that this law does not distinguish between a first-cut physical design and a tuned/optimised physical design. The reason for this is that it is not necessary. If the law is followed there is no need for a first-cut design to be followed by an optimised design—the physical design is automatically the optimised design. Where the law has been applied any physical design optimisation has been minimal or zero.

The one area where the author has found that this law does not apply is with distributed systems. The problem is the cost of sending data messages between sites across wide area network (WAN) telecommunication lines. The cost of accessing a local disk is, depending on the type of file handler and data access call, on average some 5000 path length instructions (PLI) to a single table row in a typical IBM 370 type mainframe processor. The cost of sending a data message across a WAN can, again depending on the network design, be up to 500 000 PLI. This is a difference in PLI costs of a factor of up to 100. It is clearly extremely expensive to send data messages across a WAN network.

The problem of telecommunication costs in a distributed system is further compounded in that when conducting database concurrency control and updates across multiple sites the updates must be synchronised. The concurrency control mechanism of multi-phase locking involves two remote I/O message pairs between the triggering and the remote sites. The alternative mechanism of transaction timestamping also involves multi-site intercommunication to co-ordinate the timestamps. WAN costs are yet further increased.

As will be described in chapter 4, it may be therefore necessary to break this law of logical design = physical design. Design "tricks of the trade" to "bend" the physical design from the logical requirements may be required in order to get any chance of reasonable performance. Examples of how to achieve this using enhancements to the SSADM techniques will be described.

2.2.1.2 Test logical design for efficiency

This second concept is a corollary of the first. It is based on the simple premise that, given the logical design is the physical design it makes sense to ensure that the logical design is efficient. The first two concepts combine to produce the conclusion that there should be no need to undertake physical design and optimisation! Much of the design phase could be eliminated.

This is a somewhat radical suggestion in theory but proven in practice. On two major projects all the "pearls of practical wisdom" detailed in chapter 3 to get over the deficiencies of SSADM, including testing for logical efficiency, were applied to the techniques for producing the logical design specification. The logical design specification was then used as the direct basis of database and program coding. Using the existing code generation facilities in the CASE tool for these projects the "improved" logical data model (improved because of the "pearls of wisdom") definition in the tool encyclopedia was converted into the SQL data definition language constructs of Create Table and Create Index, with the data attribute definitions becoming the table columns. The action diagrams were handcoded (the application program code generators were just appearing and were at that time regarded with some suspicion), but on a direct translation basis. No physical design was undertaken and no database or program optimisation of any kind was required.

The implications of the law logical = physical are far reaching. Firstly, it is the basis on which code generators work. A syntactically complete logical design specification is produced, a button is pressed and the physical code appropriate to the target hardware/software environment is generated. The problem is that none of the structured methods which are now widely used to produce a logical design specification explicitly recognise this law. They noticeably still produce logical design deliverables which are, in many cases, not complete, *and certainly only ensure that the logical design works* in that it can support the user business requirements, *rather than ensure the logical design works well.*

Given that the logical design should be the direct basis of the physical design it is incumbent to ensure that the logical design is efficient—if it is not then the physical design will *a priori* be inefficient and require optimisation. Currently none of the structured design methods incorporate techniques of ensuring the logical design specifications they produce are efficient. It is not surprising that the experience of those who have used code generators has been disappointing, often with extremely poor transaction response times. In some cases the disappointment has been such that some well-known companies have abandoned automatic code generation.

It may be that the inefficiencies are the result of poor code generation facilities. It may also be that the structured design method used does not recognise this law and therefore does not produce efficient logical designs.

The sooner the structured methods recognise and practise this law, the better. At least then the code generators can be judged on their merits.

2.2.1.3 Business requirements = events

A computer system responds to triggers. Triggers can be "fired" in a number of ways. A user can pose a query (Display all customers with red hair) or make an input of data (Insert a customer with red hair). The computer system can also be programmed to fire triggers automatically. For example, the passage of time beyond a certain point requires an output (List month-end accounting statistics); a certain condition occurs and the user requires to be informed or some action taken (Stock levels are below a certain threshold, therefore create a re-stocking order). A trigger is fired only because there is a business need to do something. This business need is a business requirement. Each triggering of a business requirement is an event—"do something"—display customer, list statistics, re-order stock.

2.2.1.4 Event level processing

This concept is the corollary of the previous concept. A computer system contains application programs that execute logic for a given task to "do something". The application logic is justified, written and executed because there is a business requirement. Business requirements are ultimately re-expressed as physical application programs. Given that business requirements are identified, defined and undertaken at the event level, application programs themselves should also be designed, developed and executed at the event level. Most structured design methods do not explicitly recognise (Yourdon does) that all non-realtime processing is at the event level. Certainly the event can be decomposed into sub-processes, as we shall see, but the processing itself is event fired. This simple rule is also in line with the concept of logical = physical.

Events are easy to identify, as they are objective triggers and occur at logically distinct times. The event "create customer" is identifiable and each occurrence of the event is distinct.[1] A user could insert a customer record into a database at, say, 11 o'clock and another at, say, 10 minutes past 11 or .... Event decomposition into sub-processes, by contrast, is a subjective

---

[1] A comment has been made that "create customer" is an operation and not an event. In the context above it is an event requiring the insertion of a customer record in the database and a possible set of operations for checking the attributes and the relationships with other entities.

Design Concepts for SSADM 29

skill. In line with the concept of logical = physical the author uses a "trick of the trade" and decomposes a business requirement event into its constituent problems-to-solve (an elementary process to SSADM). Problems-to-solve are more difficult to identify, partly because they all occur at the same point in time as the event and partly because a problem-to-solve can also be, in its own right, an event.

Consider the former problem. Problems-to-solve for the event create customer can be to check the customer's input documentation, check the customer's identification and check the customer's creditworthiness. These problem-to-solve tasks must all be executed whenever the create customer event occurs. Depending on the physical solution adopted, the problems-to-solve may take, particularly if computer controlled, less than a minute to complete or may each be several minutes apart if manually undertaken. Nevertheless they all logically occur together with event.

An example of the latter problem could be that the problem-to-solve check customer's creditworthiness could itself be a standalone separate event *in a different context*.

Unlike events, which are distinct and separate in time, problems-to-solve within an event are causally related. This is because they all occur at the same point in time. Problem-to-solve one in an event causes problem-to-solve two to occur causes problem-to-solve *n* to occur *et seq*. For example, validate the customer input document triggers check the customer ID triggers check creditworthiness.

Following the concept of logical = physical the logical event level business requirements produce physical application programs and the problems-to-solve produce application program modules.

Given the nature of events in batch and online processing, application programs should not, repeat not, talk to each other. Modules yes, programs no. As Yourdon correctly states, "programs are schedulable, modules are callable". The scheduling is either *ad hoc* for online or regular for batch programs. The calling is of modules calling each other as appropriate *within* an application program.

2.2.1.5 *The separation of the logical design from the physical design*

Logical design is concerned with applying a set of techniques to produce a specification of the proposed application system that pays no attention to the constraints of the target hardware/software environment on which the application will run. Physical design is concerned with taking the logical design specification and, using a different set of techniques, "re-expressing" it (not redesigning it given the concept of logical = physical) so that it can run on the target environment. Each set of techniques needs to be applied in one or more modules of the method, modules that are wholly concerned with either logical or physical design, i.e. the logical design techniques in the logical design modules and the physical design techniques in the physical design modules.

A structured method should not intermingle logical design techniques with physical design techniques within a module of the method. A design module of the method is either wholly part of logical design or wholly part of physical design. The author would look with suspicion on a method that included within the same module design techniques that are clearly logical, such as dataflow diagramming, and design techniques that are clearly physical, such as the preliminary design of a DL/1 database. This is done in the Information Engineering Business System Design method.

This seems simple enough. However, the world is not so clear cut. It will be seen in the practical case studies used to demonstrate "tricks of the practitioner's trade" that when undertaking logical design it is occasionally necessary to take into account certain physical aspects of computing. The author has found that it is sensible, and indeed most useful, *to take those physical aspects that are generic to computing as permissible considerations when undertaking logical design.* After all, the logical design is a generic design appropriate to any hardware/software environment, so why not use generic technology?

Such generic aspects could include, using file handler data storage and access facilities as an example, that direct access to a record occurrence could be by a randomiser or an index. These two facilities are generic to file handlers in general, not specific file handlers. Another generic file handler facility the author is constantly using in logical design is the blocking factor when requiring to access n record occurrences. All file handlers transmit record occurrences between disk and processor main memory a page/block at a time. It is necessary to know how many record occurrences/table rows there are in a page/block in order to ascertain whether a scan of an entity in the LDM model is efficient or not. It has been found that when undertaking transaction access path analysis the above generic but very physical facilities are very relevant to logical design. Several examples of using this generic physical technology in logical design are given in chapter 3.

What is not permissible is to consider the physical facilities that are specific to the hardware/software on which the application will run. It is not correct, for example, to take into account when building an LDM data model that the resultant database will be DL/1 and therefore build hierarchically structured LDM models, or, when undertaking transaction access path analysis, to alter the chosen path against the LDM because DL/1 does not necessarily provide an efficient or complete access path as, for example, when accessing directly on a non-root segment/record type.

Design Concepts for SSADM 31

2.2.1.6 Database design is based on total data structure and total data access

Figure 2.3 illustrates the basic components of logical and physical design and their relationships. They are relevant to all the data processing environments. In SSADM terms the data structure equates to a logical data model, data access equates to process Enquiry Process Models and the Effect Correspondence Diagrams and process logic equates to the Process Models. Data structure must be accomplished before data access can be undertaken and data access must be accomplished before data process can be undertaken. Data cannot be processed (the constructs of sequence, selection, iteration and branching) until it has been accessed (the record-at-a-time constructs of read, write, update and delete and the set constructs of select, union, intersect, difference and divide). Equally, data cannot be accessed until a data structure has been built. Intrinsically, therefore, those structured design methods that are data oriented are more soundly based than those methods that are process oriented. SSADM is a data oriented method.

Logical		Physical
Data Structure	)	
Data Access	)	)
Data Process	)	)
Message		)

Figure 2.3 Logical = Physical

At the physical level a database design must be accomplished before program design can be undertaken, for the same reasons as at the logical level.

The data structure and data access components are the building blocks of database design. Any database designer will tell you that he/she must take these two elements into account when producing the design, the data structure being the static component and the data access being the dynamic component.

This concept is based on the principle of database. A database is defined as "a single common pool of data which is structured to model the natural data, their relationships and usage which exist in an enterprise". There are two significant points. The database is composed of data that is structured and accessed and, perhaps more importantly, it is a "single common pool" of data "in an enterprise". *Therefore a database is designed for the system as a whole—the total data structure and the total data access.* The database is not designed for individual transactions.

2.2.1.7 Design the database on data retrieval business requirements

This is probably the most difficult concept to accept. When users, possibly except senior managers, are describing their existing computer systems and new business requirements they are, perhaps unwittingly, mostly talking in terms of data maintenance business requirements. "We have customers who place orders with us for products, for which we allocate stock, make despatches and raise invoices against which we receive payments." In this brief statement the user has identified six data maintenance business requirements—create customer, receive orders, allocate stock, make despatches, raise invoices and receive payment—and at least seven entities—customers, orders, products, stocks, despatches, invoices and payments. A dataflow diagram can be constructed from these business requirements and an entity relationship model can be constructed from the entities and their relationships.

In fact the only reason why this information is recorded is that the enterprise can now control and improve its operations. However, it can only do this if the information can be retrieved. "How many orders are outstanding for customer X?" "Is the stock level too low for product Y?" "How many invoices are overdue for payment?" Having retrieved this information, action can now be undertaken. *It is the data retrieval business requirements that actually drive the data maintenance business requirements, because it is the retrievals that decide what data is required in the database.* One does not put data into a computer system for the fun of it. One puts it in so as to retrieve it subsequently. Furthermore, only when the information is retrieved and presented in a valid form and timely manner to the appropriate user(s) and manager(s) can a computer system be justified.

2.2.2 Specific concepts

There are additional concepts specific to batch and online processing in a centralised data processing environment and hence of particular relevance to SSADM. They are as follows:

2.2.2.1 Events are standalone

Business requirements/events do not impact on, are distinct from and occur at different times from each other. Such events are not causally related. Event A does not cause event B does not cause C. . . .

This is particularly the case for data retrieval events. Consider two such events—one to "Display all orders for a specified salesman" and the other "List the profit and loss by month for the last financial year for a specified account". The first business requirement event could have been triggered at 11 o'clock and the second at 12 o'clock or vice versa or 10 minutes apart or.... Intrinsically data retrieval events are random, standalone and not causally related.

Data maintenance events are similar but require to be sequenced. Consider the events "Accept customer application" and "Receive customer order". They are also random, standalone and not causally related. Accept customer application could occur on January 1st and receive customer order could occur on January 2nd or March 31st or not at all. Accept customer application does not trigger receive customer order. They are, however, sequenced in that it would make no sense for the event receive customer order to occur before accept customer application.

2.2.2.2 Problems-to-solve are event based and causally related

A business requirement event can be composed of many problems-to-solve, that is, specific tasks to be done whenever the event occurs. Assume an event "Receive customer order". It is composed of a number of problems-to-solve that must be undertaken before the order can be received—the customer's credit must be checked against the price of the product on order, if the credit is OK then the availability of the product in stock is ascertained, if there is sufficient stock then the stock is allocated and the order quantity decremented .... All these problems-to-solve are undertaken together at the same time as the event itself is triggered, and in a specific and causally related order, with problem-to-solve one triggering problem-to-solve two triggering problem-to-solve three *et seq*. They are causally related because the cause of all but the first problem-to-solve being triggered is the completion of the preceding problem-to-solve.

2.2.2.3 Business requirements should be CSF and inhibitor based

Computer systems are developed to support a company's business objectives as defined in the business strategy. These objectives need to be measured so as to ascertain the company's success or otherwise. A widely accepted approach used in most information system strategy methods is John Rockart's critical success factors (CSFs). Rockart analyses the business objectives and for each identifies their CSFs—those critical measurements which, if satisfied, indicate a company's good health. A CSF could be "The proportion of the market our company has captured in the last month for our products/services". Another could be "Has the minimum % of the customer satisfaction been achieved and is the % rising?"

The CSFs are the internal measure of a company's success—the factor that makes life difficult for the company's competitors. But what about the opposite of CSFs, the inhibiting factors that make life difficult for a company— the external inhibitors, namely competitors, and the internal inhibitors, namely bottle-necks? They also need to be identified and recorded in like manner. An external inhibitor could be "the number, quality and price of our competitor(s) product(s)/service(s)". An internal inhibitor could be "the areas of the production process causing holdups".

As defined above, the CSFs and the inhibitors seem innocuous enough— the managers receive reports on their desks detailing the market share by volume and value obtained by their own and competitor companies products/services with explanations as to why their share is going up or down or whatever. That, on the face of it, is the end of it. Not so, behind these usually few and seemingly simple requests for management information is a great mass of supporting raw data. For the two CSFs identified above the information requirements are enormous—the products and services provided by the company and competitor companies, information about their quality and price, the size, type and geographical spread of the market for the product/service, any significant and total orders received this month by the company and competitors by region and type of customer, and the number of customer complaints.... Each CSF and inhibitor can thus describe a substantial part of the database data.

There is yet more to CSFs and inhibitors than voracious information needs. Each level of management requires to identify their CSFs and inhibitors and all management reports are about the measurement and causal explanation of the success or failure to meet an objective(s) as re-expressed in the CSFs and inhibitors. *All management data retrieval business requirements should therefore be reports that are CSF and inhibitor based.*

The only business requirements that need not be so based are those for the day-to-day operational running of the company. Yet even these business requirements have their ultimate origins in business objectives and their associated CSFs and inhibitors.

The information to support the management and operational data retrieval business requirements has to be put into the database. Data cannot be retrieved unless it has been previously inserted into the database. This, of course, is only achieved through data maintenance business requirements. It therefore follows that CSFs and inhibitors are also the ultimate source of both types of business requirement. Each CSF and inhibitor is thus the source of a corporation's database, are data retrieval business requirements in their own right, the father of other more detailed data retrieval business requirements and hence of all the data maintenance business requirements. *At the end of the day all data maintenance and data retrieval business requirements are the progeny of CSFs.*

CSFs and inhibitors are an excellent mechanism for tracing company strategy to its implementation. *All information in computer systems should be CSF and inhibitor based.*

SSADM is not designed to support information system strategy planning work, so it does not address CSFs. The CCTA recommends a range of strategy planning methods appropriate to a company's business. CCTA considers that strategy planning is orthogonal to the design of computer systems.

2.3 SSADM STRENGTHS

The SSADM strengths are formidable and can be summed up in one word—thoroughness. This thoroughness percolates to each of the four components of the method identified in section 1.4. One aspect of this thoroughness is the close integration of the techniques—they are an integrated set, each reinforcing and therefore adding value to each other.

The strengths and weaknesses fall into two broad classifications—general and systematic—with the systematic being further broken down into three categories—the conceptual, the structural and the technical.

When assessing the strengths and weaknesses of SSADM it has sometimes proved difficult to "pigeonhole" a facility or a technique definitively into a strength or weakness category. There are parts of the facility or technique that have strengths and parts that have weaknesses, indeed specific parts can have strengths and weaknesses simultaneously. For example, the use of semantic descriptions in the logical data modelling technique is a good thing and is therefore a strength, but the use of the facility is not as good as it could be. This part is a weakness. The approach adopted has been to pigeonhole a facility or technique on the basis of the balance of strengths and weaknesses, but still attempt to put clear-cut strengths and weaknesses into their respective sections of the chapter.

2.3.1 The general strengths

The general strengths are those that are applicable to all modules of the method. They are as follows:

2.3.1.1 *All the design deliverables are diagrammatically based*

Version 4 has become so diagrammatic that *all* the logical and physical design techniques produce diagrammatic deliverables. As far as the author is aware, SSADM is the only method that can claim this ability. It is a major improvement, for *the whole method is a set of pictures*: witness figure 2.4. The particular area of improvement is the diagrammatic representation of logic, in the use of Jackson-like structures. The earlier versions of SSADM already supported diagrams showing the structure of data in the form of the logical data model, but used documentation, the Process Outlines, to record the specification of logic. The version equivalents, the Process Models, are diagrammatic.

Inevitably there is documentation, for the reason explained in the above section. Yet the documentation in version 4 is only supportive. None of the documentation is a design technique in its own right. This is in contrast to version 3 where the Process Outlines technique for the specification of logic was only recorded documentarily and is *the* deliverable.

The much increased use of diagrams is a most welcome improvement. It means that the users are able to understand the logical design specification much more easily.

2.3.1.2 There is "bureaucratic rigour"

When giving presentations on SSADM the author has often been asked to comment on the "bureaucratic nightmare" of the method. The chief complaint is that the method is regarded as being ponderous to apply. Five aspects are constantly raised:

- Why can't the Feasibility Module be optional when certain applications "blindingly obviously" require to be computerised to enable a company to function, some obvious examples being invoicing and payroll? It is argued, with some reason, that such housekeeping type applications have nothing to do with information systems strategy.
- Why is it necessary to analyse the current system(s) before embarking on the logical and physical design of the planned required systems?
- Why do you need a Technical System Option (TSO) stage?
- The method produces vast quantities of deliverables.
- The techniques are overly thorough and wasteful—there are three data modelling techniques and all the effort in the ELHs merely produces state variables.

The author has little or no sympathy with any of these criticisms.

One of the main reasons why many computer systems have not provided the benefits anticipated is that all too often they have been developed on

SSADM Strengths

a standalone basis and not as part of co-ordinated information system and technology strategy plans supporting a well-thought-out business plan. The absence of such a plan has been identified in a number of studies as *the* single most important reason behind much of the disappointment in many of the existing computer systems. This was the prime finding of the Kobler Unit already mentioned. Their report identified that "Investing in IT on a piecemeal basis... is no guarantee for business success and in fact can slow a company down and hamper its profit performance". It was concluded that a strategy plan for the development of computer systems that supports the business objectives was one of the few factors that correlated to success in information technology.

All applications, of whatever type, be they management reporting, operational or housekeeping, must be developed and function within the business and resultant information systems strategies. Granted that companies need housekeeping systems, irrespective of whether there is an information systems strategy or not, the priority for developing *all* computer systems for a company can only be established within the framework of an IS strategy. The housekeeping applications must wait their turn along with all the other applications.

SSADM consciously does not include techniques for information system strategy planning. The version 4 manuals, however, explicitly recognise that applications need to be developed according to such a plan. And the Feasibility Module is put into such a context. "Feasibility studies [for a particular application] are increasingly focused on how the system will help achieve business objectives rather than issues solely associated with the implementation technology". Even more explicitly, "Feasibility studies may be initiated in accordance with an IS Strategy".

The Feasibility Module has been integrated with the other modules in version 4, and is no longer an add-on front-end. The advice is that the FS Module is "strongly recommended" to be carried out before undertaking the subsequent modules. This is good advice. Computer systems cost a lot of money. It is wise to conduct a quick study to minimise risk. Risk is a cost. If a project costs £1 million with an estimated benefit of £2 million but has a 70% chance of failure, then the potential of achieving the benefit is reduced. The project is therefore less valuable. Why not use a short feasibility study costing, say, £10 000, to identify that the risk % can be reduced by some 80% for a further investment of some £20 000? The benefit of the study is nearly three quarters of a million pounds.

This module should be applied even for the blindingly obvious applications—they might be necessary in business terms but the technology on which they run is not necessarily obvious; and these applications must also be seen in the context of the total picture. And feasibility studies consider applications within the overall scheme of things, the overall strategy.

Feasibility studies are responsible management   There is not the clear cut division in version 4 between the analysis of the current and required systems as there is in version 3 of the method. The Requirements Analysis Module undertakes the identification of the functional and non-functional requirements of both the current *and* the required system in the same stage, that is, stage 1 Investigation of the Current Environment. However, the identification of the requirements for the required system is only undertaken insofar as it can correct the problems and deficiencies of the current system. It is necessary to complete the identification of new requirements for the required system in stage 3, Definition of Requirements.

The author is a firm believer in making sure that the current system is well documented. Unfortunately, the reasons given in version 3 for undertaking the analysis of the current system are not restated. This is to be regretted because the reasons are still valid. The rationale behind the analysis of the current systems is that it provides:

- A firm basis of understanding the current system(s) strengths and weaknesses prior to defining the requirements of the proposed new system. Only by formally identifying the current weaknesses can it be certain that they are properly catered for in the new system. Furthermore, the real worth of a new system can only be established if it can be compared against the system it is replacing. This can best be achieved if both are properly specified and documented, preferably to the same standard.

- A clarification of the user's perceptions of his/her problems and requirements. A user may have misconceptions because of an inability to perceive or envisage the entire system. This can best be provided by the diagrammatic presentation of the current system in the form of dataflow diagrams and a logical data model.

- A possible springboard for future systems development. It could be that the current system is to be extended or upgraded rather than superseded.

It is essential, not just good practice, to have the current system properly documented. What happens if the new system turns out to be a disaster and the current system has to be resurrected—and there is no record of its capabilities?

The reasoning behind the TSO in stage 4 is equally strong. The business framework for the proposed system has been defined in stage 2 with the selection and the definition of a Business System Option (BSO). Just as the business framework must be set before the detailed specification of the proposed system is undertaken so must a technical framework be made. If the proposed system is to be in a framework of distributed database or co-operative processing, it will be necessary to include extra considerations and skills, as detailed in section 4.1, and will almost certainly require extra design effort, given the probable need to replicate the application of the design techniques per location.

Other related arguments for the "bureaucratic" SSADM approach abound. Given the clear understanding of the requirements of the proposed system obtained from stage 1, a reasonable estimate of technical complexity and risk, hardware/software options and investments can be made and planned for. *Good management requires business and technical options be considered before embarking on the major task of detailed logical design.*

Those who criticise SSADM for producing vast quantities of deliverables possibly forget that any thorough specification of a computer system, logical or physical, will produce voluminous deliverables. *Given the move towards producing a logical design specification that is the source of direct code generation, the specification must be syntactically complete down to the last dot and comma in order to be generated.* The more the specification is complete logically, the less work there should be physically and the greater the volume of the deliverables. The production of a logical design is not of itself the means of reducing the work effort or deliverable output.

Many practitioners of SSADM have said to the author that for all the effort expended in the ELH technique precious little, namely the state indicator value(s) in the entity for each event that updates the entity, is obtained. Poor SSADM! It is criticised where it produces a very small deliverable and it is criticised where it produces a large quantity of deliverables. The critics seem to confuse quantity with quality. The state indicator is the mechanism by which events that update the database out of correct sequence are identified in the Process Models. An invalid predecessor state indicator value for an event indicates it is out of sequence, so that an error message is produced and the event is rejected. This is not an insignificant issue. Without this mechanism there is no explicit guarantee that the database correctly reflects the volatile business complexities of the world that it represents. In the past this update event sequencing has been left to the "seat of the pants" approach of a skilled designer fully knowledgeable of the application—a situation not always achieved. Why increase risk when there is an excellent technique for reducing it?

2.3.1.3 *The incorporation of project management guidelines*

While SSADM is not a method for managing projects, requiring an interface guide to a specialist method, the principles of project management are detailed. Issues such as the structure of project management, the classical stages of a project, planning and control, quality planning and assurance and capacity planning and their interrelationships with the SSADM techniques and deliverables via the Information Highway (the facility for interfacing SSADM with non-SSADM activities) are all considered in detail. Indeed, they are all built into the very structure of the method, as can be seen in figure 1.5(a). There are various control points, such as the mid and end stage assessment for the project manager to take stock, and there are quality criteria defined for each and every product/deliverable being produced/delivered. A management and reporting structure is also proposed. Configuration Management procedures for the control of the deliverables are detailed. Bearing in mind that these are non-SSADM activities the project management guidance is in impressive detail.

Given the UK government sponsorship of SSADM, it is not surprising that the project management issues interface most easily with the PRINCE (Projects in Controlled Environments) project management method, also a UK government standard.

It would be helpful if SSADM identified the classical three types of "projects". The projects types are based on differing skills requirements:

- The strategy stage is one such project type, with the skills appropriate to understanding broad issues of business and information systems strategy, setting priorities, the identification of business systems requiring development and the preparing of information technology plans, all done without going into too much detail about the information that will ultimately become application systems. The "art" of interviewing is also a necessary skill. The strategy project is applied across the company.

- The next project type is concerned with producing the logical design for an application. This is done in the analysis and to some degree in the business design stage. The skills for producing a logical design do not require the knowledge of and expertise in the "bits and bytes" of the physical technology of database and application programming, but, given the move towards code generation from the logical design, do require the detailed application of the logical design techniques and the creation of the logical design deliverables to the last "dot and comma" appropriate to physical technology. Interviewing skills are still needed.

   Note—some would reasonably argue that there is a feasibility type project, where the need for and financial and technical viability of a proposed application system is assessed. In SSADM terms this is a front-end to logical design.

- The final project type is the physical design and development/implementation of the logical design specification. The skills required are those of physical technology and are traditionally divided into database and application programming.

SSADM Strengths

The boundaries between the strategic, logical and physical projects are not hard and fast. Very experienced persons may have a complete set of techniques (i.e. non-hardware/software technology) skills spanning both strategy and logical design, and others may have a complete set of design skills, spanning both the logical techniques and physical technology. Indeed the widespread use of CASE tools and their use as the logical design specification encyclopedia from which physical code can be generated is making the boundary between the logical and physical design stages increasingly imprecise.

These project types and their different skills mean that it will probably be necessary to have a different manager and team members for each project type, but with the flexibility to enable persons to span the "project" boundaries where their skills portfolio is sufficiently wide. Indeed, it is necessary to ensure that for each project type there is at least one project team member who has the skills appropriate to the next or previous project type, so as to ensure that continuity in the team for the project stage.

The point of this section is that, because of the different levels of perception of and attendant skills for the different stages in the project life cycle, from the broad strategy progressively to the more detailed implementation, it is necessary to have a permanent body sitting above the "hurly-burly" of the responsibilities of the various project stage managers. This body is the projects board (sometimes called the steering committee), which is established the moment an application area and the constituent applications that need to be developed are identified from the information system strategy study. It is the responsibility of the board to establish the business and technical direction in which the applications should go. The board therefore contains senior company business and technical directors. It must also include senior users of the application being developed, so as to ensure the specific application is what the users want. The directors are permanent members of the projects board, whereas the senior user(s) is a member for the duration of the application(s) being developed.

2.3.2 The systematic strengths

The systematic SSADM strengths are those that relate to particular aspects or techniques of the method. They fall into three categories—the conceptual, the structural and the technical. The concepts relate to those defined at the beginning of this chapter, the structure is the procedural modules, stages and steps of the method and the techniques are those used by the method for application system logical and physical design and development.

2.3.2.1 The conceptual strengths

The importance of supporting the concepts underlying the data processing environments was stressed at the beginning of this chapter. Unfortunately, SSADM fails to recognise the need to support the underlying concepts generic to computing and specific to centralised processing. It was explained that if the concepts are not supported then the method is unstable, because it is unable to provide the functionality required for the purpose of the data processing environment. Any support for a concept is because of the way the method has evolved and is not the result of a conscious decision to be conceptually sound. Any support is "accidental" rather than deliberate.

If one considers the distributed database environment and the concepts of location transparency and update synchronisation one can say with certainty that neither concept is fully supported. Few database products provide all the facilities described in chapter 4 for location transparency. For example, synchronised software recovery is supported, but no product supports synchronised hardware recovery. Until the products claiming to support distributed database do provide the requisite facilities to support the concepts they will be unstable. And so it is with SSADM. *The method will remain unstable until the concepts detailed in Section 2.2 are fully supported.*

SSADM version 4 has been improved in its support for the generic and specific concepts underlying computer systems. However, it has to be said that it is moving painfully slowly in this regard, in strong contrast to its major competitor Information Engineering. Information Engineering has adopted all but three concepts in full. All the most significant concepts are fully supported, including logical design = physical design, the database design is based on total data structure and total data access, business requirements should be CSF and inhibitor based, balance in the emphasis on the techniques and event level processing. This gives Information Engineering an inherent strength and stability. SSADM lacks this strength and stability.

SSADM now attempts to support four concepts compared to the two partially supported concepts in version 3. The new version is therefore founded on a sounder basis than the previous one. It will therefore be less susceptible to change in the future. *SSADM is becoming a more stable method*, and this can only be to the good, but it has some considerable way to go.

The concepts supported wholly or to a significant degree are as follows.

1 *Logical = physical*   There are three parts of the physical design of computer systems that this concept can support —the design of the data, the design of the applications programs and the design of the man/machine interface. This, of course, reflects exactly the component parts of a computer system identified in figure 2.3.

It was a matter of some debate in the mind of the author as to whether version 4 does or does not, on balance, support this concept. There is no explicit mention of the need for this concept, nor are identified the benefits that can be achieved if it is supported in the design techniques. There is no statement that if the logical design is efficient and the physical design is directly based on the logical design then the physical design itself will also be efficient. There is no reference to it when converting the logical design specification to a physical design.

However, the method goes a long way towards supporting the concept in practice. *Certainly it can be said to follow the concept on the data side*—for example, the entities in the logical data model become tables in relational file design and that relationships between the tables become indexes. The recognition of the concept is explicit. The manual describes "one-to-one mapping" of the logical data design into the physicals design, with the advantage being that if it can be achieved there will be no requirement to maintain redundant data or to reassemble logical records from fragmented physical records. If the physical design is not a one-to-one map it is further advised that a Process Data Interface (PDI) be created to isolate the application program's view of the database design and present the data as if it is directly from the logical data model. This is good advice.

But it has to be said that there is still no recognition that if the logical data model is efficient then so will be the physical model.

The recognition that the application programs should view the database data as if it is the logical data model is much to be welcomed. This can be done using the facilities of what are called in the relational database industry a non-procedural data access sub-language, such as SQL, or if a procedural 3GL type programming language is used and the record types/tables in the database design have been denormalised, through a PDI. Thus, as far as the accessing of data is concerned, there is only a logical view of the data. However, it should be realised that this logical = physical is *for the viewing of the data, not its design in the database*. Welcome though it is, it is not enough.

The advice on tuning the database to obtain the necessary transaction performance makes a distinction between the design "tricks" that can be applied that preserve a one-to-one mapping of the logical data model in the database design and those that do not do so. Unfortunately there is no advice that it is better to optimise by preserving the logical model rather than "bending" it. Thus, there is no advice on the need to preserve the access paths of the transactions as identified in the EAPs and ECDs. The tuning emphasis is therefore on tuning the database, whereas it may be the transaction which is the problem area. How to identify this is addressed in section 3.2.5.6.

SSADM does not, in the author's opinion, follow the concept on the process side. For example, the basis of the logical processes is the event, the business requirement. The techniques for logical process design, the EAP, ECD and their sequitur the Process Models, are based on the event. Yet applications programs are at the function level—and a function can contain many events! All computer processing is event triggered, certainly for online processing, and all is event based. There is no advice that event level processes should become application programs and problem-to-solve processes within the event become program modules. The adverse consequences of not following the concepts of logical design = physical design for process design are discussed in section 2.3.2.3.

Advice is offered that the operations detailed against the Process Models can be directly implemented in a non-procedural programming language. Within the constraint defined above this is also good advice. It is not clear why procedural code cannot also be directly generated as physical program code from the Process Models. The manual does say that a Jackson structure, with the operations and conditions added to it, may be used as the basis for defining procedural processes. Why can't it be used directly by a code generator? It is the author's opinion that but for the deficiency of the function, SSADM could support this important concept on the process side.

The other part of the design process where the concept is being explicitly followed is the man/machine interface. "There is a direct translation of dialogue controls and control elements to menus, screens and navigation facilities provided by the physical environment. The translation is aimed at a one-to-one correspondence between the logical and physical elements." The menu selection screens are based on the user roles and show the business requirements the user can trigger. The structure of the menu screens for the selected business requirements to each other is shown in the Menu Structure Diagrams. The structure of the dialogue screens is shown in the Dialogue Structure Diagrams. The advice for the design of the man/machine interface explicitly states that the logical groupings of the data elements in the dialogue designs—the LGDEs—*can* be implemented directly as physical data groups in the screen layout design. The navigation around the dialogue screens is based on the LGDEs. When this is possible then this *should* be done. Notice the "can" and the "should". Both are the correct recognition of what should be done. But, like the data side, there is still no statement as to why it is sound to follow the concept of logical design = physical design.

The reason is that if the logical design undertakes *all the aspects of the logical design specification*, that is, data structure, data access and data process, *in due proportion to their significance*, then the logical design is balanced and, if followed directly in the physical design, then the physical design will also be balanced and, *a priori*, efficient. This can be proved, and is shown to be so with examples in sections 2.4.2.3/12 and 3.1.3. *Support for the concept is therefore partial and mechanistic rather than with understanding.*

The main problem with this concept for SSADM is the total lack of support for the concept to test the logical design for efficiency, which is a natural sequitur of logical design = physical design.

2 *Event level processing* This concept is only partially supported in logical design and not supported at all in physical design. A computer system

contains application programs that execute logic for a given event to "do something", such as Create Invoice, Make Payment. These events are business requirements. The application logic is justified, written and executed because there is an event level business requirement. Business requirements are ultimately re-expressed as physical application programs. Given that business requirements are identified, defined and undertaken at the event level (one only undertakes a business requirement because of the need to "do something") and the concept of logical design = physical design, application programs themselves should also be designed, developed and executed at the event level. The event can be then be decomposed into problem-to-solve sub-processes, which become application program modules, but the processing itself is event triggered. This simple rule of event level application programs and problem-to-solve modules is also in line with the concept of logical = physical.

Given the nature of events in batch and online processing, application programs should not, repeat not, talk to each other. Modules yes, programs no. As Yourdon correctly states, "programs are schedulable, modules are callable". The scheduling is either *ad hoc* for online *or* regular for batch programs. The calling is of modules calling each other as appropriate within an application program.

Version 4 recognises this concept in logical design through the EAPs for the enquiry business requirements and the ECDs for the update requirements. Both are at the event level. Enquiries are, of their nature, event level processes (we will overlook the fact that SSADM allows enquiries to be at the function level and that a function can contain many events!) and ECDs are explicitly stated to be the effects of an update event on the logical data model. Both the EAPs and the ECDs are the inputs, in their respective conversions, into event level Process Models.

However, it has to be said that the inputs into the Enquiry Process Models are not consistent. The EAP forms the input data structure and the I/O Structure forms the output data structure for the EPMs. The EAPs are at the event level, but the I/O Structures are at the function level, and a function can contain many events.[1] No advice is given as to how to merge them when the I/O Structure represents many events.

The real problem comes when the application programs are designed in Physical Design. The programs are obtained from the functions. "Each function will become an application program or a run unit of several

---

[1] Functions are variously defined as "the processes required to respond to an input and the output generated by those processes"; "an update function may also be thought of as a group of events, which the user wishes to schedule together"; "a function is a set of system processing—all the processes that must be invoked to handle the data on the input flow". "Functions are identified from the bottom level of the Required Systems DFDs by taking each input dataflow from an external entity and tracing its path through the bottom level process or processes that must be invoked to deal with the input and to generate the output." These definitions indicate that a function can contain many events.

programs". Thus an application program can contain the logic for many events. The inefficiency this causes is discussed in sections 2.4.2.3/13.

Users work at the event level, they always have and always will. Certainly at the physical level, when the users do not require rapid response to an enquiry or update, the designer may decide to group event level processes/application programs as a batch run for reasons of system convenience and performance, but users do not think or work in terms of batches of processes. It is also true that the user can, within an event, have many interactions with the computer system. This is particularly the case with windowing technology. This technology is the latest in the ability to have a "conversation" with the computer when undertaking the processing for an event, but the initial trigger is because an event has occurred.

The author believes that the reason why SSADM has still not correctly identified and therefore does not fully support the concept of event level processing, certainly in physical design, is that no advice is given as to how to decompose processes in the DFD technique and then to use the logical event level processes as the basis of the application programs. SSADM version 3 offered not very helpful advice: "Decompose until a satisfactory level of detail is reached". Version 4 avoids this problem by not offering any advice at all in the description of the DFD technique, merely confining itself in the product descriptions to the statement that the lower level processes contain "more detail as appropriate". Equally unhelpful. How to decompose processes in a consistent manner is discussed in section 3.2.1.

3 *Events are standalone* Business requirements/events do not impact on, are distinct from and occur at different times from each other. Such events are not causally related. Event A does not cause/trigger event B does not cause/trigger C.... This is particularly the case for data retrieval events. Examples of this are shown in section 2.2.2.3.

SSADM supports this concept, but once again only implicitly. There is no advice, for example, that when drawing event and higher level process DFDs, the processes should be separated by datastores, this explicitly recognising that the processes are distinct in time and that the data they process requires to be stored "for the duration". It so happens that the techniques that define event level processing keep the processing logic of the events separate, but the reason for this is not given. The techniques are the ELHs, ECDs, EAPs and the Process Models.

Because no advice on process decomposition is given, SSADM is not able to show that if an event requires to be decomposed further into its constituent problems-to-solve then the problems-to-solve must occur at the same time that the event level process occurs and therefore are not separated by datastores. As illustrated in section 3.2.1, they all occur at the instance in time as the event, are causally related and are therefore connected directly by dataflows.

4 *Business requirements = events and event level processing*   Both of these concepts are supported in SSADM in large measure in logical design, although not supported in physical design. Support is therefore partial. If one looks at the techniques that are used to specify database processing, the EAP and ECD techniques and the resultant Process Models, they are explicitly stated to be at the event level. This is to the good. The problem area in logical design is in the I/O Structure technique. An I/O Structure Diagram is produced for each function and SSADM explicitly states that functions can contain many events. Therefore parts of logical design are at the event level and parts are at the function level—the database part at the event level and the human/machine at the function level. The problem of pitching processing at the function level continues into physical design. Application programs are pitched at the function level. All very confusing!

2.3.2.2 *The structural strengths*

The structural strengths are:

1 *The "cascading waterfall" sequence of modules, stages, steps and tasks*   All structured methods supporting a single data processing environment have adopted this top-down "start at the high level and decompose to progressively greater levels of detail" approach. With SSADM there is the classical threefold division of the stages into feasibility, logical design and physical design. The merits of the cascading approach are that it:

- enables the design techniques to be applied in an orderly sequence, so that the output deliverables of one technique can be used, if appropriate, as input to another technique. Each technique can therefore add design information value to the deliverable of the preceding technique;
- provides a firm set of procedural guidelines for the inexperienced practitioner to follow;
- provides a definitive set of points for the quality assurance function to be practised against known deliverables;
- provides boundaries between what are widely accepted as different skills groups—feasibility analysis, logical design and physical design. Few people have a fully comprehensive set of skills across more than two of the groups. It is therefore easier to delegate task responsibilities to different peoples across the groups;
- each subsequent detailed stage is within the context of the previous stage. This means that ultimately system implementation will be within the context of the organisation's business strategy.

2   *The progressive construction of the major deliverables*   There are three threads in the construction of the logical design deliverables: the data component, the database processing component and the man/machine interface component. In version 4 the data is constructed in a "big bang" approach, a logical data model being built from the combined application of the version 3 LDST, RDA and CLDD techniques. The first technique built a logical data model on the basis of the business requirements, the second built another logical data model from the analysis of the data attributes and the third technique built a logical data model as a composite of the first two models. It was all progressive. Not any more. The data side component of the logical design is weaker in this regard in version 4.

There is, in contrast to version 3, a progressive building of the other two threads of the logical design. The database processing is built up beginning with the ELH, then the EAP and ECD techniques and finally refined by the respective Enquiry and Update Process Models. The man/machine interface is begun with the I/O Structure Diagram, which is then converted into a Dialogue Structure, along with a Menu Structure Diagram.

There is thus a continuity in these last two sets of deliverables from the beginning to the end of logical design. This continuity provides both a quality assurance trace as well as a trace back to the initial business requirement the deliverable is supporting. Examples of this trace can be seen in the Do This and the Display Customer Information business requirements in figure 2.4.

2.3.2.3 The technical strengths

This is where SSADM version 4 wins. We have seen that SSADM is conceptually weak, with none of the concepts appropriate to the data processing environment for which SSADM is designed fully supported, and it is therefore an unstable method. This weakness is substantially offset by the information design techniques which, for the most part, are excellent. This is in contrast to Information Engineering, which is conceptually strong, but with some techniques not as rigorous as those of SSADM.

Certainly the SSADM techniques are rigorous. This is reflected in that, if there is a weakness, it is that *the method tends to suffer from sins of omission, not commission*, as we shall see. What is there is overwhelmingly good. *It is not what is there but what isn't there that is the problem.* The strengths of the design techniques are as follows.

1   *Logical Data Modelling (LDM)*   The technique has been improved in that the diagrammatic notation identifies additional data modelling features, but has, unfortunately, lost two important underlying strengths. The technique has been enhanced as regards data modelling, but has been weakened as a technique for producing a deliverable for database design.

SSADM Strengths

LDM is quick and easy to apply and produces a diagrammatically based deliverable that is easy to understand at an early stage in the life of an SSADM project. Credibility in the usefulness and viability of SSADM can be built up with users unfamiliar with a structured design method. The parallel application of dataflow diagramming, which also produces an easy to understand diagrammatic deliverable, reinforces the method's credibility, particularly as the two deliverables are complementary—the one showing the static structure of the data and the other the dynamic flow of the data.

The LDM technique is a substantial visual upgrade from version 3. The most noticeable upgrades are the use of semantic descriptions of the relationship(s) between entities and the diagrammatic representation of the mandatory and optional nature of the relationships through the solid line for the mandatory relationship and the dashed line for the optional relationship. LDM also includes the facility of entity sub-types.

The LDM model does not record all the semantic information about the relationships between data. Semantics is a "buzzword" meaning nothing more than the description of meaning. Semantics is becoming increasingly important to post-relational file handlers and query languages, particularly as used in expert systems and object oriented databases. This trend and its supporting design techniques and physical technology are discussed in chapter 5.

The semantics of data occur at two levels—inter-entity and inter-attribute. Inter-entity is the relationship between entities. The semantics need to describe the relationship between entities both ways—from master to detail and detail to master. Thus Customers "place" orders, which are "received from" customers. The LDM is the correct place to record inter-entity semantics, and SSADM does so. The relationships are given a descriptive title appropriate for the purpose of the relationship.

When the semantic net and class object model facilities are described it will be seen that there are, in fact, three kinds of inter-entity relationship semantics: the business semantics, the class semantics and the aggregation semantics. SSADM describes only the business semantics. All of these additional data modelling facilities are described in chapter 6.

It is the inter-attribute relationship that is not given a semantic description of any kind in SSADM. The facility is used to define facts, not data, and is used by expert systems technology. It is not suitable for traditional database technology and SSADM is quite right not to address this aspect of data modelling. Given the above, the use of business semantics to describe the relationships between entities is entirely adequate.

There are some facilities in LDM that represent a tidying up of the data modelling techniques. They were implicit in version 3 but are now explicitly included in the technique—the recognition that the attribute values may be mandatory or optional, the domain facility, the candidate key

Strengths and Weaknesses

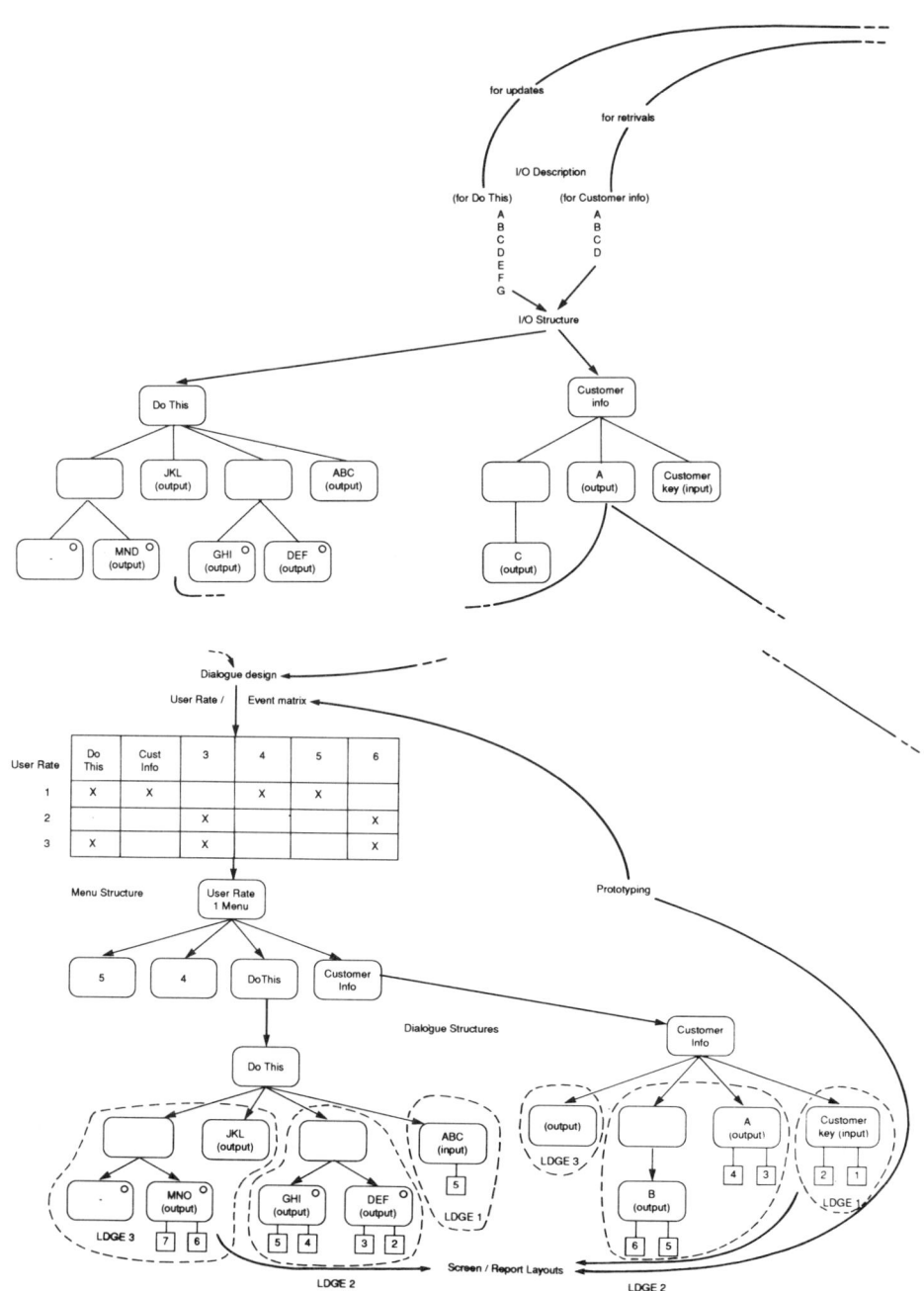

Figure 2.4 Product/technique sequence

SSADM Strengths

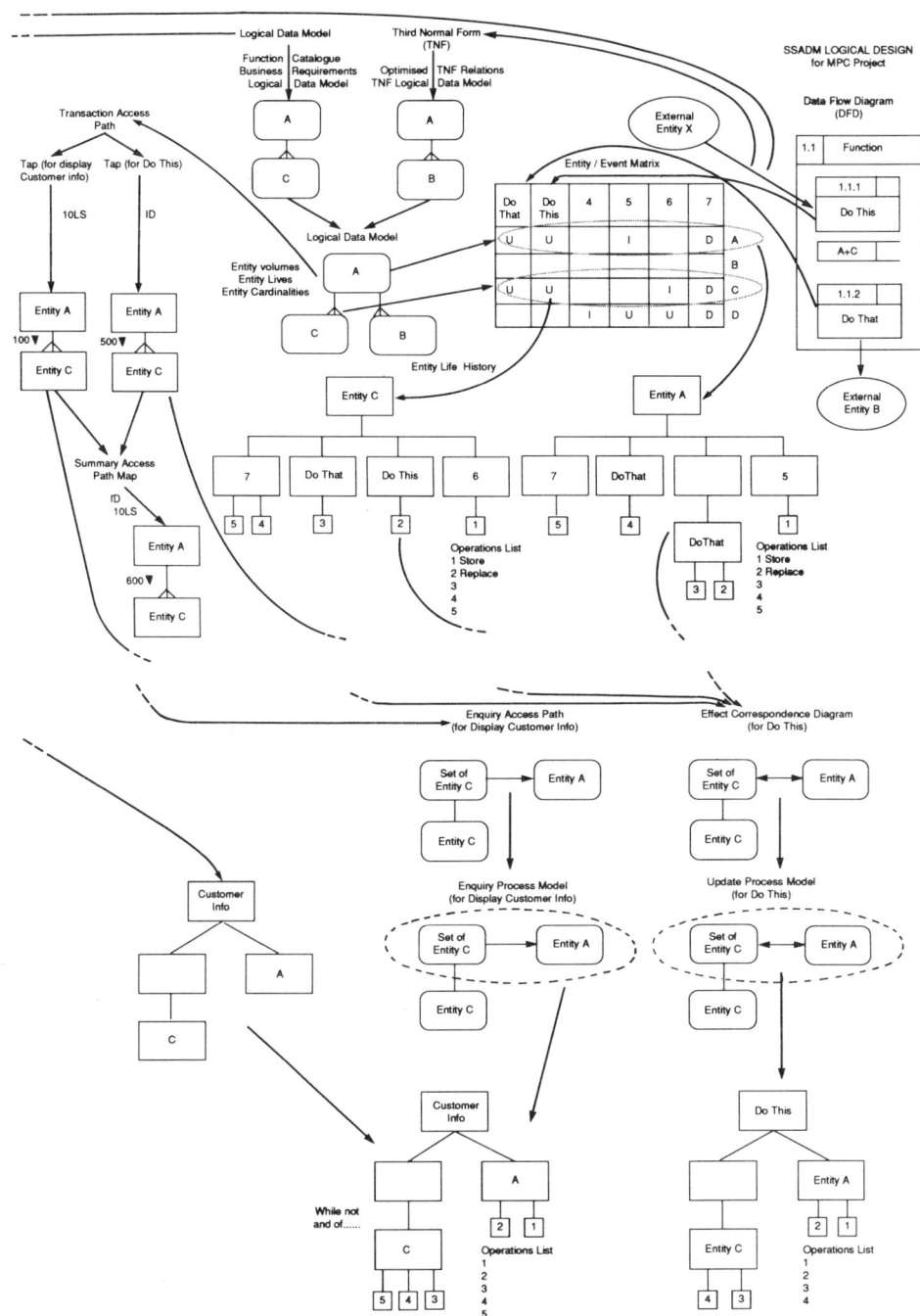

Figure 2.4 *continued*

and the transferable relationship. These are useful enhancements to the understanding of data modelling.

It has been said that the LDM model has more information than the current database technology can support, and that the technique is therefore unnecessarily complex. It is argued that the file handlers are not able to document and even less make use of the semantic descriptions and annotated conditions of the relationships between entities and cannot support exclusive relationships, so there is no need to record this information on the logical data model.

This mismatch of the logical and physical data modelling techniques and technologies is becoming less and less significant, with the ability of many file handlers to record exclusivity and inclusivity as database rules ("fired"—another buzzword—when specified business conditions occur on entity access) and triggers (fired automatically whenever the entity is accessed). New data access languages (and this does not include SQL) are now also able to make use of entity relationship semantics. Instead of having to use the = sign in the WHERE clause of SQL to access multiple tables these semantic based languages can use the relationship description. Thus a user could code (using SQL semantically) SELECT whatever WHERE Customer places Order AND "x","y". The "places" is the semantic description of the relationship Customer to Order. The argument is rejected.

The technique had two great strengths in version 3—it validated the data model against the data retrieval business requirements and, above all, made a limited attempt to test the model for logical design efficiency with the operational master entity. Version 4 is not sharp in the first strength and has abandoned the second. This is because the technique of transaction access path analysis of version 3 of the method is no longer used and the replacement techniques—the EAP for the retrieval and the ECD for the update business requirements—do not count the number of accesses and can therefore ascertain only *an* access path and not *the* access path when there are alternative access possibilities. This is most unfortunate and is discussed further in section 3.1.1.

2 *Dataflow modelling* Dataflow modelling is the technique of dataflow diagramming (DFDs) and the supporting documentation of the Elementary Process Descriptions. The great benefit of the technique is that it produces early in the method an easy to understand diagrammatically based deliverable that identifies the business processes (functions in version 4) that update the database. The entire scope of an application can be seen in the level 1 DFD. It is therefore an ideal user aid. The DFD can enhance the credibility of SSADM, in that the users can see something for their work at an early stage in the life of an application design project.

The technique is complete, except for the total absence of advice for process

SSADM Strengths

decomposition. This weakness is easy to rectify with the "tricks of the trade" proposed in section 3.2.1.

SSADM is superior to many other methods in recognising that data retrieval business requirements have no place in DFDs. DFDs monitor the flow of data around the enterprise. The flow occurs because the data changes its state in the database through a modify (insert, update or delete) of some kind. Data does not flow/change its state on a retrieval. Retrievals produce a copy of the data. The database is accessed for the required data and a copy is moved to the buffer pool. Data *per se* is not moved. Other retrieval business requirements can continue to access the database for the same data. The inclusion of data retrieval business requirements clutters up DFDs to no advantage. DFDs are intrinsically only concerned with data maintenance business requirements.

The SSADM manuals explicitly recognise this. The method advises that retrievals, particularly *ad hoc*, are best documented in the Requirements Catalogue and used to validate the logical data model, not the DFDs. By implication, the DFDs only record the data maintenance business requirements.

DFDs still play a significant role in version 4. They produce the I/O Descriptions that are the inputs into relational data analysis and I/O Structures, and are the basis for the initial identification of functions. They are also used to identify events, events being components of functions.

Extremely good advice is provided for the treatment of transient data stores. The full advice is not repeated here. One point, however, struck the author as particularly valid. A transient datastore may be valid even in the logical DFD because the required system includes some physical constraint, and may therefore need to store the data for some time. The author has never come across this, but sensibly this could surely occur.

3 *Function definition—I/O structuring* There are two reasons for the two-part title. The function plays a vital part in SSADM version 4 and much advice is offered for the identification and specification of functions. There is also a technique detailed in the section of the manual called I/O Structuring, an I/O Structure Diagram being produced for each function. This technique takes the inputs and outputs of a function and represents them in a structured way using Jackson-like diagramming. There is therefore a procedural component for the identification and definition of functions and a technique component to structure the information used by a function. The procedural component will be addressed first.

The universal function model is attractive. It makes a clear distinction between the three types of error processing: the control, the syntax and the integrity errors. The first two types of error are processed as part of the input error processing and the third is processed as part of the output error processing. This grouping is also able to support the move towards client-server co-operative processing, with front-end processing on client intelligent workstations prior to back-end server database processing. The front-end processing would be for editing that the data being input is complete (the control) and correct (the syntax) and the back-end for database correctness (the integrity).

Control processing is the checking that the input data matches the control totals specified. This is most often used for batching the input data and is a widely used mechanism for ensuring the integrity of the input data batch. Syntax checking ensures that the data values are valid, such as numeric fields containing no alphabet characters. The data editing is thus of the universal type, such as all numeric data should be just that, numeric. What cannot be done is the checking that the input data values are valid if the checking requires relating them to other data item values in the database. This requires to be done by the subsequent database processing component. This data editing is also of the application type, it being specific to the application and not universal. Neither of these two points are made in the manuals.

In a sense the input process component of the function represents the use of intelligent workstations, checking that the data is as valid as possible before being passed to the main database processing component on a back-end processor. Integrity checking relates to ensuring that the data being entered/updated against the database is correct, typically regarding referential integrity, and final data item value editing.

The I/O Structuring technique is fine. The technique is designed to show the sequence, selection and iteration of the groups of data items, the elements, being input and output to/from the function as units of information processing designed to assist the user. The elements reflect the entities of the logical data model, although this is not stated in the manuals. An element can contain more than one entity where is no selection or iteration of the data items. The data items are obtained from the I/O Description of the dataflows that cross the system boundary, from the external entity to the function process.

SSADM is correct to state that if the function is to be processed as a batch program then separate I/O structures should be produced for both the input and the output.

If the business requirement is a retrieval the I/O Structure is also used as the output data structure for the Enquiry Process Models. It is also the direct basis of the Dialogue Structure Diagram. The purpose is to take the input and output data items for a function and build them into a Jackson-like data structure diagram.

4 *Relational Data Analysis (RDA)*   This technique is perhaps more widely known as third normal form (TNF) data analysis. This technique of building a logical data model based on the analysis of the raw data items has evolved from the ideas of Dr Edgar Codd, widely recognised as the father of relational technology and the creator of the TNF normalisation rules. The technique

identifies the relationship between the data items to be used by the proposed system to selected prime key data items and progressively rearranges the data items into relations (entities in LDM) through the application of three "normalisation" rules—first, second and third normal form. Once the relations are in third normal form it has been considered that there is no further relationship between the data items in the relations other than "to the key (1NF), the whole key (2NF) and nothing but the key (3NF)".

The strength of the technique in SSADM is that it is applied up to third normal form (TNF), which many regard as the "standard" degree of the normalisation of data into relations/entities. There are no deficiencies in this regard.

In recent years it has been proposed that higher powers of data normalisation are required, the higher powers going up to sixth normal form (6NF).

The first three steps of normalisation as defined by Dr Codd are widely understood and accepted. Unfortunately there is disagreement as to what the higher powers are. This disagreement is reflected in that the different versions of the Information Engineering structured method (IEM from James Martins and Associates and Navigator from Ernst & Young) and the author have different definitions of these additional normalisation steps. IEM includes the identification of entity sub-types in the definition of 4NF. To the author 6NF identifies entity sub-types. Navigator does not include the normalisation step for entity sub-types, be it 4NF or 6NF.

From a practical point of view the first three normalisation steps used by SSADM are adequate in most application cases. Relations in 3NF also conform to 4NF and 5NF, but anomalies occur when the key of the relation contains multiple attributes. Further normalisation can be taken to decompose the 3NF relations into further optimum relations. The three additional normalisation steps are described in chapter 3.

Assuming 6NF is concerned with entity sub-types then 4NF and 5NF are, quite frankly, of more academic than practical relevance, given current file handler technology. It has been pointed out to the author that some relational file handlers, such as DATACOM/DB, can now support entity sub-types in the database schema definition, although the entity sub-types are still stored as separate tables. The author would disagree, in that there is no support for property inheritance, that is, the entity/table sub-type automatically inheriting the data properties of the super-type entity/table. It is still up to the application programmer to access explicitly the super-type of the sub-type(s) so as to obtain the super-type data. Property inheritance is discussed fully in section 6.6.1.3. Object oriented file handlers are more capable in that they do to support property inheritance.

It is pleasing to see that the technique of building a relational data model from the TNF relations is still preserved. The construction rules are the same as in version 3 of SSADM. Additional advice regarding such finer points as deep hierarchies of relations, recursive relationships and false relationships add further to the quality of the technique. Advice, although no worked example, is also provided as to how to compare the relational data model and the LDM, so as to produce what is effectively the version 3 Composite Logical Data Model.

As discussed in section 2.4.2.3/2, the results of relational data analysis often produce a LDM that is different from the one built on the analysis of the business requirements in the LDM technique. The author therefore applies the technique of building a relational data model quite separately from the LDM technique, as it used to be in version 3 of the method. *The "awesome" rigour of data modelling in SSADM identified by an expert in Information Engineering can still be preserved.*

5   *Enquiry Access Paths (EAPs)*   This technique identifies an access path for a data retrieval business requirement against the logical data model. It is therefore pitched at the event level. It shows the access path in the form of a Jackson-like structure. For the purpose of showing an access path for a retrieval event as a structure of sequence, selection and iteration of the entities being accessed the technique is fine. The technique is excellent for showing an access path in Jackson-like form.

There are certain diagramming features that aid understanding. The arrows show the direction of the access path and the accesses between entities. One can therefore ascertain the order of access. The relationships between the boxes where there is no arrow show which parts of the diagram represent Jackson-like structuring rules. There is therefore a clear distinction between access and structuring.

6   *Entity Life Histories (ELHs)*   This is the only technique specifically designed to handle the concept of time. It is therefore relevant to data maintenance events for an entity, since the events that maintain an entity require to be sequenced, inserts before updates before deletes.[1]

The technique is complete. SSADM and its compatible methods are pre-eminent with this technique for sequencing data maintenance events that update the database. Other methods, such as Information Engineering, have techniques which attempt the same task, but fail significantly to provide the full range of facilities. The Information Engineering technique, Entity Life Cycle Analysis, does not include parallel lives, quits and resumes, random event off-the-structure boxes or state variables, and is therefore a pale shadow of the ELH technique.

There are explanations in version 4 that have not been included in SSADM before and therefore further enhance the technique:

---

[1] When discussing batch application program design version 3 shows that occassions can occur where the sequence of insert, update and delete can require change. The most frequent situation is to process all the transactions for an entity together in the sequence of delete, insert and update.

SSADM Strengths

- It may be that once a random event and any subsequent events have occurred, there is a need to revert back to the main entity life. This can be done with the quit and resume facility.

- The death event of a master entity must record whether the entity is affected by the death of all its details or not. It may also be necessary to record in the detail entity ELH that the entity dies when the master entity is deleted. This is including the referential integrity rules for table deletion in the database in the ELHs. The method is the better for it.

- Effect qualifiers, that is, the same event may affect an entity in different ways depending on circumstances. When a house is bought by a public agency there may be a tenant in occupation or not. The processing will differ depending on the tenant being in occupation or not. The effect qualifiers are identified with soft brackets for the optional events Buy House (with tenant) and Buy House (without tenant).

- The major change in the technique is the adding of processing operations. The operations are those that are unique to the data maintenance task, that of maintaining the attributes and the relationships to other entities if the entity is being inserted or deleted or if a foreign key in the entity is being updated. The operations to do with database processing, and therefore also available for the retrieval events, are not included. The database processing operations are added to the Process Models, which are based on the ECDs and EAPs. The Process Models operations are therefore consistent, whatever the event type. Putting the specific data maintenance operations to the ELHs ensures this consistency.

7 *Effect Correspondence Diagrams (ECDs)* ECDs are to data maintenance events what the EAPs are to data retrieval events—they show an access path the event takes against the logical data model. The manual phrases it another way—"ECDs provide the detailed system processing by defining the effects on entities for a single event occurrence". The wording is different, the effect is the same. It shows an access path. Also like EAPs, ECDs are pitched at the event level.

The technique is fine for the purpose of showing an access path for an update event as a structure of sequence, selection and iteration of the entities being accessed. The technique is excellent for showing an access path in Jackson-like form.

The author finds the wording that ECDs show system processing strange. The database processing operations of the event are not detailed on the ECD, but on the subsequent result of the ECD, the Update Process Models (the UPMs). The UPMs show the system processing, not the ECDs.

8 Process Models The Process Models are primarily concerned with database processing, that is, the logic that can be executed before data is input to the database and after access from the database. Other processing, such as the handling of errors in the data, is also addressed. It is not concerned with the logic for the man/machine interface, that is, in dialogue design.

Process Models are constructed for both the data retrieval and the data maintenance business requirements. The technique for both types of event is essentially the same, that of taking the sequence, selection and iteration of the data groups being input and output of the data structures, merging them where there is a correspondence to form a process structure and adding operations to the sequence and conditions to the selections and iterations. However, there are underlying differences in the characteristics of data retrieval and data maintenance events. There are thus two sub-techniques: Enquiry Process Modelling (EPM) and Update Process Modelling (UPM).

The EPMs are based on the EAPs and the UPMs are based on the ECDs. Both Process Models are therefore pitched at the event/business requirement level. The EPMs have a further input component, the output data item elements of the I/O Structure. The EPM is therefore built up from two sources—the input data from the access path for a business requirement from the EAP and the output from the output data from the man/machine interface of the I/O Structure. The UPM, however, is only constructed from a single source, the access path from the ECD. There is therefore a mismatch between the sources of information for what is essentially a common technique for defining the processing logic of business requirements.

However, the mismatch can be explained, although this is not done in the manuals, which only state that if the output structure component of the I/O Structure for the update event is non-trivial then it too can be input into the UPM.

Data retrieval business requirements tend to use very little input data but produce large volumes of output data. That is their nature. Data maintenance business requirements are the other way around. They input large quantities of data and the output is in the form of an indicator saying that the input was successful or an informative error message. Any structure to this output is trivial. There is no iteration or selection and probably only one data item, the indicator or message. There is thus no point in having an I/O Structure as the output component of an Update Process Model. The explanation of the mismatch of the EPMs and the UPMs is valid.

The input component to both types of Process Model is obtained from the access paths, that is, the EAPs and the ECDs. The output for the EAPs is obtained from the I/O Structure, with the input data group elements of the structure "lopped off". There is thus one source data structure for the ECDs but two source data structures for the EPMs, which have to be merged. These two structures may not have total correspondence and the "structure clashes" require to be resolved.

The other difference between the enquiry and update events is that the operations list for the updates is more extensive. This is to support the additional tasks, such as updating the attributes and creating and deleting the entities. These distinctive features of the event types goes directly back to their different characteristics, which SSADM correctly identifies and models.

Once the Process Model for an event has been created the operations (statements of sequence) and the conditions (statements of selection and iteration) are added.

There are two types of logic—access logic and process logic. The logic can be procedural or declarative. The Process Models are only concerned with procedural logic. Declarative logic is addressed in chapter 5.

There are 12 commonly used constructs for access and process logic. Of the access logic constructs there are those for record-at-a-time processing and those for set processing. Pre-relational file handlers only use record-at-a-time access. Relational file handlers use both types. The record-at-a-time access constructs are Read, Write, Update and Delete. The set access constructs include Select, Union, Difference, Intersect and Divide. There are other constructs, such as Product, that the author has never seen used and are not included. The process logic constructs are sequence (Move A to B), selection (If...Then...Else), Iteration (Do While and Do Until) and branching (Go To).

SSADM distinguishes between operations and conditions for the specification of logic. The conditions relate to selection and iteration. By default, sequence relates to the operations.

It is pleasing to see advice being offered as regards the use of the BEGIN and COMMIT entity occurrence locking constructs for the control of concurrency/success units in database processing. This is a much underrated task. There are a few suggestions that have been found to be useful in operational systems.

- An update process may require to update many entities. It may be necessary to include several success units in the process, particularly if there are several updates, followed by a screen dialogue, followed by some more updates, followed by....

- The use of the locking facility for enquiries where there is a need to produce a total or partial database count regarding some criteria. Care needs to be taken, however, in that the other users will not be too pleased at being locked out on a X exclusive lock.

9 *Enquiry Process Models (EPMs)* These are based on the EAPs for the input component and the I/O Structures (with the input element lopped off) for the output component. It is a simple matter of merging them into a process structure, and where they do not match, resolving the "structure clash". The technique is excellent in showing the operations and conditions

to be applied to each entity to be accessed and processed for an event, the sequence, selection and iteration to be applied to the entities for the event.

10   Update Process Models (UPMs)   These are based on the ECDs for the input component for the reasons just explained. The UPMs do not suffer from the mismatch and structure clash problems of the retrieval business requirements because the I/O Structure diagram does not require to be input to the UPM. The reason for this is clear and SSADM is right to adopt this approach. The output from an update business requirement is minor. It could be merely a switch indicating that the update has been successful or an error message if the update has not been successful. Either way, the data structure of such output is trivial. The technique is excellent for the same reasons as for the EAPs.

11   Dialogue Design (DD)   Dialogue Design is in two parts: the menu selection screens and the dialogue screens. The menu selection screens are a cascade of screens for the selection of a business requirement and the dialogue screens are a sequence, iteration and selection of screens for the display of the input and output information pertinent to the selected business requirement. The menu screens are based on the user roles and the dialogues on the functions.

The technique has a number of attractive features:

- The man/machine interface (MMI) specification is based on the same technique as the specification of database processing. At the end of the day what one is doing is defining the access and process logic required for the MMI using Jackson-like structure diagrams, just as one is defining the access and process logic required for the database data. One is specifying logic in both cases and it is nice to see the same technique, albeit under two names, Process Modelling and Dialogue Design. As far as the author is aware, SSADM is the only method to recognise the similarity of what one is doing and matching the two with one underlying technique.

- The technique is extremely easy to apply. It is merely a case of taking the I/O Structures and lassoing the data group elements into the Logical Grouping of Dialogue Elements (LGDEs), the I/O Structures thus becoming Dialogue Structures.

The design of the dialogue and menu screen formats are undertaken in stage 6, Physical Design. This is entirely understandable given that it may well be that it is only in stage 6 that the screens may be identified to be dumb or intelligent terminals and designed according to MMI standards, be it *de facto* and product-specific, such as IBM's Presentation

SSADM Strengths

Manager and Microsoft's Windows/3, or open standard, such as OSF's MOTIF.

The technique is integrated with the design of the menu selection screens, which are based on the user role. The dialogues are based on the vertical access of the user role/function matrix and the menu screens on the horizontal access. The vertical axis is the function and the horizontal axis is the user role. They are therefore different views of the same information.

The choice of user role for the menu screens is sensible, *provided that the role is logically based on version 3 functions*, that is, "logically related group of events". If the user role is based on today's organisation of an enterprise there is the danger that the roles will change if there is a re-organisation. Such a role could be Section 123 head or Site Manager. This danger can be mitigated by making the user roles logical, that is, sensible to the task being undertaken and independent of organisation. For example, the role of warehousman or accounts clerk is descriptive of a task and not an organisation. It is a pity that such advice is not offered.

The recognition of the need for command structures for representing the path options the control logic can take to other dialogues for the user role when a dialogue terminates is a new feature to SSADM and is to be welcomed.

The advice in the annex on such aspects for the style guides is very helpful as a starter. A more detailed section on the ergonomics of the man/machine interface would be most welcome. Ergonomics is a skill for the most efficient use of the computer by the user and is therefore an essential part of modern systems design.

12 *Prototyping* This is now explicitly addressed in version 4. This is positive, because the more modern database products include excellent facilities for quickly developing parts of a proposed system to see how it works and performs from the designer's and developer's viewpoint, as well as what it looks like from the user's viewpoint.

However, prototyping is addressed only briefly. The full scope of prototyping is to be addressed in the Prototyping Interface Guide, which has yet to be produced. Core SSADM only considers specification prototyping, that is, the animation of the man/machine interface/Dialogue Design and report output aspects of the Requirements Specification. This will show how the system will look to the user to confirm the specification of the business requirements. As a result of the prototyping the man/machine interface design deliverables, such as the functions and I/O Structures, the Dialogue Diagrams, Menu and Command Structures, may require modification. Prototyping is not concerned with creating a working model that could be enhanced into a part or the whole of the required system.

The advice as to what applications to prototype and not to prototype, how to manage and log prototyping is very good.

13 Physical Design Version 4 recognises the changes that have taken place in application development technology with the introduction of fourth generation language (4GL) software. SSADM has long recognised the widespread adoption of relational technology but it is only with version 4 that the adoption of the new 4GL software toolset and its impact on application software development is recognised. Unfortunately the manuals do not state what 4GL software is.

Before 4GL software is described some points need explanation. There is widespread use of the terms non-procedural and procedural. Non-procedural means that one does not have to specify the logic for a particular task. For example, a menu driven query language is non-procedural in that one merely has to define what one wishes to access by defining the attributes to access, and the search argument and the data access logic is generated automatically. "For a given value of this go and get this, that and the other" is all that needs to be defined. The appropriate access logic is automatically generated. A cursor based screen painter is non-procedural in that one does not have to specify the row and column where the literal or variable data is to be displayed. Just move the cursor to the point. The screen co-ordinates are automatically generated. Procedural software requires that the logic for accessing this, that and the other and the positioning of the literals and variables would be required.

Procedural logic is, and will always be, required for process logic, that is, for the constructs of sequence (move A to B), selection (if . . .then . . . else. . .), iteration (do while and do until) and branching (go to).

4GL software contains:

- A programming language with all the logic constructs for accessing and processing data. The language thus has no limitations for data manipulation. It will also contain a large set of built-in functions for generic processing, like editing and formatting of information. The access constructs can be record-at-a-time or set access. Record-at-a-time access logic is always procedural; set access is (allegedly—see below) non-procedural. The set access is usually with the SQL data sub-language—and is regarded by SSADM as non-procedural. The process logic constructs are procedural.
- A cursor based screen painter, which can also contain data editing and formatting/display built-in functions. This part is non-procedural. The more sophisticated screen painters include the ability to define logic for the attributes on the screen. One writes the logic for the business requirement as one "paints" attributes on the screen.
- A cursor based report writer, with a more comprehensive array of built-in functions for the formatting of data. This is non-procedural.

SSADM Strengths

- A command driven query language. Invariably the language is SQL for the relational file handlers. SQL is claimed to be non-procedural. The author disputes this. One has to specify the tables to be accessed, one has to know which attributes belong to which tables and one has to specify the relationships between tables when accessing more than one table. In short, one has to know very precisely the structure of the data in the database. The claim that the user view facility is the answer to this problem is also open to question. A view is nothing more than a retrieved set of table(s) of data, and a user can have multiple user views, so still requires to know which view table contains which data. Furthermore, a view can be a very dangerous facility in novice hands. If a subset of a view is accessed the data returned is not necessarily the data requested. This can be proved. This book is not about the strengths and weaknesses of SQL and the danger of accessing view subsets is therefore not demonstrated. The ability to have a SQL command based query language should be regarded as a tool for the expert only.

- A menu driven query language. This has the same functionality as the command based query language, but the user merely has to select from a menu which information is to be accessed and presented. It is therefore non-procedural. The menu format can be a list of data items and tables from which a selection can be made and given search values where appropriate, or in the form of the screen or report format for the query, with the data items to be searched on merely filled with a value.

Both types of query language will have default screen formatting facilities, the format depending on the data being presented.

The SSADM manuals mention an application generator, but without explaining what it is. There needs to be an explanation of the difference between an application generator and a code generator. The two are quite different.

An application generator is the set of software components described above under the heading of 4GL software. As already described, this software has a range of non-procedural facilities that are able to generate some code, but it is primarily designed to aid the writing of code for an application program. Code generation is limited to the database definition, man/machine interface and database access. For example, the cursor based positioning of literals and variables on the screen automatically generates the screen column and line positions. If one uses a menu driven query language then the selection of the attributes and the search values generates the SQL DML access commands. The SQL DDL code for the definition of the database design is obtained directly from the LDM and the attribute definitions.

The part for which there is no code generation within application generators is process logic. This is much the larger portion of the code for the application being developed, and the claims about the major savings in coding effort to be achieved from application generators regarding process logic need to be treated with caution. There is no short cut to the writing of process logic. Never has been, never will be!

A code generator is more sophisticated. The operations specified in the Process Models can be used by a code generator to generate, from the single source of "logical" code, N sets of physical code. One could specify that this part of the operations in the Process Models are to be in COBOL, this part to be in PASCAL and this part to be in whatever is required. Alternatively one could specify that all Process Models for batch processing are to be in XYZ code. One specifies code once, logically.

Unfortunately version 4 addresses only application generators. There is no reason why the Process Models operations cannot be converted into code by code generators. There is not yet the technology for taking SSADM operations and generating code, but it will come.

The two technologies are not mutually exclusive. What is evolving is that the non-procedural parts of the 4GL software—the screen painter, the menu driven query language and SQL—from which code can be generated will continue to be used and the procedural component—the process logic — will be code generated directly from the logical design specification. This approach is the ideal.

SSADM requires that you:

- Recognise the large array of application development facilities available by defining a Physical Environment Classification. This identifies all the facilities available for designing and developing the database and applications systems, including data storage and access, procedural and non-procedural data processing and performance.

- Define the standards for application development.

Physical design in version 4 is based on a general classification scheme of the file handler and 4GL facilities to be used and the identification of when they will be used for which Process Models. For example, SSADM recognises that parts of what can be implemented procedurally in one design can be implemented non-procedurally in another design. The example given is the use of rules. Rules are discussed fully in chapter 5. Rules can be used in traditional application design in two ways—for the declarative definition of database integrity triggers (when an order is inserted then there must be a related customer) and of business rules (if going shopping then must take your purse/wallet). These are non-procedural statements of logic and can be defined and stored in the database schema. Alternatively they can be defined procedurally as part of the process logic for an application program inserting orders and another application program for going shopping. Which approach is adopted depends on the database product being used.

SSADM Strengths

13.1 Database Design/Physical Data Design This task is rightly separated into three parts—an initial first-cut database design that is structurally legal according to the generic facilities universally used by file handlers, followed by a product-specific design, followed by a tuning exercise against the design where the disk space and transaction response time performance objectives are not being met.

There are a number of definite positives about the SSADM approach to the design of data:

- The recognition that database file handlers use the same range of data storage and access facilities. For example, they all store data as records, store the records in blocks/pages, can index records for direct access to a record occurrence, can cluster the detail record via the master record in the same page as a hierarchical structure of optimum data storage, and can store the records in the sequence of a key. Thus, the technique produces a "universal model" database design based on "rules-of-thumb" from these generic facilities of the DBMS file handlers.

  Some five rules of thumb are listed, such as the entities being represented as record types and the grouping of the entities into physical groups. As this design is generic it does not require product-specific knowledge, and can therefore be undertaken by an analyst.

  The universal design is then converted to a first-cut product-specific design. The access paths for selected transactions are then optimised where necessary to meet space and response time performance objectives.

- The recognition that the experienced database designers can by-pass the generic approach to database design and produce a first-cut product-specific design directly or even, when skilled enough, an optimised database design.

- The recognition that the optimum way to achieve performance is to group record types in the database together in hierarchical structures, called physical groups. The technical term is *clustering* the detail record type *via* the master record type, such that the detail is in the same page/block as the master. Thus, when there is access from the master to the detail there is logical I/O to the buffer pool (you cannot tune logical I/O), but minimum physical I/O from the buffer pool to the disk. The principal basis of the physical groups of tables in the database are that the groups are hierarchical structures of data *that are accessed most frequently*.[1] This clustering enables

---

[1] There is one problem with this "utopia"—SSADM no longer ascertains the number of accesses when undertaking the version 4 equivalent of transaction access path analysis with the EAP and ECD techniques. The method therefore cannot identify which parts of the database are frequently accessed. The method's ability to achieve what it says is necessary is thus strictly limited. This deficiency and how to address it are considered in section 2.4.2.3/12 and chapter 3.

the access of the data in the database to be with as few physical I/Os as possible. The busiest part of the database is the most efficient.

- SSADM advocates the principle of one-to-one mapping of the physical design from the logical design. It goes further and says when the physical design "destroys the one-to-one mapping...it is likely to make the system less maintainable, less portable, etc.". The author could not have put this part better, but would have added that it also makes the system less efficient. *This recognition of some of the benefits of one-to-one mapping is the most attractive feature of physical design in version 4.*

  Unfortunately SSADM advocates this without explaining the reason why it is a good thing to do. It explains that there will be a saving in programming effort as there is no redundant data and no need to reassemble input and output data and that the system is less portable and maintainable. Certainly these are benefits. However, the real reason is that if the logical data model is efficient and a modern file handler with no data storage and access constraints is used and the physical design is a direct copy of the logical design then the physical design will also be efficient. Unfortunately SSADM does not show how to ensure that the logical data model is efficient. How to do this is shown in chapter 4.

- The advice for database optimisation is that only where there are physical constraints in the file handler or the transaction response time performance cannot be met should there be modifications to the one-to-one mapping of the database design. There is no problem with the physical constraints—*force majeure* is always dominant. If the file handler cannot provide the required facilities then they have to be designed around. For example, INGRES does not provide the ability to cluster the rows of the detail table(s) in the same page as those of the master table. A solution is to sequence the detail table such that the rows are stored on disk in the sequence of the key of the master table. Thus, the Orders would be sequenced in the order of Customer Number.

  However, the advice that the database may need redesign to improve the performance of a transaction response time fails to recognise that the database design should be based on total data structure and total data access, such that a transaction's accesses against the physical data structure are included as part of the whole database design. The method treats transactions as individual units of processing. In the narrow sense of transaction response times they are, but SSADM fails to realise that the transactions are but part of a whole system. They are not put into a database context. "Twist" the database design and one may improve the performance of the one transaction, but one can be sure that the other transactions that access the "twisted" part will be adversely affected. *A database is like a bowl of spaghetti: pull a strand in one part and you disturb a strand in another part.*

SSADM Strengths 67

- The advice that it is necessary to create a Process Data Interface when the database design has been "twisted" so as to preserve the view of the design as a direct match of the required LDM is very much to be welcomed. The application programmer should be isolated from the database design. The application programmer can thus design and write logic independent of the data design. *The stability of the area of the greatest development effort, the writing of logic, is preserved.* Change to the database design requires a change to the PDI only. The advantages of the PDI are major and SSADM is right to require it.

- The most excellent description of database file handler technology the author has read. Whoever wrote the material is/are extremely knowledgeable, not only of the finer points, but also of the "nuances/finessing" of database design. For example, reads of the top level record of indexes can be assumed to occur at the beginning of a transaction run such that there is only logical I/O incurred on access to these records. This is an example of a fine point in optimisation. Using memory resident tables, particularly after conducting transaction popularity polls of the tables to be accessed (not mentioned unfortunately), is the hallmark of expert finessing.

13.2 *Program Design/Physical Process Design*   SSADM version 4, unlike version 3, does not include a technique for program design. This has been replaced by the recognition of the facilities of the 4GL software toolset and the adoption of facilities to exploit it. Three facilities have been introduced—the Function Component Implementation Map (FCIM), PDI and the ability to specify certain types of logic in the database or dictionary. While these facilities are beneficial and reflect a recognition of 4GL software, they are not a substitute for a technique. The issue of a program design technique is addressed in chapter 3.

Like database design, SSADM produces a classification of the physical environment and its facilities to be used for the specification of logic, so that a strategy for using these facilities for implementing application programs can be formulated. The logical process designs, the Process Models, are decomposed onto this facilities implementation map, the FCIM.

The FCIM relates the components of logic in the Process Models onto the software development facilities of the physical environment. Exactly how this is done is not explained or shown in the manuals, but the author has worked on the basis that it is a case of allocating the logic in the Process Models, as represented by the operations and conditions, to the relevant activities in the FCIM. The activities that can be found in a function include the success units (concurrency control), syntax error handling, common processes, input/output formats and screen dialogues. There are detailed explanations of each activity in the manuals. However, if activities are to be the basis of the FCIM allocation of function logic then some that should also be included are the general data access and processing of the accessed data.

Onto these activities needs to be mapped the software development facilities that would typically be used in the 4GL software environment. Unfortunately SSADM does not detail what the facilities are and when and how they should be used. It does not say that the facilities could include command and menu based query languages, concurrency control (the BEGIN and COMMIT commands to be specified around the state indicators), report writers, 3GL and 4GL programming languages and screen painters. The purpose and functionality of these facilities is not explained. SSADM does not advise, for example, that if the business requirement is a simple retrieval of information from the database that a query language with an automatic screen formatter could be used, a common routine could be coded in a high performance programming language, some data editing could be done with a 4GL built-in function (numeric check for example) and business rules could be defined as rules in the database schema. This is strange, given the excellence of the descriptions of the database facilities.

The allocation of the logic to the FCIM facilities creates "fragments" of software, fragments that can be re-used by other functions if possible. There is a move, a move recognised by SSADM, away from the facility of the module towards these fragments of software, each fragment potentially being developed by a different component of the 4GL toolset.

The Process Data Interface (PDI) is used when the database design is different from the logical data model for reasons of constraint and performance. Either the tables do not match the entities in the logical data model or the attributes are different. The PDI creates a "mask" over the database and thereby preserves the logical data model view of the database design for the application programs. This is the view of the database as seen by the FCIM. This logical view of the physical design provides great flexibility, as it means that the database design can be changed and the only "ripple effect" is to the PDI, which obviously requires to reflect the change. All the application programs are unaffected.

SSADM explicitly encourages the one-to-one correspondence of the physical process design from the logical process design. *This is a distinct positive, and reflects the concept of logical design = physical design.* Thus the Process Models and any associated Dialogue Structure Diagrams become the basis of the application programs within the framework of the FCIM. It is a pity that SSADM does not realise the benefit to be achieved, that of efficiency, this being based on the concept that the greater the match of the physical design with the logical design, the greater the efficiency of the physical design.

There is a recognition of the use of declarative logic in the form of rules, and the use of the rules for the definition of referential integrity and business constraints. Version 4 explicitly states that the conditions for the exclusive arcs on the logical data model can be specified as rules and is much to be welcomed. What is a pity is that there is no explanation of the format of the rules and how declarative logic differs from procedural code. This is addressed in chapter 5.

Use is made of the program inversion facility, where it is necessary for a program to suspend processing for a while, while the man/machine dialogue progresses, and to invoke itself and continue processing from where it left off. Where it left off is recorded with a state indicator. What a pity that the manuals do not go one step further and, using the same mechanism of remembering its progress, fully support a conversational program. Conversational programs are more powerful than program inversion because they can "put themselves to sleep" for an indefinite period and then "re-awaken themselves" at the user's request. The additional information to the state indicator is the data on the screen that is to be redisplayed when the program restarts itself.

2.4 SSADM Weaknesses

Like the strengths, the weaknesses of SSADM fall into two broad classifications—general and systematic—with the specific being further broken down into three categories—the conceptual, the structural and the technical.

It is necessary to repeat what was said at the start of assessing the strengths of the method. It has proved at times difficult to "pigeonhole" a facility or a technique definitively into a strength or weakness category and for the same reasons.

2.4.1 The general weaknesses

The general weaknesses are those that cannot be placed into one of the specific categories. They are applicable to all the stages of SSADM. There is only one and it is a virtuous one.

2.4.1.1 An excess of enthusiasm

Notwithstanding the need for thoroughness (where SSADM is excellent) and completeness (where SSADM inevitably has some weaknesses—the perfect method does not exist) SSADM does perhaps suffer from an "excess of enthusiasm" in places. The author has found that in practice certain practices/deliverables can be ignored:

- Validating the logical data model against the data maintenance business requirements.
This assessment is based on the concept that the database design should be based on the data retrieval business requirements and is discussed more fully in section 2.4.2.1. If the concept is accepted and is properly applied (making sure that the user has properly identified their requirements to access data) then it is not necessary to use the data maintenance business requirements to ensure that the LDM can support all the business requirements. The author certainly only uses the data retrieval business to validate the LDM.

2.4.2 The systematic weaknesses

2.4.2.1 *The conceptual weaknesses*

SSADM is found wanting against many of the concepts detailed in section 2.2. *Given their importance these are significant failings in SSADM.* The concepts not supported are as follows.

1 *Test the logical design for efficiency*   Of all the concepts this is the one where SSADM is most seriously flawed and causes the author the greatest disappointment. The concept is totally ignored.

SSADM certainly and obviously recognises the concept that the logical design is the basis of the physical design. This is a major improvement on version 3. Examples of this have been referred to in section 2.3.2.3 Database Design. The method is based on this obvious and hence unnecessary to state concept. It is therefore all the more surprising that this concept, which is a natural sequitur, is not supported. If one is true, so must the other be.

The concept can now be supported with modern database file handler software, which has no data structure and few data access constraints.[1] This means that the logical data model can be represented exactly in the physical design, both on data structure and data access, the two component parts of database design. The argument that the database access paths may well be different from the logical access path does not hold true if a modern file handler is used, such as a relational product with a statistically based optimiser. Such file handlers access the tables in the sequence that provide the fewest accesses, and ascertain which paths will provide the fewest accesses based on the count of the cardinalities of the tables. The query against figure 2.5, "For a specified A and B list all the information about C", would be much

---

[1] One such access constraint with a relational file handler would be for a business requirement "For a specified Customer display the last Order". The user does not know what was the last order and neither does the relational file handler, which does not support a backward pointer, such as with a Codasyl file handler.

SSADM Weaknesses

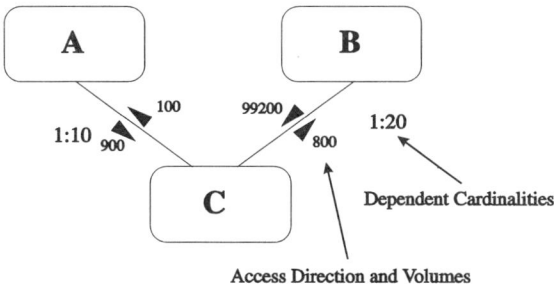

Figure 2.5 Table clustering

more efficient logically *and* physically if the access path was via A rather than via B, the logical cardinalities and the resultant physical statistics being narrower from A to C than via B to C. There is no technical reason as to why SSADM should not support this concept.

SSADM makes no support of the concept in the LDM data modelling and data navigation techniques, the prime techniques for this concept. Indeed, it cannot do so, as the techniques for ascertaining the access path against the LDM—the EAPs and the ECDs—do not count the number of accesses, and therefore cannot ascertain whether or not the LDM creates an expensive access path with a large number of accesses for a transaction. Any flaws in the LDM or the transactions cannot be identified and put right. *This is an area where version 4 of the method is in worse shape than version 3, where the technique for transaction access path analysis in the Process Outlines did count the number of accesses.* So much effort, occasionally months of effort, that is so often expended in ironing out the faults of the logical design in the Physical Design Control technique could be saved.

The most that SSADM offers in support of this concept is that the LDM should be validated against the functional requirements, and that if the required data cannot be obtained then new entities and relationships may need to be added. There is no hint that if the access path for the business requirement is terrible (the requirement is to access an A and get its Zs, and everything in between has to be accessed) the model should be altered.

As detailed in chapter 3, there are easy to understand and apply techniques to test the logical design for efficiency. They are easy to understand and apply because they are enhancements to existing SSADM techniques. Whenever they have been applied by the author they have been successful.

The area where the greatest testing for logical design efficiency should be is in transaction access path analysis against the LDM model. Logical data accesses against the LDM model convert to logical and physical I/O accesses in the implemented computer system. Database I/O is of two kinds. Logical I/O occurs when the requisite data to be accessed by an application program is located in the processor main memory buffer pool. The data is then copied into the program working storage when a data access call is issued. Physical I/O occurs when the requisite data is not in the buffer pool and must be accessed and retrieved from disk, when the buffer pool is full and some or all of the data in the buffer pool must be written to disk and when there is a COMMIT call (either an explicit call in the application program or an implicit call at the end of application program processing) and data is flushed out to disk.[1]

I/Os are expensive in machine path length instructions (PLI). Assume an IBM mainframe processor environment. The author has PLI performance figures from a number of database vendors. A logical I/O is on average some 1500 PLI. A physical I/O for a single record occurrence/table row with no referential integrity constraints or pointer linkages is some 4000 PLI. If the processor speed is 5500 machine/path length instructions per second the access to a disk would require one second of processor time. Access logic is therefore a resource-hungry process in processor terms. Physical I/O is also slow, as it occurs at the mechanical speed of the disk driver, whereas logical I/O is at the electronic speed of the processor. The disparity between the mechanical and the electronic speed is continuing to increase. Typical disk I/O times have declined from some 35 milliseconds to less than 25 milliseconds. This almost 50% improvement in physical I/O times has been vastly outpaced by the enormous increase in processor speeds, from millions of instructions per second to over 500 million instructions per second and growing.

In contrast to access logic, process logic is relatively cheap in PLI and, because it is executed in processor main memory, functions at electronic speed. Advice the author has received from a number of independent sources and database vendors is that, again on average, the PLI process logic overhead between data access calls in an application program is some 500 PLI for a 3GL and 700 PLI for a 4GL programming language. Programs that do not have to massage data prior to or after input or output to a terminal or database can effectively ignore process logic overhead.

The important point is that access logic is much more expensive than process logic. Fortunately, if the application is well understood and fully and correctly specified, the number and type of logical accesses to the data, given known assumptions about the application—this online business requirement is executed 100 times a day and a customer has on average 1000 orders outstanding—can be measured very accurately. Using "tricks of the practitioner's trade" to be described in chapter 3 for the technique of

---

[1] Chapter 4 describing distributed databases shows that there are two other types of I/O to database I/O.

SSADM Weaknesses

transaction access path analysis, it is possible to ascertain for each business requirement, *inter alia*:

- if a business requirement has multiple entry point possibilities to the LDM data model which is the best entry point;
- if a business requirement has multiple access path options which is the best option;
- where, when and how to modify the LDM or transaction if either is inefficient;
- when to "cheat" and incorporate derived or redundant unnormalised data in the LDM.

All this will keep data access overheads to a minimum.

Unfortunately the standard SSADM data access techniques of EAPs and ECDs do not support any of these tests to ascertain if the LDM model or business requirement data accesses are efficient. They can't—they don't count!

Given the failure to test for logical efficiency, it is not surprising that the author has found in many consulting assignments that much database optimisation/tuning using the Physical Data Design Optimisation technique has been needed. Many hours of unnecessary effort could have been saved if SSADM supported this concept.

2 *Events are standalone and problems-to-solve are causally related* The two concepts are nowhere explicitly explained in the SSADM manuals and neither is it shown in any of the design techniques for data processing. The only technique that can diagrammatically represent these two concepts is dataflow diagramming. DFDs can show the standalone nature of event level processes by having data stores separate processes, as shown in figure 3.22. Such processes must be separated in time and therefore stand alone. The simultaneous and hence causal relationship between problem-to-solve processes can be represented by dataflows between processes, as shown in figure 3.23, where process 4.1.1.1 triggers 4.1.1.2 *et seq*. All the problem-to-solve processes for the event Allocate Fork Lift truck occur at the same time as the event.

Unfortunately DFDs do not play the full role in SSADM the author believes they should, partly because they do not show the correct way to decompose processes and are therefore unable to support this concept.[1]

---

[1] The role of the DFDs will decline further with the adoption of object oriented design—see chapter 5.

3 Design the database on data retrieval business requirements This is nowhere addressed by either version of SSADM. It is not appreciated, as argued in sections 2.2.2.4 and 2.2.2.5, that *it is ultimately the data retrieval business requirements that actually drive the data maintenance business requirements*, because it is the retrievals that decide what data is required in the database. Data is only in the LDM, and hence in the database, because there is a need to retrieve it according to user requirements. The data maintenance business requirements do not add any new information in the LDM model that has not already been identified as required by the data retrieval business requirements. They do not enrich the model, or indeed the company's business. One does not put data into a computer system for the fun of it. One puts it in so as to retrieve it subsequently. Data maintenance business requirements are a necessary evil.

Only when information is retrieved and presented in a valid form and timely manner to the appropriate user(s) and manager(s) can a computer system be justified—only then does data have value.

Clearly, the data maintenance business requirements must be part of the overall system design. The concept of database says so. As business requirements they will ultimately produce database update application programs. The data accesses of the data maintenance business requirements are not forgotten in database design because they are included in the summary access path maps. The design of the database will therefore automatically give due weight to the data maintenance accesses. As business requirements they generate accesses to the entity model and are therefore part of the overall database design. *However, the structure of the final logical data model, and hence in the fullness of time the design of the database, should be based on data retrieval and only the data retrieval business requirements.*

It is a pity therefore that the data maintenance business requirements as described in the Elementary Process Descriptions are used to validate the LDM. If the data retrieval business requirements have been properly identified this task is not necessary.

It should be noted that *ad hoc* retrievals cannot specifically be catered for in the database design process. The best that can be done is to ensure that the LDM model, and hence the database, supports all conceivable data retrieval needs.

4 Business requirements should be CSF and inhibitor based As identified in section 2.2.2.5, CSFs are the ultimate basis of an organisation's information. One of the great weaknesses of SSADM is that it does not include a stage for information systems strategy planning, where the CSFs are identified. Other methods do take advantage of the CSFs as the basis of the design process. Both versions of Information Engineering, IEM from James Martin & Associates and Navigator from Ernst & Young, pay close attention to this concept. The IEM Information Systems Planning stage explicitly requires the analyst

SSADM Weaknesses 75

to identify the CSFs, inhibitors and objectives of each organisation unit, along with their performance measures. Indeed IEM includes a suggested questionnaire with specific questions regarding the issues for business objectives and their likely CSFs and inhibitors. Navigator explicitly demands that CSFs are considered when designing application systems—"focusing on CSFs when designing systems" is emphasised in the Monograph series manual. There is a specific chapter in the techniques manual on executive information needs analysis. It also explicitly requires external CSFs to be identified, and is therefore outward looking, presumably in relation to the company's competitors. Information Engineering therefore fully supports this concept. This is a major strength and one that is missing from SSADM.

One point, perhaps niggardly, is that both IEM and Navigator could be more specific in their support for this concept by stating explicitly that data retrieval business requirements for management must be CSF and inhibitor based. This is not done and has to be gleaned by implication by the intelligent reader of the manuals. It would also be useful to state the need to create a trace of each data retrieval and data maintenance business requirement back to a CSF or inhibitor as a quality assurance exercise.

5 *Database design is based on total data structure and total data access*   SSADM supports this concept as regards data structure but not as regards data access. The LDM is the total data structure, but nowhere does SSADM produce summary access paths showing total data access.

This concept is an extension of logical = physical. As stated in the description of this concept in section 2.2.2.3, a database system is based on the total of these two components and is therefore designed for the application as a whole and not for individual transactions. If the database supports all applications then the database must be designed for all applications, for the enterprise as a whole. The database should not be designed and even less tuned for individual transactions (a controversial statement!).

SSADM version 3 recommends that the transactions with the highest interactive peak, those requiring fast response times, those with long batch runs and those with high volume should be checked against the database and modifications made to the database or transactions where the performance objectives are not met. Version 4 merely states that the major functions should be measured for their "usage of resources". Major functions are most likely to be those that maintain the entities with high volumes in the database. That would match only one of the criteria in version 3.

Until recently all leading structured methods, including SSADM, have not produced summary data access information. Database designers have therefore not been able to see the advantages of ascertaining total data structure and total data access in the logical design specification. Information Engineering is the first structured method, as far as the author is aware, that makes any significant attempt to provide summary data access information. Unfortunately SSADM does not. In fact, it cannot. Version 4 no longer includes transaction access path analysis. The version 3 Process Outlines include this technique, but Process Outlines are not included in version 4. The replacement techniques, the EAPs for the enquiries and the ECDs for the updates, are inadequate for the job of access path analysis, as they do not count the number of accesses to the logical data model. *The EAP and ECD techniques merely define an access path, they do not analyse it.*

One administrative but not insignificant point is that, with the summary access path maps, these two crucial inputs to database design are combined onto one sheet of paper (or at most five). Secondly, and most importantly, *a balanced database design can now be produced.* It is balanced because it is based on a total data access map drawn on a total data structure diagram that, being combined, automatically and simultaneously *takes these two basic components of database design* (see figure 2.3) *into account in due proportion to their significance to each other.*

Because total data access is included in the database design no one transaction is favoured, no one transaction is excluded. Each transaction is automatically given their appropriate significance in the total design, based on their proportion of access to each entity. If one transaction incurs 50% of the database accesses logically, this will be directly reflected in the physical design. Transactions are therefore automatically given their due significance in the overall application performance. *Balanced designs are a priori efficient.* This is proved in section 3.1.4. Wherever and whenever the author has applied this concept the databases have never required re-tuning.

Where performance in the balanced design was below the objectives set it was because of the nature of the application. Where this occurs nothing can be done, so long as the business remains unchanged.

Figure 2.6 illustrates the relevant portion of a logical data model the author produced for an airfreight company. The purpose of the application was to record the rates charged to customers for the shipment of goods by air across the world. The application was developed and implemented at the time of high inflation in the United Kingdom in the late 1970s and early 1980s. The rates were therefore under constant change. In total some 80 000 rates were added and changed per year. Each rate was mandatorily related to 12 master entities. This was the widest network data structure the author has ever seen. Thus, whenever a rate was inserted the referential integrity constraints demanded that each of the 12 master entities to the rate were checked to ascertain their presence in the database.

This referential integrity access inevitably generated extremely high disk I/O, because each master entity was spread around like "grass seed" on disk, with the rate being able to be clustered in the same page/block on disk to only one of the master entities. The client had installed IBM's high performance DL/1 database management system and yet was suffering from disastrous

SSADM Weaknesses

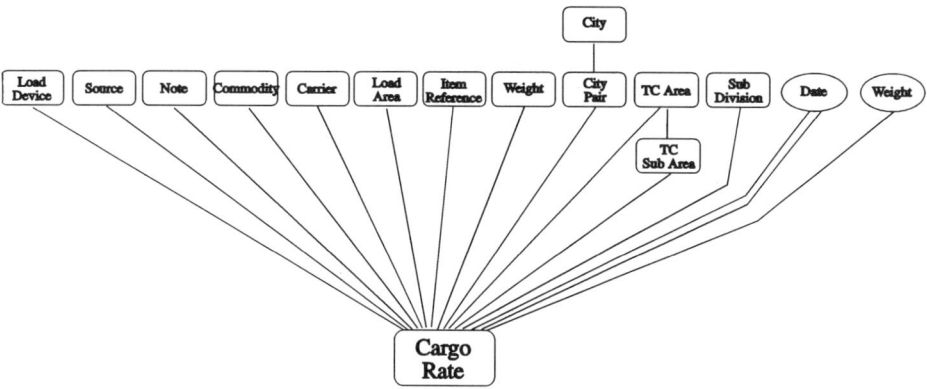

Figure 2.6 A "bad news" data model

response times on many transactions, particularly on inserts of rates. The fact that DL/1 only supports implicitly a hierarchical data structure and a single breadth of network data structure was, given the broad network, immaterial to the overall problem. Any file handler would have faced such referential integrity performance problems.

The purpose of the study was to find the cause of the problem. All traditional approaches to database tuning had been tried. However, no logical data model had been produced. Once a model had been constructed and the breadth of the data structure finally understood it became readily apparent that the file handler was not the problem. It was the nature of the business. The airfreight company had no option but to purchase a much larger computer.

The point of this example is that the solution to the problem was not found by expensive physical transaction performance monitoring and tuning as advised by the SSADM Physical Data Design Optimisation technique but by a quick exercise to build an LDM model and review it. It so happened that the cause of the problem was found in the total data structure component of database design rather than in the total data access component. A budget of some three months' work of physical database tuning had been planned by the client. The solution to the problem was found *logically* in two days.

A perfect example of how total data access on total data structure, i.e. summary access path maps on the LDM model, can provide many physical answers occurred at an insurance company. The company was using a relational file handler that was providing an appalling performance on response times. The LDM model contained some 60–70 entities. The model was not particularly significant—there were no deep hierarchies and, most importantly, no broad networks. The total data structure was therefore not the problem. It was not until the various summary access path maps were produced that the problem was spotted and a solution was devised. 85% of all the data accesses were to six entities and of these accesses 80% were data maintenance. The six entities contained the bulk of the application data with many thousands of entity occurrences.

The file handler was ORACLE, an excellent product in the round but flawed for this particular application. ORACLE up to version 5 locks data at the entity/table level. The six tables were continuously in a state of being X locked (the exclusive lock preventing any other access until lock release) in their entirety, even though only a few table rows at a time were actually being modified. There was inevitably a permanent queue of transactions waiting to access these tables. The other problem, minor by comparison, was a slow processor. Whenever the transactions could be processed the queue backlog cleared only slowly. There was thus a mutually self-reinforcing "traffic jam". The solution was simple—adopt ORACLE version 6 with the teleprocessing sub-system and lock the tables at the row level.

The point of this example is that, in contrast to the airfreight example, the total data access component of logical design as mapped against the total data structure component was the crucial element in ascertaining the reason behind the physical performance problems. Again it was the logical design of total data structure and total data access that found the problem and provided the answer, not the optimisation technique. *Physical design and tuning was not the solution—in fact it was part of the problem.*

Unless all the information necessary for summary data access is ascertained, presented and interpreted the full benefit of this concept will not be obtained. Given that the accesses of individual transactions cannot be counted, there is no way that SSADM can produce summary access path maps. SSADM cannot support this important concept.

2.4.2.2 The structural weaknesses

1 *No small project version of the method* The manuals implicitly assume that all application design projects require all the stages and tasks be applied in full. No account is taken of the fact that many projects may be for minor and *ad hoc* application systems for running on a mini or PC/PS type processor. Other structured methods have produced a small project version of the full-blown method. Information Engineering, for example, advises that where the project is small (advice as to what is small is given), and the risks are low, the logical specification need not be to a high level of quality. Various tasks, such as the analysis of technical options and relational data analysis, can be omitted.

Designers of applications that span data processing environments should "cherry pick" the appropriate satellite specific technique(s) on an as required basis while continuing to "cascade" use the core generic techniques.

SSADM Weaknesses

2 The absence of advice on software packages There is little consideration in SSADM of application software packages and how much of the logical design specification should be produced before assessing the suitability of the package(s) for the application. All that is mentioned is one half page in the section on Technical Systems Options that a package solution is one of the technical options that can be considered when there is a satisfactory level of functional match between the package and the logical requirements specification. There is no advice as to the issues that need to be addressed, such as the contractual aspects regarding support and changing the package, how to measure functional match and aspects of software portability and scalability.

It needs to be appreciated that, if a package solution to the required system is chosen, there is no need to undertake stage 5 of the method.

2.4.2.3 The technical weaknesses

Like the strengths of SSADM, the weaknesses are considered from two viewpoints—the techniques for information systems strategy planning and the techniques for application systems logical and physical design.

1 Logical Data Modelling (LDM) The technique is excellent. Notwithstanding this, there are points of improvement that can still be made.

- The technique now includes not only data modelling on the basis of the business requirements but also modelling on the basis of the analysis of the data attributes through relational data analysis (RDA). In version 3 these techniques were separate, even to the extent of being applied in different stages. The benefits of this separation are discussed in the next section.

- It seems strange that, given the LDM technique now includes entity sub-types, the application of RDA stops at fifth normal form. Sixth normal form tests for conditional dependence and thereby identifies entity sub-types. One part of the technique does not match another part of the same technique. The full application of relational data analysis is addressed in section 3.2.3.

- The necessity for the crow's foot is, in fact, an explicit recognition of poor diagramming practice. In over a decade of drawing entity models the author has never found it difficult to position the detail entity below the master. There is also the benefit that the model is much easier to interpret. One can quickly appreciate which entity is the master and which is the detail just by looking at the model rather than following a potentially labyrinthine maze of relationships to ascertain via the crow's foot which entity is the master and which entity is the detail. It is also much easier to ascertain the impact of the model on database performance, broad networks being bad news on disk I/O. It can readily be appreciated that the entity model in figure 2.6 will create performance problems, with heavy referential integrity overheads.

- Introduce operational entities. These have been lost in version 4. The nearest equivalent is the entry point type on the access paths in the EAPs and ECD where there is access via non-key attributes of an entity. The problem with this approach is that:

  — The database designer has to look at all the EAPs and ECDs (probably not these as they will require entry via the primary key), one for each business requirement, and this is a laborious task.
  — The ability to generate schema DDL code for the indexes from the LDM will be lost. Many CASE tools provide this facility when the LDM is recorded in the encyclopedia.

  Operational entities occur where a data retrieval business requirement requires to access an entity on other than the prime or relationship key. Assume there is a requirement to retrieve all customers with red hair. The prime key of customer is customer number. It would therefore, as the model stands, necessitate a scan of the entity table to ascertain customers with red hair. It would be a better approach to create an operational entity of colour of hair as an entry point to customer and access directly on colour with a value of red and from there access the requisite customers. Such operational entities contain only the key attribute to be searched on.

  Version 4 requires that where there is a business requirement accessing an entity on other than the prime key then create another master entity. This is a most strange approach. Such master entities would contain, just like the operational entity, only a key attribute. There would therefore be in the database design single attribute tables of data, to be maintained by an application program. The database design rules would recognise these as standard entities and hence standard tables of data, whereas in fact they are not and should be indexes. By contrast, operational data groups would be recognised for what they are (they have a different shape), become indexes and be automatically system maintained. The LDM should represent operational entities, as in version 3.

- Although entity sub-types are now included, LDM fails to mention that the entity sub-types inherit the properties/attributes of the master entity. Thus if the Customer entity has the entity sub-types Creditworthy and Non-Creditworthy Customers then these two entity sub-types will also inherit the properties/attributes of the Customer. The phrase in the SSADM manual "the super type entity has attributes which are common to its sub-types" is not the same thing as property inheritance. Property inheritance

SSADM Weaknesses

is automatic. That is not implied in the manual. The significance of the property inheritance facility is discussed in chapter 6.

- Advice on one-to-one relationships. There isn't any. All that is said is that should there be a one-to-one relationship between entities it needs to be made a one-to-many or the two entities merged. There is no advice as to the dangers of merging the entities with different keys, only that it shouldn't be done—they are different things serving different information purposes and the attributes of the entity whose key is not used in the merged entity would not be in third normal form. Neither is there advice as to how to ascertain which of the two entities needs to become the many of the new one-to-many relationship. Advice that the author gives is to ascertain how the entities would be as if there is a need to support a history of the two entities. Which entity would repeat over time? Assume that the one-to-one entities are Member of Parliament and Constituency. At an instance in time there is a one-to-one relationship, but over time it is obvious that a Member of Parliament repeats for a given Constituency. Member of Parliament becomes the detail of the Constituency.

- No advice is given regarding the definition of set cardinalities. The logical design volume documentation happens to make reference to the minimum, average and maximum dependent volumes (set cardinality) of the detail entity(s) to the master entity. However, as discussed in paragraph 3 below, these cardinalities are inadequate.

2 *The failure to have three distinct data modelling techniques* SSADM version 4 has one technique to produce the logical data model—LDM modelling. Having built the LDM model it is then "checked" by confirming the correctness of the allocation of the attributes to the entities by applying the technique of data normalisation/relational data analysis (RDA). RDA is applied against the LDM. A perfectly valid approach, were it not for frailties of *"homo sapiens"*. The designers would not be human if they did not view the data items in the entities in the context of the entity model. There is therefore the danger that applying the RDA technique in such a manner would tend to "confirm" the entity model, rather than "test" it.

SSADM version 3 deliberately has three separate and distinct data modelling techniques. Version 3 of SSADM first built a logical data model based on the analysis of the business requirements (the LDST model), then as a separate step analysed the data items in the I/O Descriptions using RDA, built a RDA logical data model using a set of four data modelling rules, and finally combined the LDST and RDA data models into what was called a Composite Logical Data model—the CLDD. The author listened to a lecturer on SSADM describe the data modelling techniques of methods like SSADM (version 3) and LSDM as "awesome". It was said with respect.

The reason for this separation is that each technique has different strengths and weaknesses. Where entity modelling is strong is different from where relational data analysis is strong. Entity modelling builds a logical data model based on business requirements, preferably the data retrieval business requirements, works at the entity level and with skilled designers and knowledgeable users is quick and easy to build. RDA, by contrast, builds the data model based on the mechanistic grouping of the raw data items into relations by the application of the data normalisation rules. Crucially, however, both techniques provide different perspectives to the data (entity modelling being top down and based on business requirements and RDA being bottom up and based on data) and thereby may produce different results, even for the same application. This may sound surprising but it has very often occurred.

In one application for a railway company the users were quite explicit in defining a thing called a train. A train entity was correspondingly identified in the LDM entity model. However, when RDA was applied it was found that there was no such thing as a train. What was identified were things such as engines, coaches and guards' vans that were joined for the purposes of a journey. Another example was for a major international oil company, where the users kept on talking about oil wells. In fact analysis of the data identified no such thing. What there was was a derrick (they were exploration wells), a hole in the ground into which were inserted conduit pipes, which branch out and hopefully hit oil and gas accumulations. Together all these things formed the concept of an oil well. Yet again, in an insurance application the users spoke about brokers and underwriters. Relational analysis of the data items showed that there were no such things. There were account holders, and a holder could be simultaneously a holder for broking accounts and a holder for quite separate underwriting accounts.

These examples are but a few of the wide variations between the data model results that can be ascertained from these two data modelling techniques. SSADM version 3 is therefore entirely correct to keep these two techniques distinct and applied at different stages of the method.

The first two techniques are now brought together as a continuum and there is no concept of a CLDD logical data model, merely the LDM model. Keep the two techniques separate and apply them at different points in the method by different designers. This assists the independent application of the techniques. It adds an extra element of rigour.

The above adverse comments need to be put into context. The techniques of building separate data models based on the business requirements and the raw data items and the advice for combining them into what was previously called a Composite Logical Data Model (CLDD) are still preserved in SSADM. It is just that the techniques are not regarded with the same degree of distinction and separation. The LDM can be made a CLDD. Why not enforce it?

3 *The paucity of LDM model and data navigation statistics* The LDM model requires by-product information about the entity volumes, lives and cardinality ratio statistics. The volumes and lives of the entities are fully covered by SSADM.

The cardinality statistics that are required are minimum, maximum, average and working average. Assume that the customer entity has 100 occurrences in the application and the order entity has 1000 occurrences. The minimum cardinality is zero as a customer may have placed no orders; the average cardinality is 10 orders; the maximum value, established in discussion with users, is that a number of customers have up to 150 orders. The working average is the average that *actually exists* when a customer has placed orders and *is the statistic that is used by relational file handlers* with statistical query optimisers for ascertaining the optimum access paths to the database. For example, assume that 50% of customers have no orders and the remaining customers have, more or less, an equal amount of business. The working average would therefore be 20. This is information needed by the database administrator.

If it is known that certain customers have fairly high volumes of orders then these can be explicitly identified in the documentation. For example, it is known that all customers located in Birmingham have on average 20% more orders than customers located elsewhere. When it comes to undertaking transaction access path analysis if it is known that the vast bulk of accesses are to those customers that place orders, which seems not unreasonable, then clearly the working average is the value to go for. If the customer is based in Birmingham then the number of orders for that customer is 24. The only structured method that addresses working average cardinality, as far as the author is aware, is the Navigator version from Information Engineering.

Given the thoroughness of the LDM technique, the tardy SSADM description of the need for this crucial statistical information about the data is surprising. The statistics are as important for logical design as they are for physical design. The entity volumes and cardinality ratios are crucial information for transaction access path analysis. Without this information the technique cannot be applied properly and certainly it could not be established whether the LDM model or transaction access path is efficient. It is crucial information for the second concept, that of testing the logical design for efficiency.

4 *Relational Data Analysis (RDA)* The latest thinking on data normalisation has moved beyond the first three normalisation rules. This is recognised by Dr Codd in his acceptance of the Boyce/Codd normalisation rule. Others, such as Dr Fagin, have developed further normalisation rules, that of fourth, fifth and sixth normal form. This additional normalisation is valid because it

has practical application and should, in the author's opinion, be supported by SSADM.

The SSADM manuals describe the further normalisation steps of 4NF and 5NF. It seems strange that the currently final step of 6NF "functional dependence" is not taken, as this relates to entity sub-types. Yet entity sub-types is included in the LDM technique. There is therefore a mismatch between the LDM data modelling technique based on the analysis of the business requirements and the RDA data modelling technique based on the analysis of the raw data items. The first goes the full length to sixth normal form and the other stops short.

5 Dataflow Modelling (DFM) Notwithstanding the increased role that the technique has within version 4, which the author welcomes (dataflow diagramming is a much underrated technique), there are a number of concerns:

- Advice is provided that the level 1 DFD is used to show the scope of the system, yet in the same paragraph the manual states that the system boundary should not be drawn on the level 1 DFD. Why not? The boundary line represents the scope of the system and is a clear and visual means of illustrating the point. The author very decidedly draws a systems boundary around the level 1 DFD.

- The manuals seem to imply that the Required System DFD should contain only D type datastores. The advice is that "M" datastores are only to be found only in the current system. Why cannot the required system include parts of the business and its associated data that are manual? The author can recall advising a client that the required system was so slow moving that there was little point in computerising it and that it should remain manual, although enhanced. The datastores were manual and were recorded as type M.

- The dataflows should only be one way at the lowest level of process decomposition. This seems strange advice. What about updates to the datastore—should there be two dataflows? If this is the policy then there will an unnecessary number of dataflows in the DFD for the updates. Updates to the database, which the datastores represent, is a two-way process. Updates require to read access before an update access, so why not represent it logically as a two-way flow? The author continues to use two-way dataflows to show updates to a datastore.

- The area where the author feels the greatest improvements can be made is in process decomposition and decomposition consistency. The advice offered in the version 3 manual is not the unhelpful decompose "until a satisfactory level of detail is reached", but it is no better—"the lower level diagram contains more detail in terms of the ... processes that belong with the higher level process". Equally uninformative.

The approach towards process decomposition adopted in the manuals is confusing. The manuals mention decomposing up to three levels of decomposition ("a third level should be the lowest required"), but there is no advice as to how to undertake process decomposition. The lowest level processes should be at the function level. "Decomposition stops when functions can be identified. Finer detail is not required ... since this is defined using later techniques within SSADM." But there is no advice as to what levels the two upper-level processes should be. And, adding to the confusion, the manuals also say that process decomposition should also identify events, and a function can contain multiple events and is therefore more detailed than a function.

The manuals state that when checking the DFDs for completeness and consistency each process should have a strong active verb and a single object, such as Amend Tenant Details. This is very much a "do something title", which, as explained in section 3.2.1, indicates that this is a "do something" event. Given the advice in section 3.2.1 and in the SSADM manuals, the processes at the lowest level must therefore be at the event level. If the lowest level is a function and contains many events, which "do something" title is to be used for naming the process?

Other advice on decomposition is that if the Elementary Process Description is lengthy the process should be decomposed further. But what is lengthy?

Readers of this book would do well to read the James Martin Associates version of the Information Engineering method to obtain some useful tips on process decomposition. The tips have been used in section 3.2.1, along with some further suggestions.

- DFDs are used as the basis of identifying functions and "business" (author's quotation marks) requirements and events. This is confusing. Business requirements, be they data maintenance updates or data retrievals, are the same as events. A function can contain many events. The manual says so. Functions therefore cannot be the same as business requirements/events. DFDs are therefore used to identify two levels of business activity—the function and the event. But processes are decomposed to only the function level, that is, potentially above the event level, so apart from intuition and skill, where in the DFDs are the business requirements/events shown and how are they to be identified?

SSADM still mentions update events and data retrievals requirements. Why is there this need for distinction? The only difference between a "do something" update event, such as Create Invoice, and a "do something" retrieval event, such as List Customers with Red Hair, is that the update

events need to be sequenced. Why is a retrieval business requirement also not an event in the same way as an update business requirement?
- There should be only one "driving" input to a process. Why? Why cannot there be multiple user roles, each represented by an external entity, which can trigger a given process? Assume there is an event level process Allocate Stock. At a manufacturer of advanced aero engine equipment the author found that several user roles could allocate stock. Clearly the storeman could trigger the event, but so could the store manager and so could the stock audit clerk. No one trigger was the driver.
- The SSADM manual reflects inconsistency of decomposition in the diagrammatic examples. Consider figure 2.7, which is typical of many DFDs examples illustrated. It shows the decomposition of process P1. Process P1A occurs at a point in time, let us say 11 o'clock. At 11 o'clock it is triggered by the external entity EE1 with a dataflow DF1, updates datastores D1 and D1/1 and triggers process P1B, which must also therefore execute whenever process P1A executes. P1B is therefore causally dependent on P1A. It is only triggered by process P1A and never triggered by an external entity. These two processes therefore occur logically at the same time. In the real world it could be that P1A takes two seconds to execute, so that P1B is triggered within two seconds of 11 o'clock. P1B updates datastores D1/2 and finishes. The data in the datastore stays there indefinitely until retrieved at a later point in time, let us say at 2 o'clock, by process P1C. P1C is therefore time independent of P1B and is in no way triggered by or causally related to P1B. There is inconsistency—P1A and P1B are causally related and are executed together at the same logical point in time; P1C is a distinct and separate process occurring at an arbitarily different point in time. These two groups of processes are therefore at different levels of decomposition. P1A and P1C are at the event level and distinct. P1B is a sub-process within and is not distinct from P1A and therefore at the problem-to-solve level.
- SSADM DFDs do not support the concept of time—process A occurring before process B. Given that events in batch and online processing are standalone, processes at the event and higher function and business area levels of decomposition should be separated by data stores. Data between such processes is stored for an indefinite period of time. Such processes are not therefore related in time. The manual states that the process decomposition should go no further than the function level, such that time is implicitly excluded, as a function can be higher than the event level. It is a pity that SSADM then illustrates DFD examples with processes not separated by data stores and connected by dataflows. Where processes are connected by dataflows they are intrinsically causally related and not distinct and standalone in time. Such processes are connected in time in that they occur together.

SSADM Weaknesses

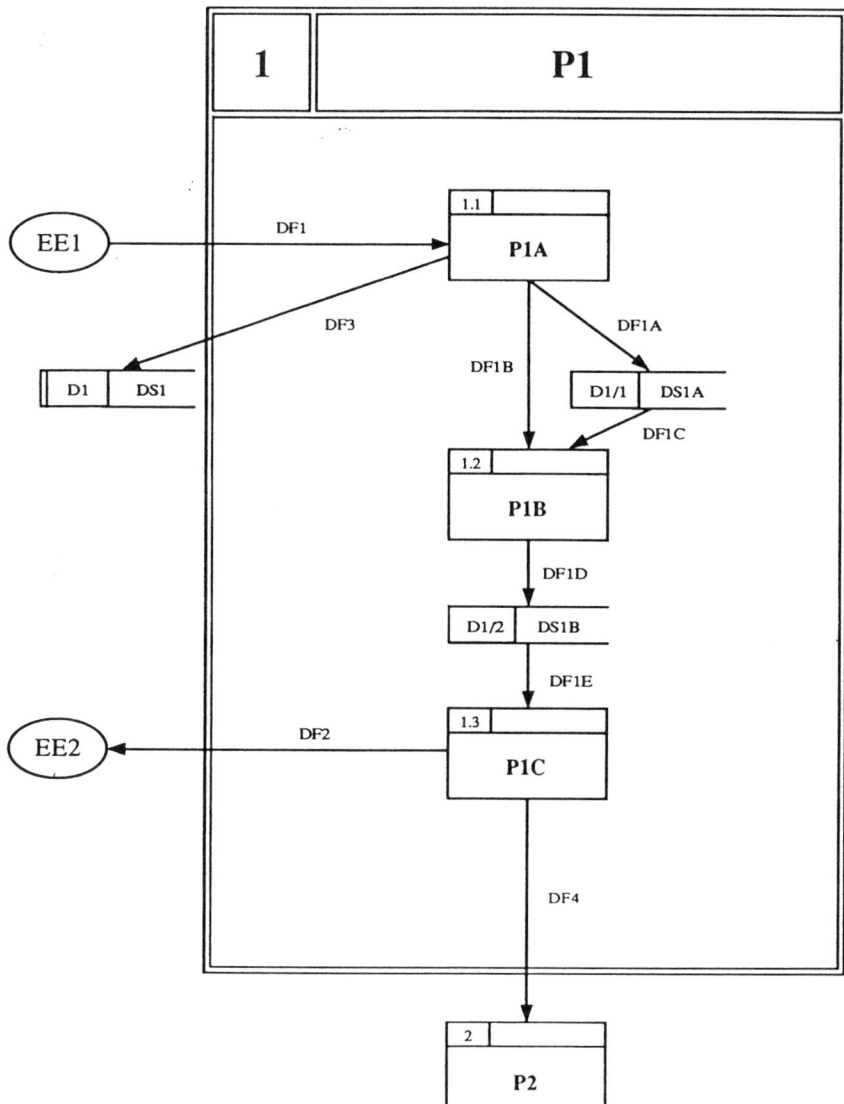

Figure 2.7 Inconsistent process decomposition

All in all, the author is not convinced that the technique has been improved in version 4.

6 *Function Definition—I/O Structuring*   These two are considered together, as the I/O Structure is a sub-technique within Function Definition. The prime problem is the function. The author feels there is little to commend the concept of the function. Nevertheless, its role in SSADM is explicitly recognised as important. "Functions definition is a central analysis and design technique in SSADM." It identifies "units of processing ... on which the physical design is based". "It identifies how to group system processing to support the user's task." "The identification of functions is concerned with ascertaining which events and/or enquiries the user wishes to be processed together." "Functions are units of processing designed to assist the user."

Functions are the user role view of processing and the basis of application program design. Each function becomes an application program or a run unit of several programs. The problem of the function being the basis of application programs is discussed in section 2.4.2.3/13. The issue of the user role view of processing is discussed here.

The concept of the user role view of processing is new to the author, and in his opinion is highly dangerous. Relational technology has the facility of the user view in the SQL data sub-language, but this is a user view of data, not a user view of processing. The SSADM user view relates to processing. In version 4 the definition is "a set of system processing which the user wishes to schedule together, so as to support their business activity". This is not necessarily logical, with the user role being subject to change through reorganisation. Version 4 does not require the grouping of processes on the basis of their "logically sensible" groupings, but on the basis of the company's current perception as exemplified in the user role of how the business should be run. If processing in the physical design is to be organised on the basis of user roles and their functions, what happens if there is a re-organisation of the company and user roles and the attendant functionally based processing are changed? Assume there is a user role of invoice clerk, who can receive invoices, send them and cancel from the system. But the invoice clerk is not allowed to deal with money and therefore cannot process the other events dealing with invoices. This is left to an accounting clerk. The system is designed accordingly. Then there is a company re-organisation. The invoice clerk is allowed to process receipts of money but not repayments. The events which the invoice clerk now requires to be processed together are different. On the basis of version 4 the system needs to be redesigned. Re-organisation is not uncommon. Are the application program designs to be changed as well? Surely not! Surely what should be done is to identify processing on the basis of what is logically correct, and that is at the event level. Users work at the event level—they "do something", like allocating stock, sending invoices and receiving orders. They do not work at the function level. It is noticeable that all the examples of functions in the manuals are "do something" processes, such as enquiries, which by their nature are events, such as Produce Report, and updates, such as Terminate Tenancy.

The user role view of processing should be built with the menu selection screens for the user to select which event level business requirements to trigger. Section 3.2.2 discusses how to model this.

Now put this in the context of the version 4 definition of a function.

The author believes that it is most unwise to design an application system on the company's organisation. It should be based on a logical system specification. Any variation from this will lead to instability. *The design of system processing on the basis of a company's user roles is as unstable as the company organisation.*

Given the weakness of the function, in order to make the risk of redesign as small as possible, it is important to ensure that the user roles are logically sensible. Such roles could be warehousman and invoice clerk. Such roles are in line with functions *as defined in version 3 of SSADM*, where the definition of "a logically related group of events" is excellent. There could be a function of Stock Control and Invoicing. Within the Invoicing function there could be a set of events, such as Create Invoice, Send Invoice, Receive Invoice Payment, Adjust Invoice Amount, Clear Invoice and Delete Invoice. All these events are logically related to the function of Invoicing. They have nothing to do with the user role and are independent of any company organisation, unless both are also logically organised. How to handle functions with many events is discussed in section 3.2.2.

7 *Enquiry Access Paths (EAPs)* For the purpose for which it is designed the technique has a serious deficiency. It can show *an* access path in a Jackson-like structure—it cannot show *the* access path in a Jackson-like structure.

There is a basic problem with the Jackson-like approach—it does not count. If the business requirement has multiple entry point possibilities—"For a specified A, B and C go and get all the details about D"—which entry point do you enter on—A, B or C? The one that provides the fewest number of accesses to the logical data model. But one cannot tell with EAPs, because EAPs do not count the number of accesses.

The manual advises that when the business requirement has several entry point possibilities ("If an entry point for an enquiry is several entities...") then all the entities and relationships should be included in the access path. No! This is the one thing you should not do! This is not the way a file handler generically works and surely SSADM should be representing logically what the file handler does. One of the entry points will provide a better access path than the other entry points. The best entry point should be ascertained through a transaction access path and the access path traced from it. The EAP should then be based on this access path. How to do this is described in chapter 3.

There is also the problem that the access path, whichever entry point is used, has too many accesses. It is an online business requirement but the number of accesses is, let's say, 5000. There is no way that the response time will be within a few seconds. The reason for this large number of accesses could be that the logical data model supports the business requirement but does not do it well/efficiently. It needs modifying. But it is impossible to ascertain that the access paths have too many accesses unless there is a mechanism of counting the accesses.

Both of these problems are particularly the case with data retrieval business requirements and can only be solved if there is the ability to count the accesses. *EAPs are therefore only able to validate that the LDM supports the business requirements, but not that it supports them efficiently.*

The only advice offered on this issue is that if the steps of the EAP technique do not produce an access path and the required attributes are not in the LDM then the LDM should be modified "to match the enquiry requirement" by adding an new master entity entry point, creating new entities or relationships, or some process oriented solution, such as a sort. No examples are provided. The nearest to checking that the LDM is efficient is where there are alternative access paths "some initial thought may be given to the number of logical reads entailed in the alternative routes". But there is no technique or even advice to ascertain the number of read accesses and what these numbers mean.

With data maintenance business requirements it is very much a case of accessing a specified occurrence of A. The user knows which entity/table occurrence to maintain. The problem of ascertaining which access path to follow is therefore not often a problem. The issue of long access paths is also not serious, it usually being an entity/table occurrence being maintained and possibly some relationships to other entities being checked for referential integrity purposes..

The understandable statement that the physical access path may be very different from the logical access path should be discounted with the modern file handlers. The relational file handlers with statistical optimisers count the cardinality ratios between entities/tables and between indexes and tables and therefore have the same physical access paths as those against the logical data model (ORACLE version 6 does not have a statistical but a rule based optimiser and therefore the physical access path may well be different from the logical access path). The pointer chain technology file handlers, such as the Codasyl type, will also have the same access path as the logical access path. They follow the relationships between the record types in the same way as the logical access paths do.

The technique for counting is transaction access path analysis. This technique was applied in version 3 in the Process Outlines. This has no longer been continued in version 4 and *is definitely one area where version 4 has taken a retrograde step*. Fortunately, it is easy to put right. Re-introduce access

SSADM Weaknesses

path analysis and then base the EAPs on them. This is what the author has done on a recent very large project.

The access types that can be used in an EAP is of the direct access to an entity occurrence, accessing the master entity of the detail and accessing the next detail from the master. There is no mention of accessing in the reverse order—why cannot access to the detail entity be in a backwards direction as well as a forward direction?

8 Entity Life Histories (ELHs) It is very difficult to find fault with this technique. ELHs support different types of update events. There is the type that is the same no matter when it occurs or how many times it occurs. This is the usual type of event. There is the type, known as effect qualifiers, where the processing of the event depends on circumstances. The example given in the manual was for the event New Property which had different effects depending on whether there was a tenant resident or not. ELHs also support entity roles, where an event occurrence can effect more than one occurrence. This could occur, for example, when a person closes one account and opens another. This event type is not new to version 4.

The reason why these event types have been mentioned is to put them into the context of the next type of event, an event type that is not the usual, and is neither an effect qualifier nor an entity role. Assume the following. The entity is a loan. The loan is repaid with the event Make Repayment, which, of course, iterates. All the repayments are the same except the last, which is different in that a congratulatory letter requires to be sent to the borrower asking if he/she would like another loan. The ELH would look as in figure 2.8. It seems as if this type of event is like an effect qualifier, except that in the examples effect qualifiers are treated as mutually exclusive and hence optional, whereas with the loan example the differing effects are a sequence. It does not seem that the loan situation type of event is addressed in SSADM.

It is a pity that an example of the difference between an event and an effect is not given. Consider marriage for a woman in a Christian society. The event is getting married, the effect is that her surname changes. Change Surname is an effect, not an event.

9 Effect Correspondence Diagrams (ECDs) The ECDs suffer from the same problem as the EAPs—they can show *an* access path but not *the* access path— and for the same reason. This is not such a problem with the data maintenance events as with the data retrieval events, as there is usually only a single entry point—the entity to be maintained—and one access path—from the entity being maintained. And data maintenance events usually have short access paths, so that the number of accesses is not large as compared with the data retrieval access paths.

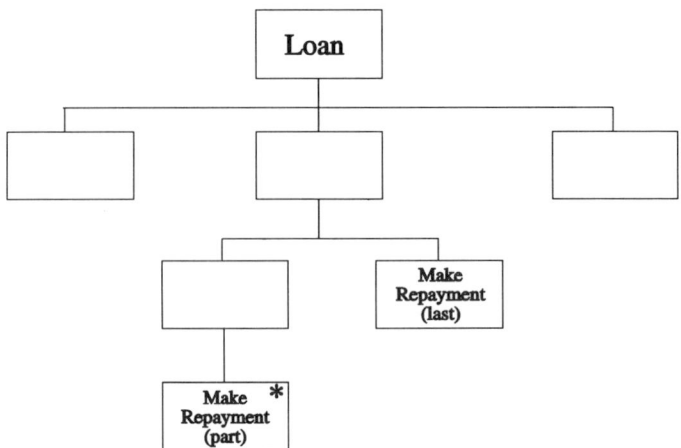

Figure 2.8 ELH user roles

The ECD shows which entities are accessed for the event and one can follow the access path from the entry point entity. However, the clarity of the access path is lost to some degree by not using the arrow as in the EAPs to show the direction of access. The use of the double-headed arrow gives no indication of the direction of access.

The double-headed arrow is intended to show one-to-one correspondence between effects between entities. The author finds this strange. They are added to the ECD where access is from a detail entity to a master entity. But this is no different from the EAPs, which do not support double-headed arrows or the concept of one-to-one correspondence when accessing from detail to master. It seems superfluous for data maintenance events, which only differ from data retrieval events in that they require to be sequenced. We shall see that this one-to-one correspondence aspect generates further problems when creating the Process Models.

10 Process Models The Enquiry Process Models and the Update Process Models are considered together as they are similar techniques. Both show the access path against the LDM for a business requirement.

There is a general concern for both the EPMs and the UPMs. It is the lassoing of the one-to-one correspondence on the EAP and ECD access paths, that is, where one occurrence of an entity access is to one occurrence of another entity (such as accessing from the detail entity to its master entity) or accesses of a Jackson iteration "set of another entity" box (such as from a master entity to its detail). This one-to-one correspondence is represented by the arrows in the EAPs and the double-headed arrows in the ECDs. Several problems arise because of this lassoing:

SSADM Weaknesses

- The sequence of access between the entities within the lasso is lost. Consider the logical data model in figure 2.9 and the business requirement "For a specified C list all the information A and B and all Ds and Es where …". A, B and C are lassoed into one process box for the Process Models as if there is one access to this box called C. Yet in reality there is a sequence of accesses to C, then B, then A.

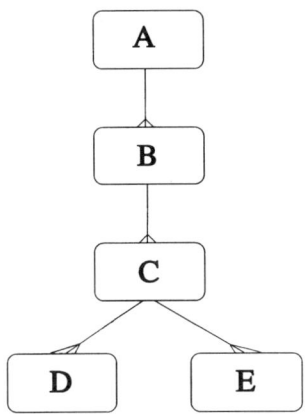

Figure 2.9 A logical data model

- The operations which are at the entity level in the EAP/ECD are also merged by the lasso facility. It may be that there are a considerable number of operations appended to each of C, B and A. They are merged to form a potentially overly large group of operations. Excessive size reduces maintainability.

There are other problems as well. For example:

- If one takes the lassoing facility away the operations are added at the entity level within the event. This is fine if the logic is appropriate to the entity. If the entity is Customer and the logic is "Customers with red hair get a bonus of 10%" then the operations are correctly associated with the Customer entity. But if the logic is not appropriate to an entity what then? The logic could be a declarative statement "If Tuesday do XYZ otherwise do ABC". This logic is not appropriate to a single entity. It could also be that it is necessary to obtain some information from a number of entities and then process the findings in summary form. Where on the Process Model should these operations be added? They are, of course, appropriate to the event and not the entity, and they should be added to the root box at the top of the Process Model. The problem with this is that the root box is itself an entity—see Enquiry Process Models below.

10.1 Enquiry Process Models (EPMs)

There are several problems with the technique:

- There is a mismatch as regards the I/O Structures and the EAPs that are together input and merged to create the EPMs. The EAPs are pitched at the event level and the box at the EAP process model (that is, after the EAP has been converted into a Process Model structure with appropriate lassoing) is an entity/lassoed entity. The root box at the top of the converted I/O Structure process model is the process name at the function level. The input data structure is therefore at the entity level and the output data structure is at the function level, and a function can contain many events. The input data structure from the EAP and the output from an I/O Structure to an EPM are therefore unsynchronised. What if the I/O Structure has many events? And what happens to the data for the events in the I/O Structure that have nothing to do with the EAP event? Which part of the I/O Structure should be eliminated? Presumably the data for the events that are not appropriate for the EAP input to the Process Model should also be "lopped off", in addition to the input data group elements. How is that to be merged with the EAP? No advice on this is given in the manuals. Certainly there will be a structure clash, but which kind? The ordering, boundary and interleaving clashes do not seem to be suitable.

  The solution adopted by the author has been to pitch the I/O Structures at the event level. *Anything to do with database processing should be pitched at the event level.*

- There is a further mismatch in the input and output data structures for the EPMs. If one considers figure A7 on page LS-LDPD-25 of the SSADM manual one can see that the root box of the EAP is for an entity and the root box for the I/O Structure is for the function. They are joined by a correspondence arrow, yet they do not correspond. The correspondence arrow from the EAP data structure should be to the Application Details entity in the I/O Structure diagram. It is noticeable that the first of the leaf boxes on the output data structure is always an entity, which one hopes matches the entity (or lassoed entity) as the root box of the input data structure. The SSADM manuals cause the mismatch by requiring that the correspondence between the two structures for the merging task to be root box to root box!

  It is a pity that there are no examples of how to recognise the different types of structure clash just by looking at the input and output structures. Consider figures 2.10 and 2.11. When drawn on a project it was at first considered to be an ordering clash. When it was presented as such to an audience of SSADM and Jackson experts there was *substantial* disagreement—it's a boundary clash, it's an interleaving clash, it isn't a clash at all!

SSADM Weaknesses

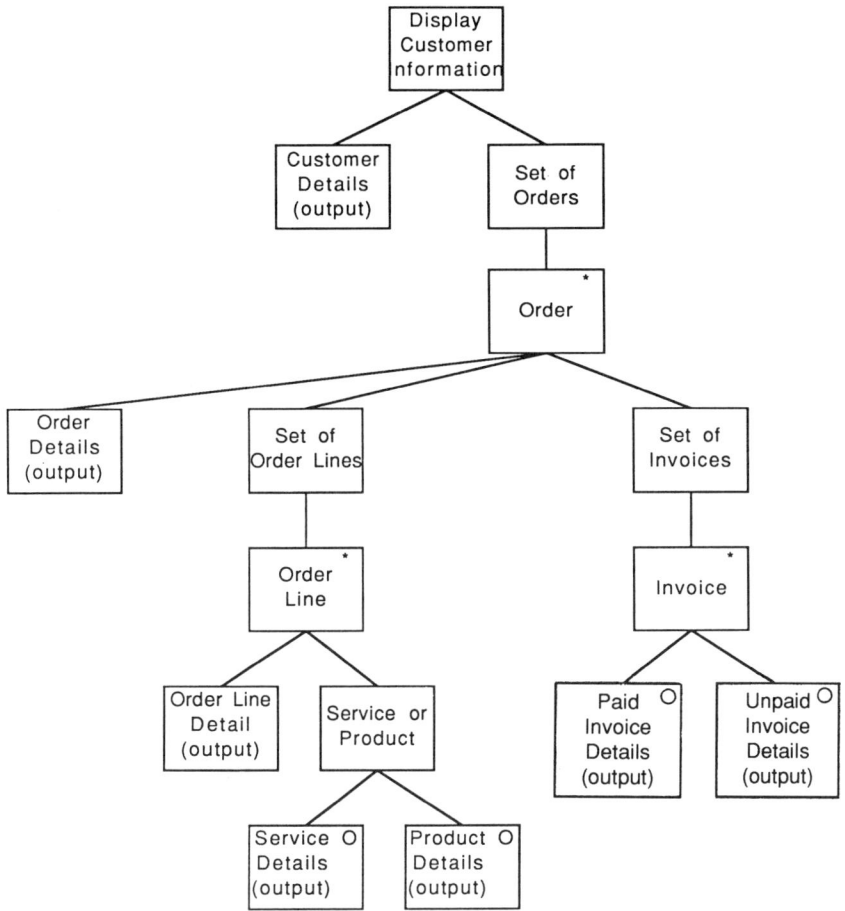

Figure 2.10 Enquiry output data structure for display customer information

10.2 Update Process Models (UPMs) There are no deficiencies that are specific to the UPM technique.

10.3 Dialogue Design The problems that relate to dialogues are:
- They are pitched at the function level, this reflecting their origins in the I/O Structures. Once again there is this problem that processing can be regarded at more than the event level, even though users work only at the event/business requirement level. There is no advice, where the function contains many events and only one event is triggered by the user, what to do with the residue of the function data in the I/O Structure that has nothing to do with the event/business requirement being "dialogued".

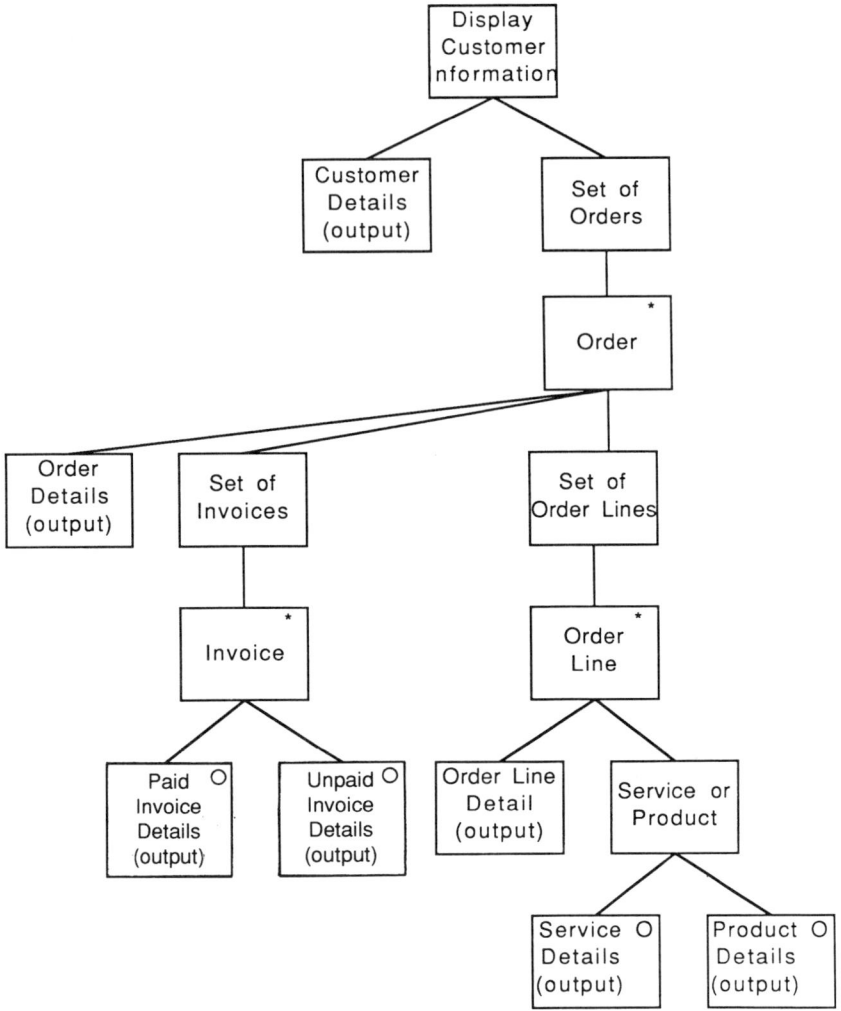

Figure 2.11 Structural clash for display customer information

- The operations for the processing of the dialogues are not added to the dialogue structure diagrams. Indeed there is no indication within SSADM where the operations for the man/machine interface processing should be defined. The traditional way that online screen processing used to be done is with dumb terminal screens, where the input screen was sent as an input message to the application program running on a back-end mainframe type processor, the message was processed against the database and an output message sent back in response to the screen. No processing of the input or the output was done at the screen. But technical developments have moved on, and intelligent screens can now undertake screen processing where it should be sensibly be done, at the screen. This use of intelligent front-end screen processing is best exemplified with the client-server architecture that is proving very popular with the latest relational file handler offerings, such as from SYBASE and the latest version of INGRES.

The client is a front-end processor/intelligent screen and the server is a back-end mini/mainframe type processor. Each can request a service of the other, where the service can be "Can you send me some data?" or "Can you process me some data?" The great advantage of this architecture is that processing can be put where most appropriate. Such client screen processing includes word processing functions, graphics displays, spreadsheets and the man/machine interface through screen formatting and presentation and editing the input data, such that only valid data, as far as possible, is passed to the server processor for database processing. The back-end server processor is then left with the processing that is "rightfully" its, such as database management and the production of bulky, typically corporate, reports.

The optimum is to define the server processing as operations on the Process Models and the client processing as operations on the Dialogue Structure Diagrams. *The trouble is, by not including operations to the Dialogue Structure Diagrams, SSADM is not able to support such client-server type processing.*

- The manuals do not consider the increasing use of windowing technology in the design of the man/machine interface. They do not identify that the LGDE could be used to support windows. Indeed the manuals explicitly state that a LGDE is not a window on the screen. SSADM seems to envisage the old pre-windowing technology where the format of the screens is fixed in the design process. There is no recognition of the pick and point technology of windowing.

The LGDE is a grouping of data items for presentation on the screen. Their sequence, iteration and selection reflect the navigation within the dialogue. A screen can contain multiple LGDEs. All that needs to be put dynamically against the LGDE in physical design is such information as its position on the screen, whether it is a front screen or not, whether it is minimised or not and one can easily envisage the LGDE as a window in the screen canvas area. The problem with the technique is that the structure diagram assumes a sequence from left to right in the LGDE dialogue navigation paths, and there is no ability to reflect that there might be parallel use of multiple LGDEs in a totally random manner.

11 Prototyping The major aspect that is disappointing is that prototyping is not used for the rapid development of applications of low business and technical risk. This is an area which Information Engineering has exploited with the Rapid Application Development mechanism, thus achieving tangible results for parts of the total application within a short time period. Credibility with users can be achieved with this approach.

The concern is not that prototyping should be regarded as a positive development, which it is provided the proper tools are available and used in a proper manner, but that SSADM only advises on the good points about prototyping and does not identify any shortcomings. Prototyping cannot provide a total answer to systems design and development. The prototyping shortcomings are:

- It cannot be a substitute for the application of the logical and physical design techniques and the production of their deliverables. Furthermore, prototyping tools as represented by 4GL software, by their nature, can only produce physical design deliverables, for the simple reason that the tools function in a hardware/software environment. They can produce:
  — a database schema;
  — the table definitions of the attributes;
  — the menu and dialogue screen formats and their dialogue sequence;
  — the data access logic implicit in the selection of menu options.

  This is fine for computer technicians but, apart from the man/machine interface of menu and dialogue screens, are totally unsuited to untrained users. One of the major reasons for the move towards the use of techniques of logical design is that the methods produce deliverables that are logical and mostly diagrammatic and therefore, by and large, understandable by the users. This means that prototyping only has limited practical use for the user for the full specification of the system. *As envisaged by core SSADM, prototyping can be used as a validation mechanism of the logical design, but not for its specification.*

- Another problem is that prototyping cannot represent all the design information contained in the logical design deliverables. The deliverables that cannot be supported are:

  — The DFDs. Prototyping cannot represent the decomposition of processes or the relationship of processes to each other. These diagrams are most useful tools for increasing the user's understanding of the system as well as being, if used properly, a useful logical design deliverable.
  — The ELHs. Prototyping cannot represent the sequence, selection and iteration, the grouping of iterations and selections, the abnormal quitting from the usual sequencing of events and the state variables used to re-express the sequence etc. in which the data maintenance events for a particular entity occur.

SSADM Weaknesses 99

— The access paths as represented in the EAPs and the ECDs.

12 Database Design

Some of the advice offered in the eight activity steps in Physical Data Design needs to be treated with some caution:

- There is at times a failure to reflect the latest file handler technology. For example, it is advised (Activity 1) that some of the information contained in the logical data model does not translate into the physical database design, such as the exclusion arcs and the relationship names. It should be borne in mind that the exclusion arcs are in fact business rules about the entity relationships, and can be defined as rules in the database schema. There are file handlers that can support the definition of rules in the schema, such as INGRES and DATACOM/DB. The rules are as much part of the data design as the SQL DDL definition of the tables, attributes and indexes.

- The latest file handler technology also very much makes use of the semantic description of the relationships between entities in physical design (Activity 1). Object oriented file handlers need these descriptions to identify property inheritance. However, inheritance is not recognised in SSADM, even though there are entity sub-types which require inheritance. The semantic required to support property inheritance is of the "is a" type and not the more free form type described in the manuals. The relationship link phrase rules of the LDM technique define database semantics in the form of the "business" description of the relationship. The role and importance of property inheritance is described in chapter 5.

   Relational and pre-relational file handlers do not make use of the semantic descriptions of the entity relationships, so it is understandable that neither does SSADM, which is designed for this type of technology. Relational file handlers are now supporting entity sub-types as separate tables with a common key (the key of the super-type), but without property inheritance. It is a pity that the deficiencies of the technology have been carried over in the design techniques. This will require revision when SSADM is enhanced to support expert systems and object oriented technology.

- The advice on secondary indexes (Activity 2), that is, indexes on non-key entry points, implies that there are prime and non-prime keys. Such is not the case in physical design. Keys can be used for three purposes—to ensure that each row of the tables is unique, to sequence the table rows in the order of the chosen key and to act as an entry point to the table. The attributes chosen for these purposes can each be different. There is no concept of a primary key in physical design,[1] and thus an index is an index.

- Because SSADM does not recognise the concept of testing the logical data model for efficiency nor the ability to produce a balanced database design from the summary access path maps, it offers advice that is most unfortunate, which is the database designer should "only optimise to meet essential requirements". Don't tell the users whose business requirements are deemed not to be essential. They will not be pleased. The advice is also, in the author's opinion, wrong. With the total data structure and total data accesses encapsulated in the summary access path maps *the significance of individual transactions in the overall database design is automatically taken into account in due proportion to the total data structure and data access*. All transactions are given their due attention. No one transaction is favoured, no one transaction is ignored. There is a failure to recognise that a balanced database design based on a logical data model that has been tested for efficiency and total data access will be automatically efficient.

SSADM only takes total data structure into account when producing the initial unoptimised database design (Activity 1). But Activities 4 and 5, which finesse the initial design, take no account of data access. Activity 4 states that a record with more than one mandatory master entity should be placed in the physical group of the master whose key is part of its own key. This is a data structure rule. But since when has the placement of a detail table been based on a key? Activity 5 says that the detail entity should be placed in the physical group of the entity to which it has the least number of dependent occurrences. Another data structure rule. Both of these rules ignore data access.

The placement of tables into physical groupings is for the purposes of minimising disk I/O (you cannot minimise logical I/O). Since disk I/O is because of data access then data access should be the deciding factor, not keys or least number of dependent occurrences.

Consider figure 2.5. Should table C be clustered on disk via table A or table B in order to reduce physical disk I/O? Some structured methods the author has used state that a detail table should be clustered on disk next to the master table to which it has the fewest dependent occurrences. Table C should therefore be cluttered via table A and not via table B. But

---

[1] Post-relational file handlers are now able to support what are called "surrogate keys". These are system generated keys that are composed of a system generated number concatenated with the time that the table row was created. Rows are never deleted, only archived. The benefit of the surrogate key is that users can now pose historical type questions, such as "What was the information about X as of a point in time?", that point in time being stored in the surrogate key.

The surrogate key is also going to be one of the prime facilities to be used in object oriented file handlers for the linking of objects belonging to the same class. How this is done is discussed in chapter 5.

this rule takes no account of data access. What if the summary access path map showed that the number of accesses between tables A to C were 1000 per day and between B to C were 100 000 per day as illustrated? Clustering C to A as recommended would save, assuming all dependent occurrences of C were in the same page/block as A (this can be easily calculated), 1000 physical I/Os but at the cost of 100 000 physical I/Os between B and C. Table C's rows would be spread around on disk like "grass seed" as far as table B is concerned. If table C was clustered via table B instead, because it is much the busiest access path, then, again assuming all dependent C's are clustered on the same page as the master B, 100 000 physical I/Os would be saved, at a cost of 1000 physical I/Os between A and C—a hundredfold improvement in performance based on taking total data access into account. It is only when data access to the master entities are more or less equal that the database set cardinalities should be considered.

The failure to take data access into account in this manner in the data structuring rules means that extra work may, probably will, be subsequently required in the database design optimisation technique. Why not take data access into account in the initial database design on the lines indicated above and described in chapter 3, rather than later in design optimisation? It is easier to accomplish and potentially would save considerable effort in the much more difficult and laborious task of database tuning.

The failure to take data access into account also adversely affects Activity 6, determining the block size to be used for a table/file of data. The advice provided is good, that you require to ascertain the most commonly accessed physical groups and size the block so as to be able to accommodate the largest of the commonly used groups. The trouble is that you do not know which are the most commonly used/accessed groups because SSADM does not count the number of accesses and therefore cannot produce the total accesses. Determining the commonly used groups from the Function Definitions and their ECDs and EAPs is a very inadequate mechanism. First of all there are n Function Definitions, ECDs and EAPs, and no database designer wishes to read n documents. Furthermore ECDs and EAPs do not count the number of accesses a function takes. Granted the Function Definitions document the frequency of the function, but that does not tell you the volume of accesses, and it is the volumes of access that one requires.

- Having undertaken transaction timings using the design optimisation technique SSADM does not show how to interpret the result if performance is not met—is it the transaction, the processor or the database at fault? How to do this is addressed in chapter 3.

Additional points of concern are:

- The transaction timing form has a number of features missing that should be included. An example of the form that the author uses is illustrated in figure 2.12. The additional features for the processor overhead are:
  — The front-end and back-end processing required if a client-server file handler architecture is used;
  — The operating system processing overhead (some vendors such as IBM and ICL separate the TP monitor from the operating system, others such as DEC and Hewlett Packard do not);
  — The processing required for the database I/O, that is, the logical I/O to the buffer pool and physical I/O to the disk.
- There should be a reasonable assumption that these control software actions are triggered as a result of the database record being accessed/actioned. Information Engineering estimates this overhead as some 20% on the processor. The additional features for the disk accesses include the I/O incurred for journalling/logging and for the paging of application program, file handler and operating system/TP monitor software.
- The final overhead to include is the telecommunication line times.

In total these missing features represent a substantial portion of the processing overhead contributing to transaction response times. Because of these additional points of concern, estimates of processor speed and size based on the SSADM calculations should be treated with caution.

Record/Table Name	Call Type	No. of Calls	Processor Time					Disc I/O				Comments
			File Handler	Application Program	Interrupt Handler (OS+TP Mon')	Disc I/O Overhead	FE/BE; Multi-Proc	Index	Data	Journal	Virtual Paging	

Note: the data disc I/O includes I/O to support referential integrity

Figure 2.12 Transaction timing form

SSADM Weaknesses

13 Program Design/Physical Process Design   There are a number of concerns:

- The first and most significant is that there is no technique. This is the only part of the logical and physical design process that is not addressed with a design technique, and seems all the more extraordinary in that there is a technique in version 3 of the method. No reason for the absence of a technique is given. And it seems strange, with no plausible explanation in view. For the major area of systems development there is no technique. The toolset of database facilities has not led to the demise of the database design optimisation technique. Why should there be a demise of the version 3 program design technique, which is excellent, in particular for batch program design? Information Engineering has not abandoned program design because of 4GL software. Why should SSADM? *This is the second area where the author believes that version 4 is the poorer over version 3.*[1]

- The manual states that while there is considerable certainty about how to convert a data design into a database design there is uncertainty about how to implement a logical process specification in a specific programming environment. SSADM continues to reflect that uncertainty. What is provided is a set of guidance as to what to do. In the author's opinion "what to do" is not an adequate substitute for "how to do it". A substitute for a technique, albeit a poor one, would have been worked examples of the products/deliverables to be produced, but again there is nothing shown. This section of the manual is therefore difficult to grasp, and some of the deliverables, such as the Function Component Implementation Map (FCIM), are still unclear, even though the author has asked trainers of the method to describe them.

- There is yet again the problem of the function. SSADM says that the functions become the application programs. Given that functions can contain many events, an application program can contain the logic for many events. The problem with this is that the application programs are triggered at the event level, but because they can contain code relevant to many events can pull unnecessary code into the processor main memory, which can therefore become overloaded, requiring increased paging of the application software and a lowering of performance. The sooner SSADM drops the concept of the function being the basis of any physical design, the better. It is very much a "red herring". Application programs are triggered and executed at the event, the business requirement, level. They always have and always will be at the event level. Their grouping into batches for offline running is purely a programming and operating design convenience—it does not alter the fact that the logic is still organised by event into individual

---

[1] A reminder that the other area is the absence of a numerically based technique of access path analysis.

application programs. The only real difference between offline and online programs is that the user triggers the online programs whereas the operator triggers the offline programs and the offline triggering requires careful batch design. There is still event level processing, either way.

- SSADM physical design does not recognise the increasing adoption of client-server architecture, where much of the processing of logic is now being done on intelligent workstations/front-end processors. As indicated in section 4.1.2.2 the advantage of this is that the front-end processors are able to undertake the processing that is "sensibly" theirs. No attention is paid as to how to split the logic between the client and the server processors.

  Given that operations are not added to the Dialogue Structures then there is no way that SSADM can support front-end processing. The operations have to be defined against the Process Models, such that the processing is implicitly done on a back-end processor.

  Surely, what is needed is that operations be specified logically where they are sensible, that is, on the Dialogue Structures for the man/machine part of the logical design specification, and against the Process Models for the database part of the logical design specification. At least the logical design specification can then explicitly support the not so new client-server technology. If the physical design assumes dumb terminals the operations on the Dialogues Structures can be readily and easily transferred to the Process Models prior to program design.

- The manuals attempt to describe what can be specified non-procedurally, but give no examples. *The components that can be non-procedural are access logic and the man/machine interface.* There are two types of business requirement, those that retrieve and present and those that retrieve, process and present. An example of the first could be "For a specified salesman display all orders won in the last six months". The request is purely access logic, which can be specified by the use of a menu based query language, to be presented in a screen format already painted or to be default created dynamically according to the data retrieved. All this is explicitly non-procedurally specified, with the data access and presentation code automatically generated, the former as SQL and the latter is screen painter code. *Non-procedural logic can be code generated.* An example of the second business requirement could be "For a specified salesman display all orders won in the last six months and calculate the bonus". The calculation of the bonus is process logic, which has to be specified procedurally. The programmer's pen must come out for this part of the program. *Process logic will always be procedural.*

- Although solutions are discussed, no examples of the resolution of three types of structure clash for the Enquiry Process Models is given. A pity. The advice for resolving an interleaving clash is not very helpful. The clash is resolved by the applications of two techniques, entity-event modelling and logical database process design. Yes, but how? Entity-event modelling includes the ELH and ECD techniques and logical database process design includes creating EPMs and UPMs. No explanation of how the techniques are to be used is given. This problem will be addressed fully in the to be released shortly (late 1991) SSADM/3GL Interface Guide.

There is no recognition of the scope of code generators. One of the great advances made by version 4 is that it now contains its own command language and syntax, not totally complete as we shall see in chapter 3, but a major start nonetheless. This language can become the basis of a code generator, such that where there is a need for procedural coding then the operations and conditions in the Process Models could be used as the basis of direct physical code generation.

3

"PEARLS OF PRACTICAL WISDOM"

This and the following three chapters provide advice as to how SSADM can be applied:

- more effectively for the data processing environment to which it is targeted;
- more widely in data processing environments for which it is not targeted.

All the advice is based on practical experiences and is correspondingly supported by real world or representative examples from case studies on which the author has worked.

For the new technology trends of knowledge based expert systems and object oriented systems the practical experience is still patchy. The policy followed for this book is that where there is an absence of practical experience, so that worked examples based on real life cannot be described, advice on the application of SSADM is explicitly caveated in that it is based on the author's knowledge rather than experience.

This chapter describes "pearls of practical wisdom" the author has learnt over a decade of applying a number of structured methods and found useful and practical in the specific application of SSADM. The pearls have found re-expression as "tricks of the practitioner's trade" that ensure that the concepts identified and described in chapter 2 are fully met when producing the logical design deliverables. The tricks have been tried and tested and have not been found wanting. They are proven. All the examples of the tricks are based on real world experiences. None of this chapter is theory.

The chapter is in two parts:

- Tricks that are generic to logical and physical design and not specific to any method or data processing environment. Given the purpose of this book, they are considered in the context of SSADM. The generic tricks fall into two categories—conceptual and systematic.

  The conceptual tricks reflect the generic concepts that logical design = physical design and test the logical design for efficiency. The tricks are:
  — ensuring that the entity relationship diagram (ERD) model is efficient;
  — ensuring that the transaction access paths are efficient;
  — producing and interpreting summary access path maps;
  — ensuring a balanced database design.

  The systematic tricks are:
  — ensuring that process decomposition is consistent and sensible;
  — ensuring that the process decomposition/dataflow diagram processes are used correctly in application program design;
  — appreciating the different kinds of keys to an entity;
  — building a composite LDM;
  — taking relational data analysis to sixth normal form;
  — ensuring that events that update the database occur in the correct sequence;
  — integrating the different aspects of the man/machine interface;
  — applying additional points of detail in the application of the design techniques.

- Tricks that are specific to version 4 of SSADM. Problems have occurred on a number of projects where version 4 has been applied and it has been found necessary to make enhancements to the method. The enhancements relate to:
  — the full adoption of Jackson standards rather than the production of Jackson-like diagrams;
  — the adding of operations to the Dialogue Structure diagrams, these reflecting the logic for the man/machine interface;
  — modifications to the Enquiry Access Paths (EAPs) and the Effect Correspondence Diagrams (ECDs) to cater for "jumping" access paths;
  — modifications to the Process Models to cater for common procedures and getting rid of the lasso facility where there is a one-to-one correspondence between entities accessed, so as to preserve the sequence of data access to the entities;
  — stopping the combining of the element boxes in the I/O Structure diagrams, this in effect being the equivalent of the lasso facility in the Process Models.

There are two ways in which enhancements to anything can be done—addition or modification. Addition is much the best approach as it does not put what is being enhanced at risk, and skills and experiences are preserved. Throughout this book the policy has therefore been to enhance SSADM by addition wherever possible. *Fortunately the SSADM techniques tend to suffer from sins of omission rather than sins of commission—what is there is overwhelmingly good and requires enhancement by addition of "tricks of the practitioner's trade" to what is already there rather than modification by redesign.* Only where there are sins of commission is there any attempt at modification to SSADM.

3.1 CONCEPTUAL TRICKS

3.1.1 Testing for logical design efficiency

The testing of the logical design for efficiency is a crucial concept to which SSADM, along with all other leading structured design methods, pays no attention. Given the concept that logical design = physical design, a concept that SSADM version 4 now explicitly supports, and the inexorable move towards using the logical design as the source for code generation, it is incumbent on us to make sure that the logical design is efficient. At least the physical design has a chance of being efficient from the start and may not need any subsequent optimisation at all.

The two aspects that need to be tested are the logical data model (LDM) and the transactions that access it. The testing, however, is not equal for all the transactions. There is a substantial difference in the nature of transactions that update the LDM and those that retrieve data from it. The data maintenance transactions *tend* have short access paths, mostly to only one entity occurrence (create customer, change order quantity, delete stock). There is not much testing for efficiency to be done there. By contrast, many of the data retrieval transactions have much longer access paths (display orders for customers where ... and products where ... and invoices where ...) and potentially several possible entry points (customers, products or invoices). As will be seen, it is against the data retrieval transactions that the LDM most needs to be tested for efficiency. All the case study examples are for data retrieval transactions. Never once has the author found it appropriate to test data maintenance transactions against the LDM.

Any inefficiency in the LDM and the logical transaction accesses to the model can be ascertained during the process of data navigation/transaction access path analysis. (The use of the term "transaction" here is a logical transaction, not a physical transaction. A logical transaction is the same as a business requirement. A physical transaction occurs in an application program and is a logical unit work (LUW). A LUW is a unit of database and program recovery and is defined in the program between a BEGIN and a COMMIT statement. Any database modifications made between these two points is flushed from the buffer pool to the log file and possibly to the database when the COMMIT is reached.) Note that the word "technique" is not used when describing transaction access path analysis, for the simple reason that the SSADM manuals do not describe one. Transaction accesss path analysis was identified in chapter 2 as *the* area of weakness in SSADM.[1]

Testing the LDM and transactions is undertaken together because it is difficult to separate them in practice. If a transaction is found to be extremely inefficient against the LDM it could be either the transaction or the LDM that is at fault. The "tricks" in this section show how logical inefficiency can be ascertained and how the LDM and transaction can be optimised.

The following aspects of efficiency are considered:

- Business requirements with multiple entry points.
- Balancing data retrieval versus data maintenance overheads.
- Data redundancy.
- Key sequence.
- Minimising entry points.

3.1.1.1 Business requirements with multiple entry points

The issue of logical access path efficiency occurs where a business requirement has multiple entry points. A typical example in figure 3.1 occurred in an application system for a local authority. Store items were held in stock and the policy followed was that whenever a store item was issued an order was immediately placed to re-stock the store, so as to maintain a constant stock level. Thus, whenever an issue was raised, money to pay for the order was involved, such that an issue related to an account. Analysis of the LDM had shown that for an account there was on average 1000 issues and for a store item 70 issues, with 10 issues matching an account and store item.

The business requirement was "For a specified account and store item list all issues". The entry point options were on the specified account or store item. Entry on the account generated 1000 accesses to all its issues, 10 of which related to the specified store item. There were therefore 10 further accesses to the store item in order to obtain the description of the item. The total logical accesses when accessing via the account entry point was therefore 1011. The reverse access path via the store item generated only 81 accesses.

---

[1] The EAP and the ECD techniques are not classified as transaction access path analysis techniques. Although they draw a diagram of the accesses against the logical data madel for a business requirement they do not count the number of accesses and therefore cannot ascertain that the access path is efficient or the best.

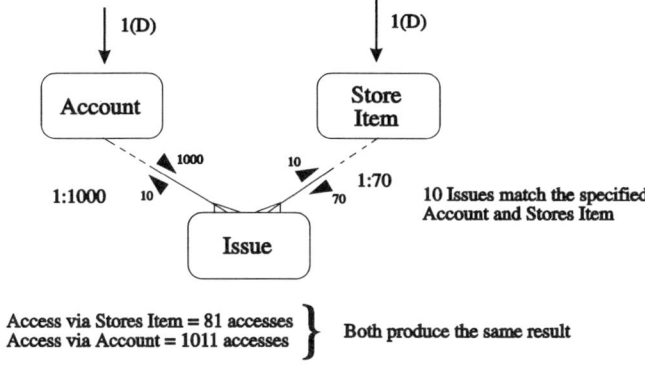

Figure 3.1 Logical access path efficiency (1)

Thus accessing via account produced an access path more than 12 times less efficient logically than access via store item.

The author concedes that a physical design could negate the above conclusions. However, the author works on the principle that, as a rough rule-of-thumb, one logical access against the data model will produce on average one physical I/O access to disk. A quick estimate of the number of physical I/Os that a typical relational file handler would incur against figure 3.1 would be one physical I/O to access the index pointing to the account, which would generate a further physical I/O to access the appropriate account table row. Assuming there is an index on the issues table by account number and the bulk of the index is in main memory the number of index accesses to issues would be, say, 10 physical I/Os. Ten index pages at the bottom level of the index contain 1000 pointers to the 1000 issue table rows relating to the specified account. The issue table rows, assuming they are stored in entry sequenced order (the most usual mechanism used by relational file handlers), would generate a further 1000 physical I/Os, as the issues relating to the account would more than likely be spread around like "grass seed" on disk. Direct access to the 10 occurrences of the specified stores item relating to the accessed issues would generate one physical I/O via the index and 10 further physical I/Os to the stores item table rows. The total number of physical I/Os is therefore 1022. This compares with the 1011 logical accesses to the data model. The assumption of one-to-one logical access to physical I/O is therefore not unreasonable.

The only occasion in practice where the basis of direct conversion from logical to physical has ever been significantly awry was where a very small database (some 8 megabytes) was pulled in its entirety into processor main memory. This virtually eliminated the physical I/O overhead on the processor, which was only incurred on the initial load of the database tables, but in no way affected the logical I/O to the buffer pool—data has to be accessed. In all other cases there has been a remarkable similarity of the physical I/Os to a balanced database design taking into account total data structure and total data access overhead and the number of logical accesses to the LDM.

Where the disparity between the accesses between two entry points is as great as in the example then the logical conclusion—access via store item rather than via account—is valid at the physical level. One does not need the database optimisation techniques to tell you this!

3.1.1.2 Balancing data retrieval versus data maintenance overheads

This situation occurs when the logical data model appears to be inefficient at supporting a business requirement and potentially requires improvement by modification. The questions are—how to ascertain if the LDM is inefficient and will modification result in improvement?

Both examples in this section relate to a client in the Third World. The author's terms of reference were to minimise any unnecessary complexity in application programming.

The first example in figure 3.2 shows the situation where the LDM was inefficient but the proposed modifications proved to be the reverse of an improvement. The application was a personnel system, with the database

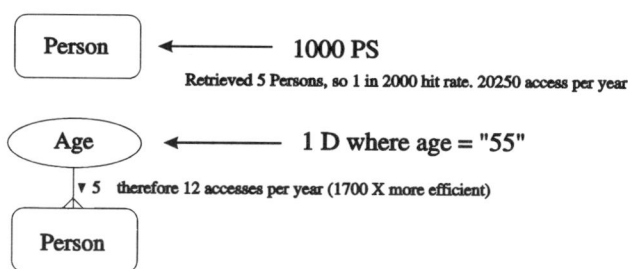

Figure 3.2 Logical access path efficiency (2)

containing 10 125 persons. The business requirement was "List all persons who are 55 years of age" and ran twice a year. The users requesting this requirement were the pension support staff. People retired at the age of 60 and the staff required to know the pension cash flow requirements in some 5 years' time. There was no requirement to list the persons in person number order. The entry point access type was therefore physical sequential.

The first access path chosen was to scan the person entity to ascertain who was 55 years of age. Each scan generated 10 125 accesses. The client company was, however, composed of young people, so that out of 10 125 accesses only 5 persons were retrieved—a hit rate of 1 in 2000. Even with a large blocking factor it was clear that this was a very inefficient access path. The installation used IBM hardware and it was known that the largest block size could go up to the full track on a disk, which would equal some 20K. However, it was also known that the person entity contained many attributes and would require some 600 bytes of data. (Notice here that the boundary between logical and physical is extremely imprecise. *Physical considerations that are generic to all hardware and software requirements can quite legitimately be considered as part of the logical design process.* After all, one is producing a logical design specification that will be physically implemented. What is not permissible at the logical level is to consider specific hardware/software considerations. This at times is difficult to achieve if one knows the target physical environment.)

It was crystal clear that the initial logical access path was going to be inefficient, whatever the target physical environment. The LDM was therefore modified with an operational entity of age above person. The data retrieval access path was now extremely efficient, with direct access on age where age = "55", which in turn related to the 5 persons of that age. The total logical accesses via age was therefore 6 as opposed to 10 125 accesses via person—a data retrieval improvement by a factor of more than 1700!

However, this improvement was more apparent than real. Nobody has the secret of eternal youth. We all have birthdays. Assuming 250 working days per year, the need to keep people's ages up to date and the resultant requirement to scan the person entity each working day to ascertain who has a birthday, the cost of maintaining the second access path was found to be 2.5 million logical accesses per year. The data maintenance access costs on the second access path were therefore 123 times greater than the data retrieval benefits!

It was therefore clear that the first access path was the one to choose, even though it was not very efficient.

Many people have pointed out that a simple alternative solution would be to write the application program to test for date of birth. However, this would have involved the users or the application programmers in some mathematics. Given the terms of reference, it was not followed.

The second example is the reverse of the first—that is, where the LDM was inefficient and the proposed modification proved to be a substantial improvement. The example is based on the same personnel system. It is illustrated in figure 3.3. The business requirement was run once a month and analysis had shown that a person changed his/her job seat twice a year and their section once every two years. Given that the business requirement wanted to list the staff movements during the last two years then the cardinalities between person and person/job seat for this transaction is 4. (Note the concept of transaction cardinalities introduced. As far as the author is aware, the Navigator version of SSADM is the only structured method that supports transaction cardinalities.) Person/job seat was stored in time order.

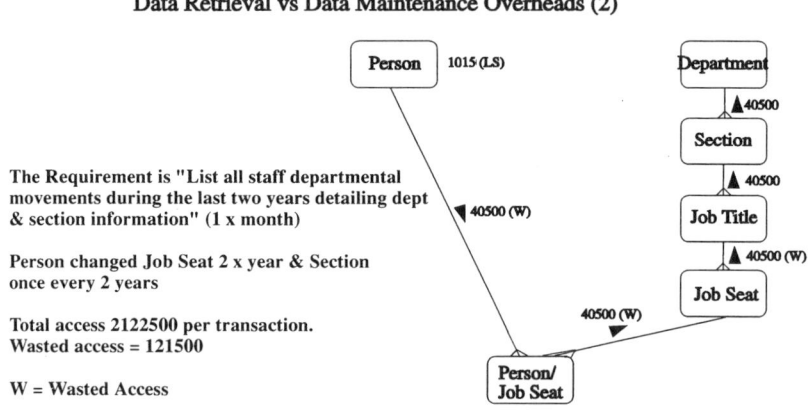

Figure 3.3   Logical access path efficiency (3)

Assuming that the dependent person job/seat is related to person in time order then the access path and the number of accesses was as illustrated. The entry point type on person is logical sequential, as it was necessary to list the persons in person number order. The total accesses for each running of the transaction was 212 625 logical accesses. Given that the only entities holding data of relevance to the business requirement were person, section and department, the accesses to person job/seat, job seat and job title were totally wasted—a total of 121 500 accesses. Assuming on average that one logical access generates one physical disk I/O (as indicated earlier, a not unreasonable assumption) and that each physical I/O is some 30 milliseconds then about one hour of dedicated I/O time would be wasted per transaction execution. The LDM was therefore inefficient for this transaction.

The relationship between a person and a section was relatively stable—to be precise, four times more stable than between a person and a job seat—and would therefore have low data maintenance overheads. A link data

Conceptual Tricks 115

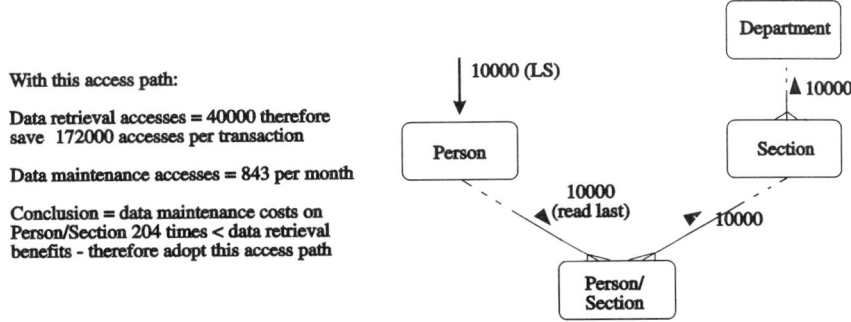

Figure 3.4 Logical access path efficiency (4)

group between person and section was therefore created—see figure 3.4. The transaction set cardinalities between person and person/section, given the time span of the business requirement, was unity, so the total data retrieval accesses was now 40 000, a reduction of some 170 000 accesses. This was a saving of some 86 minutes of dedicated disk I/O time, of which only one quarter—the accesses to the person/section link entity—were wasted. The data maintenance costs of this new data group was found to be only 843 accesses per month, i.e. some 204 times less than the savings on data retrieval. It was clear that, unlike the first example, the LDM required to be modified.

As with the multiple entry points example, the logical conclusions based on the above clear-cut evidence would be valid in any physical design.

3.1.1.3 Data redundancy

The example as to when it is profitable to introduce redundant data into a logical data model is based on figures 3.5 and 3.6. Redundant data can either be derived data as a result of a computation or a duplicate copy of data stored elsewhere in the database. The client was an oil company. The oil wells were exploration wells, so that tests, injections, flow rates and hours off per day were constantly monitored, often at second intervals. This information was required to be stored for many months in order to obtain a pattern. The set cardinalities between conduit and test, flow rates, injections and to a lesser degree hours off per day were therefore enormous—over 1 million, for example.

The data retrieval business requirements were typically of the kind "Display the average flow rates for a specified well for the last month" or "Display the average down time for a specified conduit for a specified week". The number of accesses to the detail entities to conduit were far too high, particularly for online transactions. It was clear, even at this logical design

Constant need to know such things as:

* Average flow rates per week/month

* Average conduit down time per week/month

Very wide cardinalities of conduit to detail entities, therefore expensive to provide answer

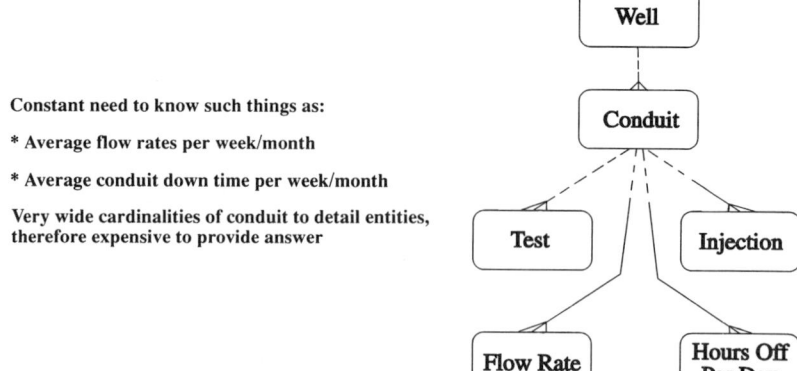

Figure 3.5 Logical access path efficiency (5)

Created a Weekly Summary Entity:

high data maintenance logical access but low physical cost:

* Conduit not an entry point entity
* Hierarchical data structure, therefore can cluster Weekly Summary via Conduit
* Have to access through Weekly Summary anyway to insert dependent entities

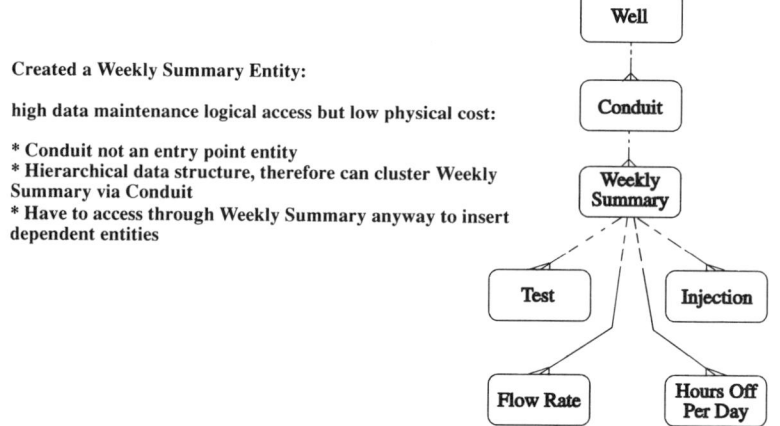

Figure 3.6 Logical access path efficiency (6)

stage, that the physical transactions would fail the response time objectives. The solution was to introduce a weekly summary entity as a detail of conduit. The summary data group would hold summarised statistical data of each of its detail data groups. Most data retrieval transactions now finished at weekly summary and were highly efficient.

Clearly there was a data maintenance cost involved. It was ascertained that the extra data maintenance costs were high at the logical level but low at the physical level. This example again illustrates that the boundary between logical and physical design is often blurred. The reason why the physical

Conceptual Tricks 117

cost of inserting weekly summary on to the LDM was low was that the data structure is a hierarchy and weekly summary could therefore be clustered in the same page/block as its conduit, which was the usual entry point entity. The set cardinalities between conduit and weekly summary was a fixed 52 (i.e. one year of historical information) so that with a large block/page size a conduit and all its associated weekly summaries could fit in one block/page. The weekly summary entity could convert into a physical table of data, with each row requiring some 30 bytes of packed decimal data storage. Fifty-two rows would therefore fit easily within a 2K block/page, even including the master conduit table row to which weekly summary was related.

It was clear that the proposed logical design would be physically efficient. Accessing weekly summary would generate one logical I/O access to the buffer pool for every insertion of test, flow rate, injection and hours off per day and therefore appear to be expensive. Crucially, however, physically no extra disk I/O would be involved whatsoever. As explained in section 2.4.2.1/1 physical disk I/O is slow and expensive and requires to be kept to an absolute minimum. The transaction would therefore be physically efficient. The redundant data was therefore highly profitable.

3.1.1.4 Key sequence

This trick is illustrated in figure 3.7. The client was a holiday travel company and the requirement, *inter alia*, was to monitor the locations at which persons had taken their holidays. Analysis had shown that a person takes on average three holidays a year and that a location has about 1000 holidays a year. It was necessary to hold 20 years of holiday history. The set cardinalities between person and holiday was therefore 60 and between location and holiday was 20 000. A typical business requirement that was frequently triggered was "For a specified person display all holidays taken during the last 2 years". Assuming that the holiday entity is unsequenced then the access would be to enter on person and to read all 60 holidays for that person to ascertain which occurred during the last two years. If the relationship between person and holiday was sequenced on the keys of date of holiday and person number then reading reverse from the specified person to the holiday entity would require only six accesses.

It was clear that sequencing holiday by time in relation to person would provide major physical design optimisation benefits, irrespective of whether the file handler was first generation hierarchical, second generation Codasyl or third generation relational. Sequencing the relationship by time improved the retrieval access efficiency by a factor of 10. There was also no data maintenance overhead, as the holidays were entered in time sequence anyway.

Clearly if other business requirements want to access holiday via person on another search key the date sequence of holiday would not be appropriate.

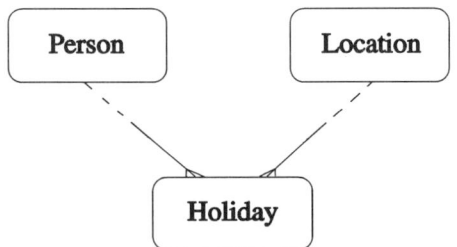

Business Requirement:
"For a specified Person list all Holidays taken in the last two years"

(With key sequence read reverse Holiday for 6 records - 3 holidays per year)
(Without key sequence read all 60 records - 10 years of holidays)

Figure 3.7 Logical access path efficiency (7)

The search key most frequently used requires to be the chosen sequence key of an entity. The choice of key sequence is, of course, the usual design balancing act.

3.1.2 Producing summary access path maps

The need for summary access path maps is based on the concept of database. As explained in section 2.2.2.3 a database should be designed for the enterprise as a whole and is based on two of the three major components of a logical design specification, namely data structure and data access. Given that the database is for the whole enterprise then it is necessary to ascertain total data structure and total access.

By designing a database on the basis of total data structure and total data access the author believes that the design *a priori* will be balanced and that a balanced database design is inherently efficient. "Tricks" to support this concept are discussed in this and the next two sections.

The mechanism for producing summary access path maps is to add up the access paths of each individual transaction. Five different summary access path maps are required:

- *Overall System*. This is the total data accesses for all the applications at an enterprise and is the access component of a balanced database design.
- *Online transactions*. This map shows those portions of the LDM that require high performance.
- *Batch Transactions*. This is the least critical map and has not proved of value in practice.

Conceptual Tricks 119

- *Data Retrieval Transactions.* This map shows the total retrieval accesses against the LDM.
- *Data Maintenance Transactions.* This map shows the total modification and referential integrity accesses against the LDM.

The two last summary maps are essential for distributed database design and are discussed in chapter 4.

The sum of the batch and online and the data retrieval and data maintenance transaction summary access path maps each equal the overall system summary map.

Consideration of the summary access path maps is in two parts—how to produce them and how to interpret them. The production of the summary maps is a slow and repetitive but vital task. The steps are as follows:

1 *Decide a time span.* The author tends to use a time span of one month. Any time span is acceptable. On a highly volatile online system the author has used a time span of one day.

2 *Obtain the transaction volume and frequency.* The frequency of a transaction could be, for example, daily or weekly. The volume is the number of times the transaction runs per frequency, for example 500 times per day.

3 *Assuming a one month time span:*
   - If the frequency is daily multiply each transaction access to an entity by 21 (working days) or 30 (calender days) as appropriate.
   - If the frequency is weekly multiply each transaction access by 4.2.
   - If the frequency is monthly multiply each transaction access by 1.
   - If the frequency is quarterly divide each transaction access by 3.
   - etc.

4 *Multiply access volumes to each entity in each transaction access path by the result of 3.*

5 *Draw the monthly access volumes on the LDM for each transaction.*

6 *Do the above for each transaction.*

7 *Combine all the accesses to produce the total summary access path map.*

A walked through example of the above is given in figures 3.8 to 3.12. The time span assumed is one month.

Once the summary maps have been produced the database administrator has the two components that go into database design—total data structure and total data access—on a single sheet of paper. Contrast this with the current situation with most structured methods. The database administrator will have a logical data model and a whole set of transaction access path maps, one for each business requirement. If there is a grand total of 200 data maintenance and

Data structure against which the summary access path map will be built

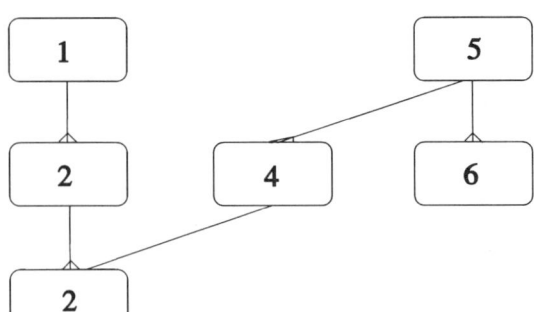

Figure 3.8 Summary access path maps (1)

data retrieval business requirements then the database administrator will have 201 sheets of paper, one of which shows total data structure (the LDM) and 200 of which show individual transaction access paths. In order to get an overall "feel" of the total accesses the database administrator will have to read and absorb the information from 200 individual transactions—a virtually impossible task.

3.1.3 Interpreting the summary access path maps

Figures 3.13–3.17 illustrate five representative summary access path maps for an enterprise. They require to be interpreted in order to achieve a balanced database design, balanced in the sense that the design takes total data structure and total data access into due account. Note that the version 3 facility of the operational data group has been reinstated. This is necessary so as to show key only data groups that will be system maintained as indexes.

The interpretation is as for a typical relational file handler. Starting with figure 3.13:

1 Each entity becomes a table.

2 Each relationship between master entity and detail entity becomes an index.

The key of the index is the key of the master entity and the index points to the detail entity.

3 Each entity not at the bottom of the LDM (i.e. an entity with detail entities) requires an index on its prime key.

This is to ensure efficient support for referential integrity.

4 Decide entry point types.

Conceptual Tricks

SYSTEM: ANY

(1 x day)

FUNCTION NO: 123				FUNCTION NAME: EXAMPLE 1		
Data Group	Acc. Type	Read Type	Acc. Via	No. Acc.	Data Items	Conditions and Comments
1	R	D	-	1		
2	R	P	1	10		Read entire set forward
3	R	P	2	10		Read last
4	R	C	3	10		

Access Type

I = insert   D = delete
M = modify   L+ = add to optional link path
R = read   L = remove from optional link path

Read Path

D = direct
PS = physical sequential
LS = logical sequential
C = via child
P = via parent

Example 1

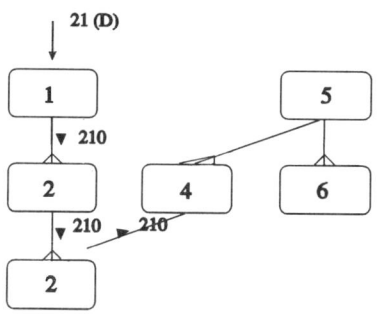

Figure 3.9 Summary access path maps (2)

Entity A is accessed only directly. It is a highly stable data group (see figure 3.16) with the few data maintenance accesses being updates rather than inserts and deletes (for the sake of brevity the summary maps do not show a breakdown of the type of data access, but can easily be detailed to do so). The message from the summary maps is to choose a randomised direct entry point access mechanism. Indexed access would be more expensive in disk I/O and processor overhead because it requires to be accessed and maintained. If the file handler does not support randomised access an index

SYSTEM: ANY

(150 x month)

FUNCTION NO: 321					FUNCTION NAME: EXAMPLE 2	
Data Group	Acc. Type	Read Type	Acc. Via	No. Acc.	Data Items	Conditions and Comments
5	R	D	-	1		
6	R	P	5	3		Read reverse last three occurences
4	R	P	5	10		Read last
3	R	C	4	100		Read entire set
2	R	C	3	100		

Access Type

I = insert   D = delete
M = modify   L+ = add to optional link path
R = read   L = remove from optional link path

Read Path

D = direct
PS = physical sequential
LS = logical sequential
C = via child
P = via parent

Examples 1 & 2

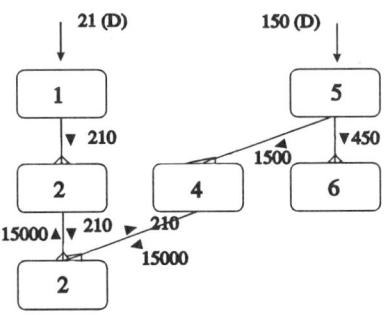

Figure 3.10 Summary access path maps (3)

would be required to provide direct access entry. The disadvantage that randomisers incur, that they do not provide logical sequential access unless a specific type of algorithm is used and the randomised key values are in a known continuous ascending or ascending sequence, is not a concern here as there is no requirement to access on entity A in logical sequential order.

There are no entry points on entity B so cluster entity B via entity A on disk. The data group volumes and set cardinalities document shows that the

Conceptual Tricks

SYSTEM: ANY (1 x week)

FUNCTION NO: 456		FUNCTION NAME: EXAMPLE 3				
Data Group	Acc. Type	Read Type	Acc. Via	No. Acc.	Data Items	Conditions and Comments
1	R	D	-	1		
2	R	P	1	10		Read entire set backwards
3	R	P	2	10		Read last

Access Type
I = insert   D = delete
M = modify   L+ = add to optional link path
R = read   L = remove from optional link path

Read Path
D = direct
PS = physical sequential
LS = logical sequential
C = via child
P = via parent

Examples 1 & 2 & 3

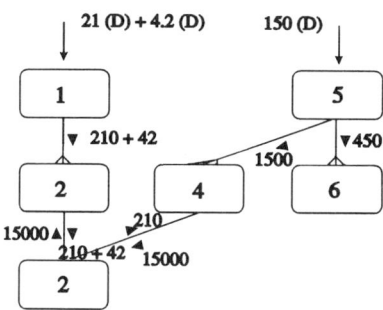

Figure 3.11 Summary access path maps (4)

working average set ratio is low—let's say 10—and that 50% of entity A have detail entity B. The entire cluster of entity A and its related entity Bs will therefore fit on one page/block, such that the 100 logical accesses between entity A and entity B will not generate any physical I/Os in the implemented system. Clustering here would be very advantageous.

Entity C is a problem as regards what entry point mechanism to choose. 90% of all entry point accesses are direct, so that randomised access initially looks a good choice. However, the remaining 10% of entry point accesses require logical sequential retrieval. If the accesses were batch then an extract and sort routine would be acceptable and randomised entry point access still a good choice, in order to avoid the cost of maintaining and accessing via an index. However, reference to figure 3.14 shows that the logical sequential

System Summary Access Path Map

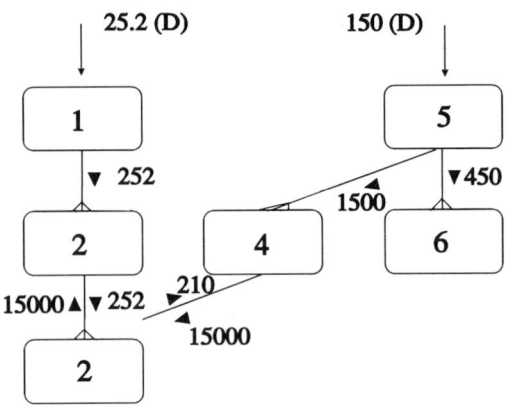

Figure 3.12 Summary access path maps (5)

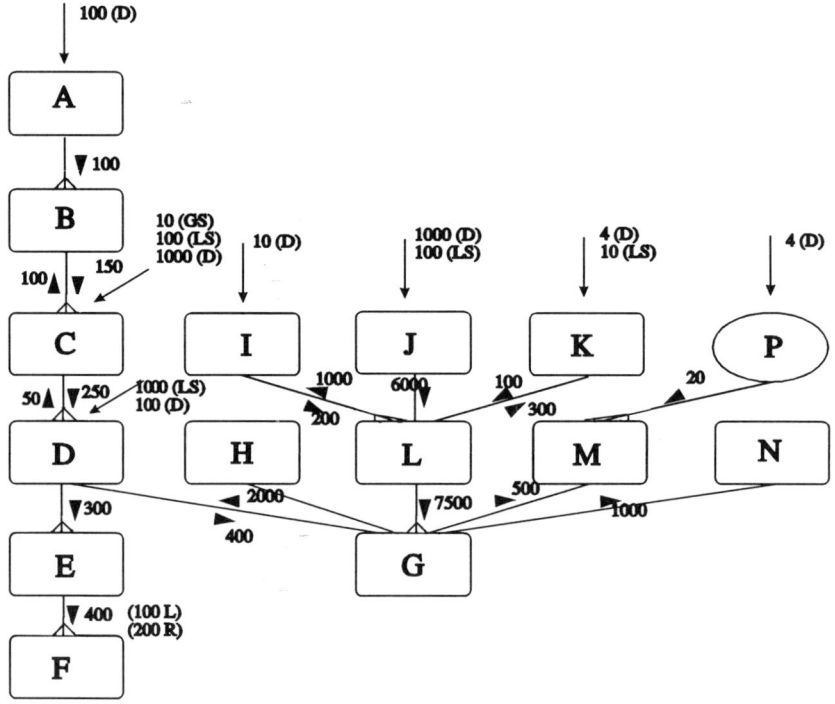

Figure 3.13 Total summary access path map

Conceptual Tricks

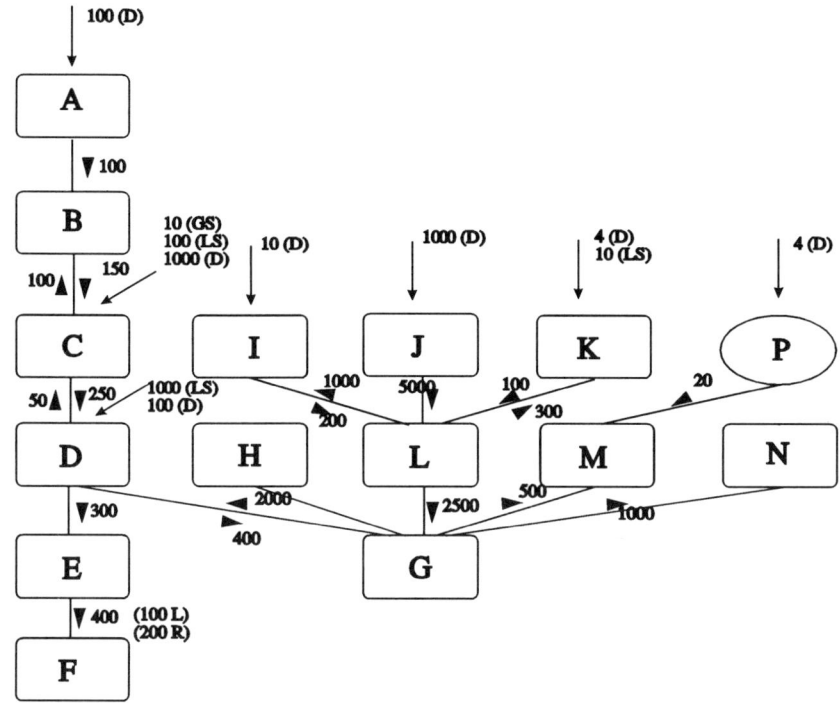

Figure 3.14 Online summary access path map

entry point access is online, so high performance is required. Dynamic online sorts, particularly with large volumes of data, are bad news and should be avoided wherever possible. It is clear that entry point access on entity C needs to be via an index.

Entities M and N appear not to require any entry point access. This is not the case. They require either randomised or indexed access (preferably randomised as all the entity accesses are direct) to support access from entity G.

5 Cluster the non-entry point entities.

Entities E and F are not entry point data groups, so cluster them via a master entity. In this case there is only one master entity, so cluster them via entity D. A hundred accesses between E and F are last, i.e. for a given E access its last occurrence of F where the sequence of F is known, and 200 accesses are read reverse from the last occurrence of F. Given that relational file handlers do not support pointer chain technology, a scan and sort routine for these

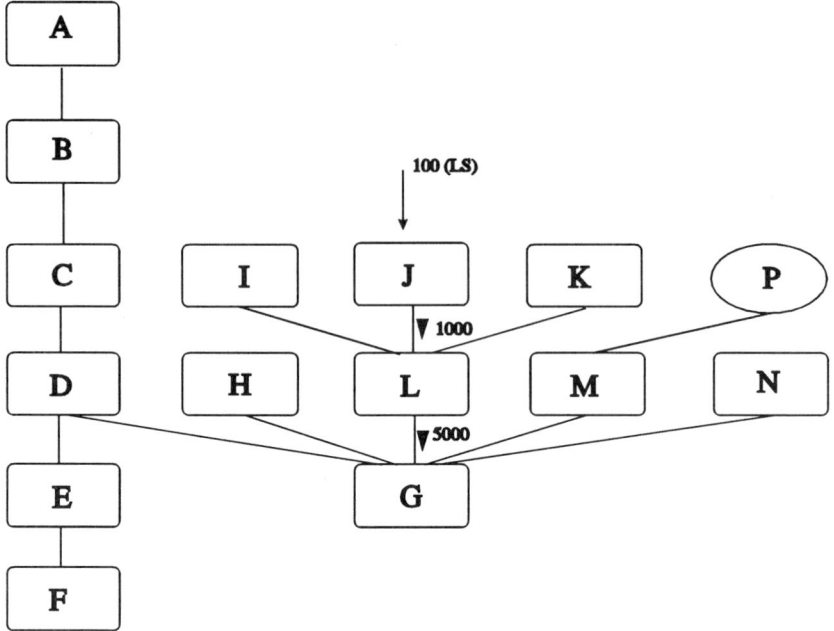

Figure 3.15 Batch summary access path map

transactions requiring last and first access will be necessary, followed by access to the desired entities. If the key sequence on which the last and reverse access is different from that of the foreign key relationship key between E and F then access to F can be optimised with an index on the last and reverse key(s). For example, if the last and reverse key is date and date is not part of the foreign key in F to E then index F on date.

Entity D is the opposite of entity C. 90% of the entry point accesses are logical sequential, the data maintenance overheads are low—see figure 3.16—and the access is entirely online—see figure 3.14. Indexed rather than randomised entry point access is therefore required. In addition, it would be worthwhile, based on the above evidence, to store the data on disk in logical sequential order as a "key sequenced data set".

Cluster entity G via entity L and entity L via entity J. Both entities relate to multiple master entities, so to which master should they be clustered? The total summary access path map provides the answer. Entity G is accessed more than three times more frequently via entity L than via any other master entity and entity L in turn is accessed five times more frequently via entity J than via any other master entity. Clustering as advised above would therefore have the maximum benefit as regards reducing physical I/O on access between these three entities.

Conceptual Tricks

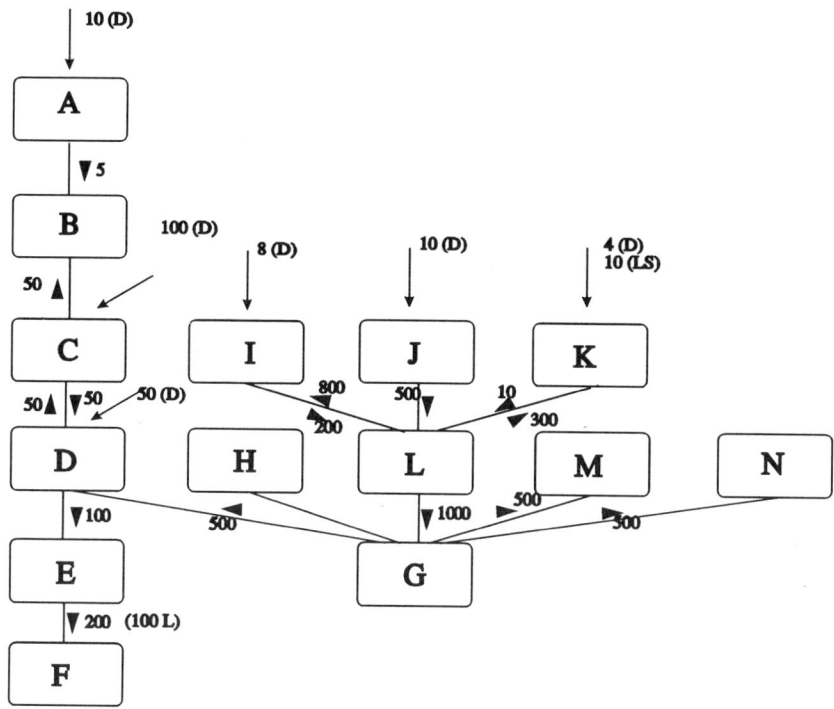

Figure 3.16 Data maintenance summary access path map

A KSDS has one possible and one definite advantage. First the prime index on the key data item on which the file is now sequenced is much smaller (by a factor of the number of record occurrences per page/block) than the secondary indexes or any indexes if the file is not key sequenced. For example, if 10 record occurrences could fit per page/block then the prime index would be 10 times smaller than the secondary indexes. This is because an index on the key sequence data item only requires one pointer per page/block—see figure 3.18. The example here is the VSAM KSDS file handler from IBM. A control interval equals a page/block. The control area is all the control intervals in a cylinder on the disk. It illustrates that the lowest sequence pointers in the index point to the last table row in a control interval block and that there is only one pointer per control interval, not per table row.

By contrast, all secondary indexes and indexes on non-key sequenced files require one pointer per table row occurrence. The key sequence index is therefore smaller and more likely to be held in the processor main memory buffer pool and thereby reduce disk I/Os.

But care needs to be taken. The strategy of one pointer per page/block has a significant disadvantage—the block offset position of each table row in the

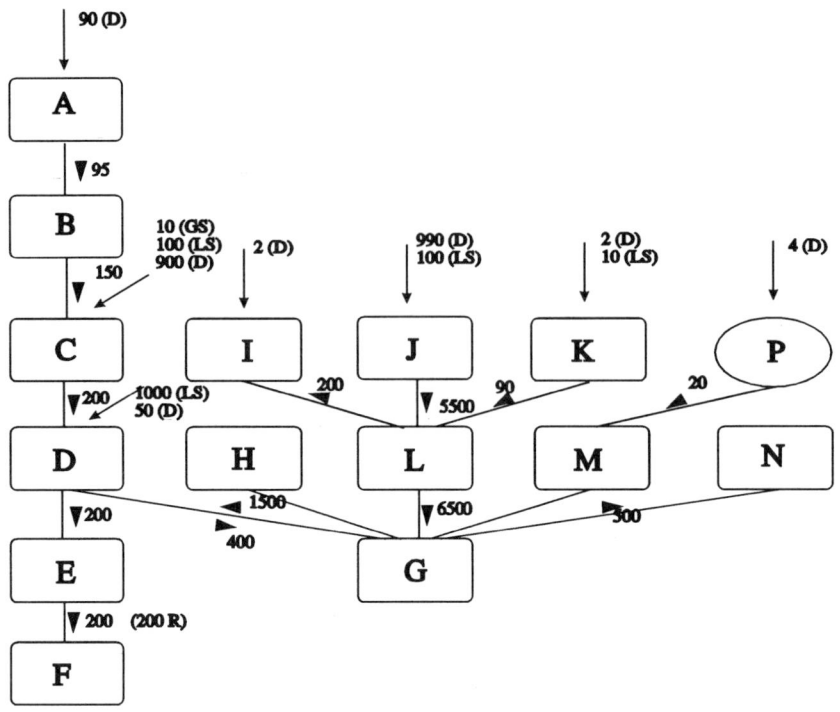

Figure 3.17 Data retrieval summary access path map

page/block is not given. The file handler therefore has to read on average half the rows per block to find the desired row. Logical I/O to the buffer pool is thereby substantially increased. If the blocking factor is 20 the logical I/O overhead on the processor would be increased by a factor of 10!

The second and undoubted advantage is that, by storing the table rows in logical/key sequenced order, 10 key sequenced accesses to 10 records stored in key sequence would generate only one physical I/O. The main disadvantage of key sequence data storage is on record insertion. If the block into which the table row requires to be inserted in order to maintain key sequence is full then overflow of some kind, such as block splitting, is required. However, figure 3.16 shows that the data maintenance accesses per month to entity D are low, of which 70%—let's say—are updates. Block splitting through insertion is therefore low risk. Make entity D a KSDS.

6 Delete non accessed entities.

Delete entity H from the database. It is never accessed.

Conceptual Tricks

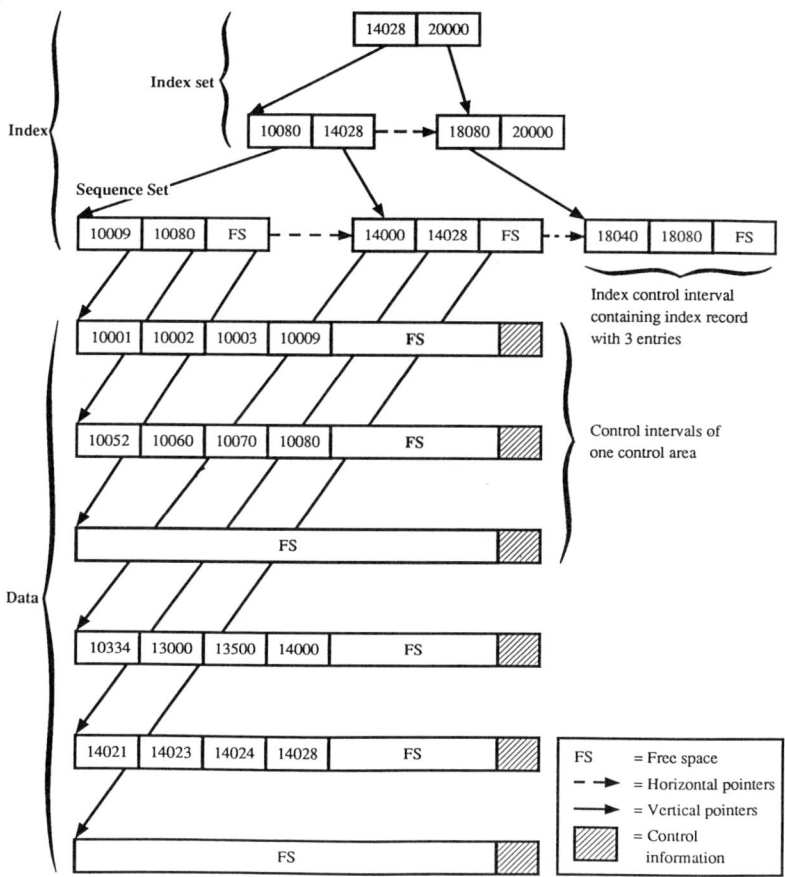

* VSAM is an IBM trademark

Figure 3.18 Index structure for a VSAM key-sequenced file

7 *Make operational entities indexes.*

Entity P becomes an index.

3.1.4 Producing a balanced database design

This trick reflects the concept that the database design is based on total data structure and total data access. While SSADM produces a total data structure LDM and total data access matrices the method does not guarantee to produce a balanced database design because, as explained earlier, the EAPs and the ECDs do not count the number of accesses in the access paths. SSADM therefore is liable to produce a tuned design—tuned for the transactions that happen to be selected for optimisation. There is also a further problem—the data structure component of database design is at the enterprise level whereas the data access component is mainly at the transaction level. An SSADM database design may not therefore match the definition of database as detailed in section 2.2.2.3.

By accepting the concepts of logical design = physical design, testing the logical design for efficiency and producing and interpreting the summary access path maps as detailed in section 3.1.3, the basis of balanced database and program designs that are inherently efficient can be easily achieved.

SSADM accepts the first but not the second or third concepts. It is this failure that prevents SSADM from ensuring that the database and program designs are balanced and *a priori* efficient. Once that advice is provided then much of the need for subsequent database and program optimisation will be eliminated. Much effort could be saved.

Two SSADM design techniques need to be enhanced to take total data access into account as well as total data structure. The first is the production of the Physical Data Design from the required LDM. This "first-cut data design" is obtained by making a number of transformations to the LDM that are based on five "rules of thumb" generic/universal DBMS data structure facilities concerned with the placement of data, to be carried out in eight activities. DBMS product-specific rules are then applied to the data design to produce a more implementable design.

But the rules of thumb are not meaningfully based on data access, the other component of database design. Such advice as to data access for the placement of the data into physical groups (rule of thumb 3) is somewhat ritualistic, for the EAP and ECD techniques that show the access paths against the LDM do not count the number of accesses, such that the number of accesses that is the basis of the groupings cannot be ascertained. In fact the groupings are actually based on the least dependence occurrence rule—activity 5. The problem of this approach is discussed in section 2.4.2.3/12.

Notwithstanding that the author believes physical database optimisation should not be necessary if a balanced database design is produced, the real world is likely to require some physical optimisation. No file handler currently gets "ten out of ten" so that compromises to and a certain amount of "tweaking" of the design for certain types of transaction access may be required. Of course, given the concept of logical = physical the fewer, the better. Database optimisation is the task of the second technique, namely Physical Design Control (PDC).

The PDC technique currently produces a tuned database design based on transactions the database administrator decides to test. The problem with this is that the administrator cannot be sure that he/she has chosen the best

Conceptual Tricks

mix of transactions. The manual states that the functions that consume the highest resources should be measured for performance, these functions being those accessing the entities with the most number of occurrences. The author cannot agree with this. Clearly where there are a large number of occurrences it is likely that there is a heavy data maintenance overhead, but that is no indication that there will be a corresponding overhead for the data retrieval business requirements. There are other things to consider as well, for example when inserting with a large number of referential integrity checks to make. The performance overhead of this was well illustrated when considering figure 2.6.

The conventional wisdom offered by a number of structured methods is to select those transactions that:

- require fast response times;
- occur at the highest interactive peaks;
- occur in high volumes;
- run for a long time.

This was the advice in version 3 of SSADM.

All good stuff. But what if the following occurs? The database is as drawn in figure 3.19, with the vast bulk of the data accesses occurring in only a small part of the database. Summary access path maps drawn against the LDM give a good visual impact. From the summary maps in figure 3.19 it is clear that the small part of the database requires the greatest optimisation. Yet it so happens that the "flashy" transactions selected for testing are to be found in the 20% area. There is no guarantee that the transactions selected on the basis of the above criteria will always fall within the small 80% area. Probably yes, but certainly no. *It is only by having summary access path maps as described earlier that the part of the database requiring greatest optimisation can be properly ascertained.*

By having summary access path maps the transactions to be selected for timing can be chosen on the basis that they access those parts of the database that have the greatest accesses and therefore require, without question, the greatest design optimisation.

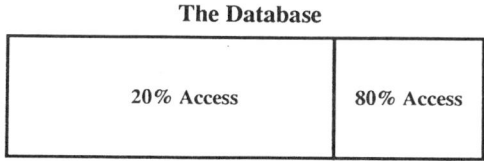

Figure 3.19 Database access distribution

By using the LDM and summary access path maps as the basis of choosing where the design optimisation technique will be applied the inherent efficiency of a balanced database will be reinforced. The mechanism by which the technique should be enhanced is the interpretation of the summary access path maps as detailed in section 3.1.3.

3.2 Systematic Tricks

3.2.1 Ensuring process decomposition is consistent and sensible

This is a notoriously difficult task to achieve, primarily because so much of it is subjective. The processes in the DFDs can be drawn at multiple levels of detail appropriate to the user specifying their requirements. This is one of the great strengths of the DFD technique. It can be pitched at whatever level is appropriate. Although SSADM allows decomposition of all the components of SSADM the author has only ever decomposed the processes. He has found it appropriate to decompose the processes to four levels of decomposition— business area, activities/functions,[1] events and problems-to-solve.

The problem with process decomposition is that the identification of business areas, functions and problems-to-solve is subjective and therefore open to interpretation. A hypothetical but realistic port authority is used as an example to illustrate the four levels of decomposition. The decomposition is illustrated as dataflow diagrams (DFDs).

The island port company has six major business areas as detailed in figure 3.20. These business area processes very often match the high level management structure of the enterprise, but it is not necessary for this to be so. Indeed, it is extremely important that the decomposition is undertaken "logically", that is, without consideration of any physical constraints, such as management organisational structure. Any match between the organisational structure and the decomposed processes is because the management structure is also "logically" organised to reflect the natural breakdown of the company's business operations.

The cargo handling business area is decomposed into a set of four functions—see figure 3.21. Business area and function processes are higher than the event level processes. A trick to identify processes that are higher than events is to entitle the processes using either a gerund ending with an "ing" suffix for the title (for example, accounting and passenger handling) or a double noun for the title (for example, general cargo, harbour maintenance).

The boundary between business areas and functions is not precise. Using a paraphrase of the official definition of a function in IEM, a function is a group

---

[1] Functions in the context of this section are as defined in version 3 of SSADM, that is, a "logically related group of events".

Systematic Tricks 133

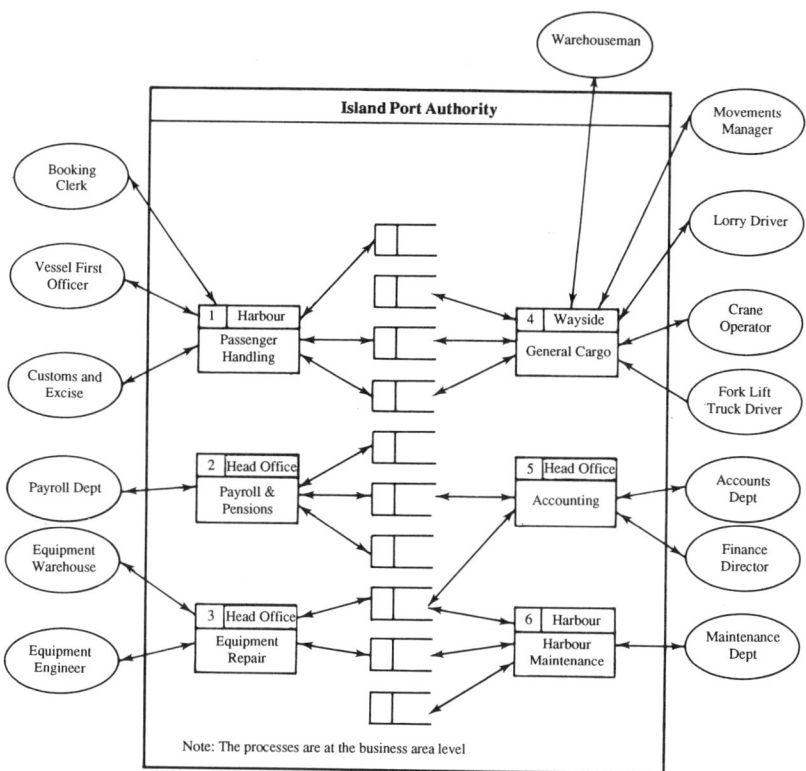

Figure 3.20 Dataflow diagram: level 1

of business activities which together support one aspect of the company's business. But what is an aspect? If one takes the function of cargo loading then, as illustrated in figure 3.22, there are six events, all of which are "sensibly" associated with the function in that they are concerned with the aspect of loading cargo onto a vessel. One could follow this argument further and say that a business area is nothing more than a sensibly related group of functions. For example, the functions in figure 3.21 are obviously associated with the business area of general cargo.

Event level processes are much easier to identify, as they are objective triggers to which the computer must make a response. A naming convention trick that the author uses to identify event level processes is to give them a "do something" title. In figure 3.22 the processes are so named—Load Pallet, Allocate Crane, Move Pallet (to Crane). This style of naming convention is also appropriate to problem-to-solve processes, as illustrated in figure 3.23. This is because problem-to-solve processes can, depending on circumstances, also be events. In the example the problem-to-solve Advise Movements Manager

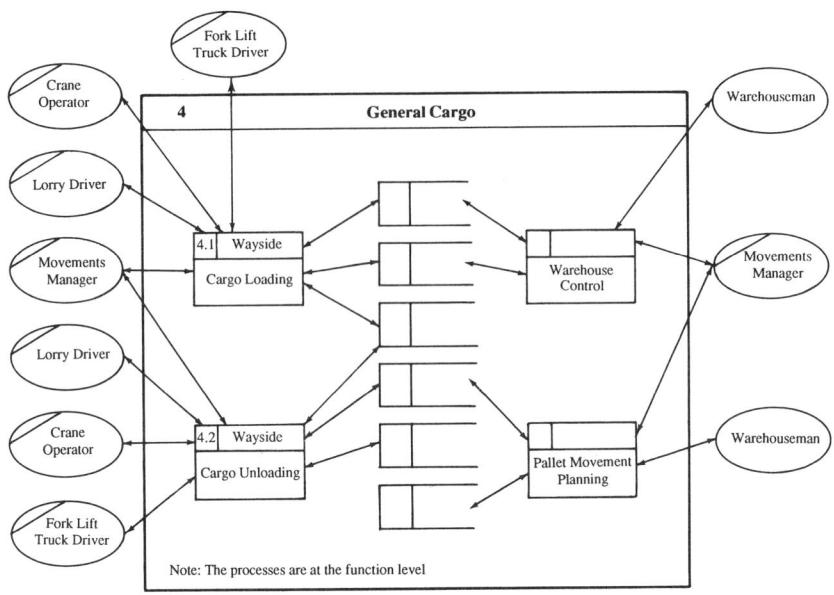

Figure 3.21 Dataflow diagram: level 2

so happens to be a problem-to-solve within the event of choosing/allocating a fork lift truck. In another business context it could well be a standalone event apropos anything to which the Warehousman, the Fork Lift Truck driver and the Movements Manager require to be advised on.

Decomposing processes on the lines described above has a number of advantages:

- It automatically pitches the processes at the two highest levels of decomposition to a level that is readily appreciated by the senior user/manager. The naming conventions of using gerunds and double nouns to describe the processes also facilitates an understanding that the processes are higher than events. The processes on the level 1 DFD/PDD is a deliverable that can be presented to and understood by the users of the major business areas in an enterprise, namely the senior managers. The functional processes at level 2 are logically grouped within business area and are processes that typically reflect middle management. The port company had a middle manager responsible for warehouse control at the main port and the organisation of all pallet movements within each of the ports and a middle manager responsible for cargo loading and unloading within each of the ports. Notwithstanding that the business areas and functions are subjective groupings of business requirements/events for

Systematic Tricks

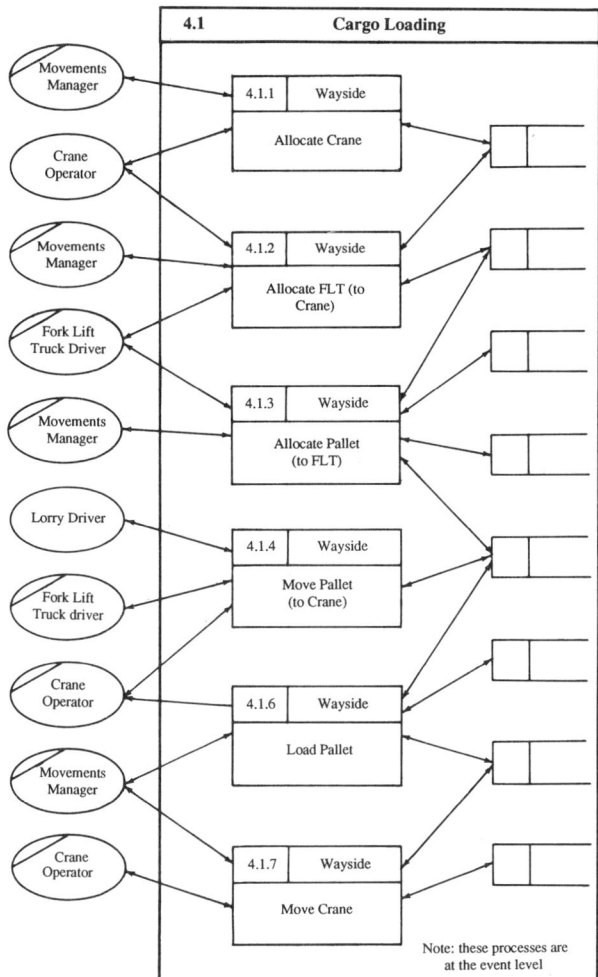

Figure 3.22 Dataflow diagram: level 3

users/senior managers to understand, the processes identified are logical to the business and not the management structure. It so happened that the port management was "logically" structured.

- It explicitly recognises that events are triggers on the computer system by using the "do something" naming convention. The fact that events in batch and online processing are standalone is explicitly reflected in that the processes in the DFDs are separated by datastores. Process A does not trigger process B.

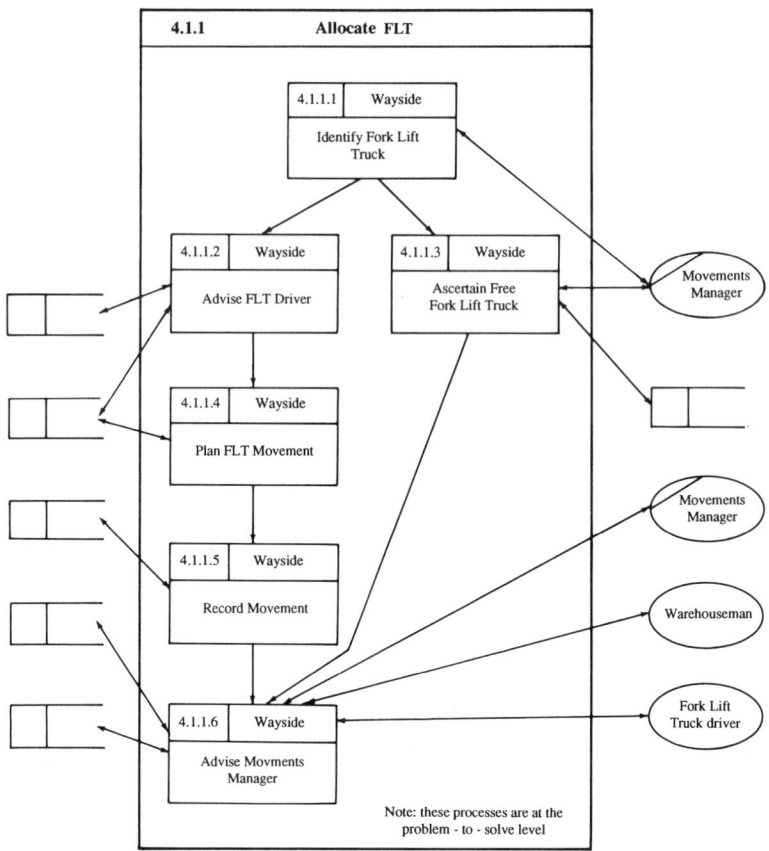

Figure 3.23 Dataflow diagram: level 4

- The DFDs are able to recognise explicitly that problem-to-solve processes in an event are causally related and occur logically at the same point in time as the event by linking the processes by dataflows. Process A triggers process B. It also recognises that problems-to-solve can also be events by naming them with a "do something" title.

- The processes at the two lowest levels can be used directly as the one-for-one basis for identifying application programs and their constituent modules. Given the concepts logical = physical, business requirements are events, event level processing, events are standalone and problems-to-solve are causally related it is natural that the events become programs and the problems-to-solve become modules. It also has the advantage of recognising that in centralised processing application programs should not talk to each other. It confirms Yourdon's statement that "application programs are schedulable, modules are callable". This approach to using the two lowest level DFDs as the direct basis of program design also supports the concept of logical design = physical design.[1]

- The DFD processes can now be used much more soundly for cross referencing to the other SSADM deliverables. The event level DFD processes will now match exactly the events on the ELHs, the ECDs, the EAPs and the Process Models. One can take this matching of the DFD process decomposition with other SSADM logical design deliverables. How to do this is discussed in the next section. One can therefore match the direct statement of "what is required" defined in the Elementary Process Descriptions written for the event level processes with the "how it is to be achieved" specification in the Process Models. The author therefore does not, as recommended in SSADM, write the elementary descriptions at the lowest level of process decomposition. If the SSADM advice is followed and one stopped process decomposition at the function level there would be a mismatch.

3.2.2 Matching the DFD processes, the I/O structures, the dialogue structures and the menu structures

On a project the author has just completed, a set of new SSADM version 4 modelling standards was adopted whereby the cross-relationship of the design products was much tighter than the standard SSADM requirements. There is no SSADM requirement, for instance, that the decomposed processes in the dataflow models match the input and output screens of the application system. If the decomposition of DFD processes is done in the consistent manner described in the previous section then this matching can be achieved. The integrity of the design deliverables is thereby increased.

A set of DFDs with the processes rigorously decomposed at the activity, function and event levels were produced at the client site. These were then used as the basis of the I/O Structures, Dialogue Structures and Menu Structures. Typical examples of the DFDs are shown in figures 3.24–3.26 and the logical data model in figure 3.27 with appropriate data attributes.

---

[1] The author is beginning to have some reservations about using the lowest level DFD processes as the basis for the identification of application program modules. The allocation of the Process Models into software "fragments" of the Function Component Implementation Map (FCIM) as advocated in the Physical Process Specification attractively reflects the facilities for software development that are now available with 4GL software. The straight coding from the logical operations in the Process Models as with 3GL programming languages is becoming a thing of the past—see section 2.3.2.3/13 for the use of the FCIM.

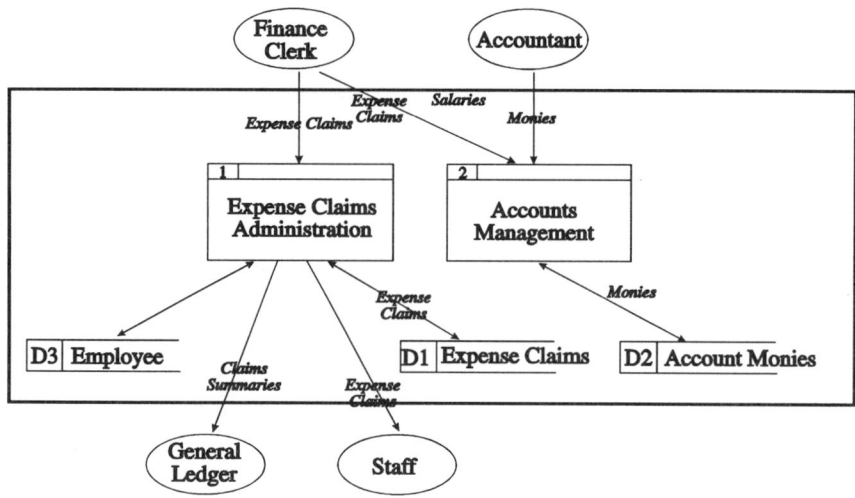

Figure 3.24 Level 1 DFD (with activity level processes)

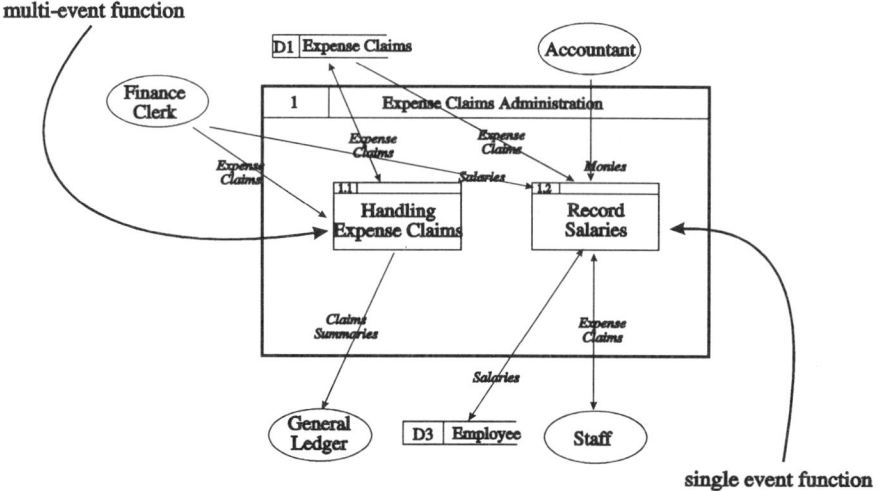

Figure 3.25 Level 2 DFD for finance (with function level processes)

3.2.2.1 *The I/O and dialogue structures matching the DFD processes*

An example I/O Structure for function "Handling Expense Claims" is given in figure 3.28. This function has three events as can be seen in figure 3.26. Wherever there are multi-event functions the events require to be marked as options in the I/O Structure are illustrated, the options being Receive

Systematic Tricks

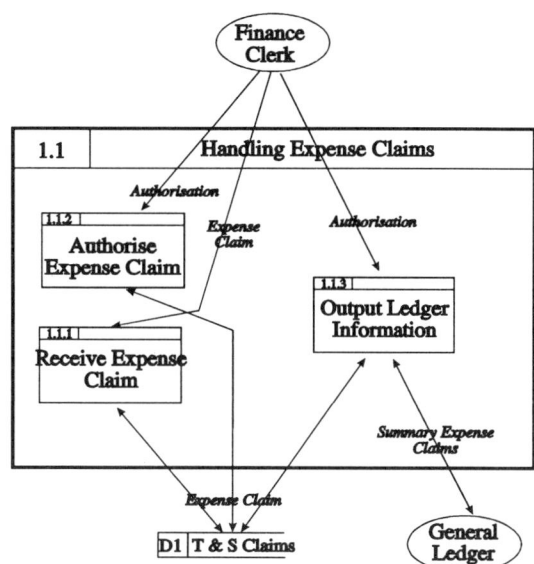

Figure 3.26 Level 3 DFD for finance (with event level processes)

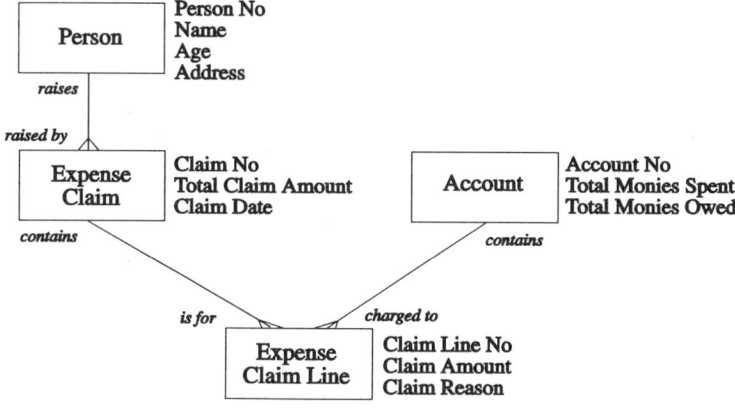

Figure 3.27 The logical data model (with normalised data attributes)

Expense Claim, Authorise Expense Claim and Output Ledger Information. As can be seen the I/o Structure matches exactly the processes in the DFDs. The diagram also shows the groupings of the data attributes, whether they are input or output and the sequence, selection and iteration of the groupings.

Where a function is a single event function the I/O Structure would be modelled at the event level. Such a single event function is shown in

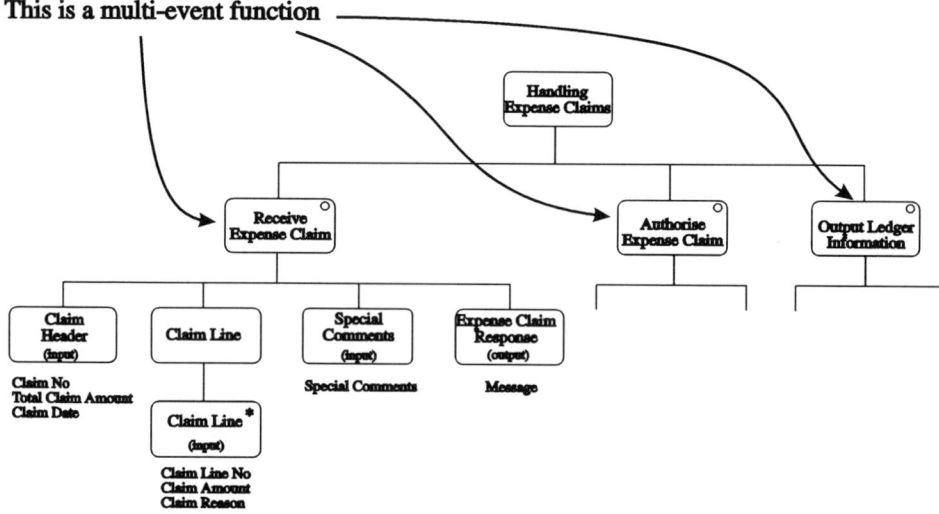

Figure 3.28 I/O structure for function 1.1 "Handling Expense Claims"

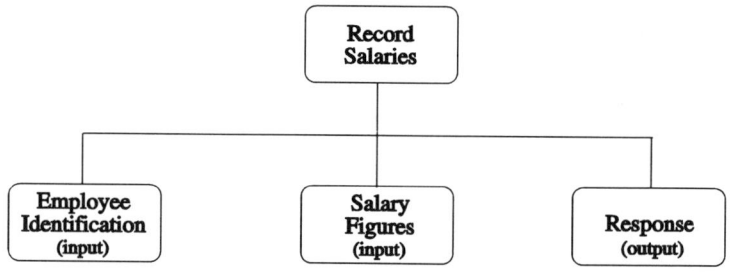

Figure 3.29 I/O structure for function 1.2 "Record Salaries"

figure 3.25 with the "Do Something" process Record Salaries. This event level function was not decomposed any further. The matching I/O Structure for this could be as in figure 3.29. As can be seen there is no need for the optionality of the events to be built into the I/O Structure.

There is no need to show the Dialogue Structures, these being merely "LGDE'd" I/O Structures.

3.2.2.2 The I/O menu structures matching the DFD processes

The structure of the menus for the user role of Finance Clerk—the external entities in the DFDs defining the user roles—can be seen in figure 3.30. Again the structure matches the decomposition of the DFD processes. The main menu would show a selection of the two activities the Finance Clerk is allowed to use (DFD in Figure 3.24), the next level menu for the selected activity Expense Claims Administration being two function options of Handling Expense Claims and Record Salaries with the next level menu screen for the selected multi-event function Handling Expense Claims being the event options of Authorise Expense Claim, Receive Expense Claims and Output Ledger Information.

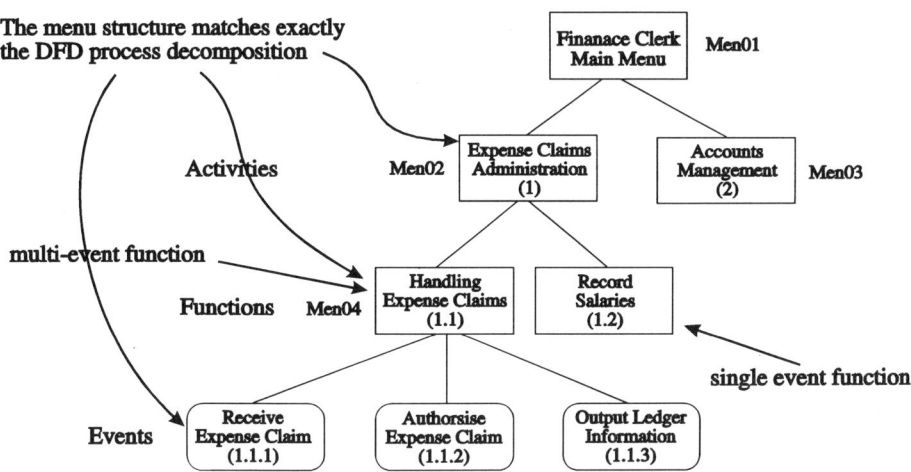

Figure 3.30 Menu structure diagram for user role finance clerk

The sharp eyed reader will notice an error in the above I/O Dialogue Structures and the Menu Structures. Given that, where the function has many events there is the use of optionality in the I/O Structures, then there is no need to repeat the optionality in the Menu Structures.

The solution to this, of course, is not to have functions with many events. The reasons for this are discussed in section 2.4.2.3/6. Optionality in the selection of what the user wishes to trigger should be in the menu screen, not the dialogue screen. The dialogue screens are there to support online business processing.

3.2.3 Taking relational data analysis to sixth normal form

SSADM now includes advice to decompose data to fifth normal form (5NF), although stating that relations in 3NF also conform to 4NF and 5NF.

For the last decade 3NF has been widely accepted as the ultimate in data normalisation/data decomposition. It was believed that relations in third normal form have no further hidden relationships—the non-key data items are dependent only upon the entire prime key of the relation and there are no relationships between the non-key items.

Within the last few years it is being increasingly argued that there are further stages of data decomposition, three more to be precise. The full six normalisation rules are:

- first normal form (1NF)...test for repeating groups;
- second normal form (2NF)...test for inter-key dependence;
- third normal form (3NF)...test for inter-data dependence;
- fourth normal form (4NF)...test for multi-valued dependences on multi-key key-only relations;
- fifth normal form (5NF)...test for derived information;
- sixth normal form (6NF)...test for conditional dependence.

Examples of the first three normalisation rules are described in the SSADM manuals and are not repeated here. Normalisation from fourth to sixth normal form is discussed below. The above definition for 4NF agrees with that in SSADM, but 5NF in the method is defined as no loss of information when the relation is decomposed into three or more relations. This is not as defined above. The SSADM manuals also do not provide examples of 4NF and 5NF. Furthermore, it is strange that the method includes entity sub-types in the LDM but does not include the normalisation rule for entity sub-types, that of 6NF. This section therefore shows examples of these three higher powers of normalisation.

Assume a third normal form relation composed of subject, teacher and textbook with all three attributes marked as key items—a key-only relation. A trick that the author has learnt is that if a multi-key key-only data group appears in first normal form then it is likely to be retained as part of a final optimised set of relations, because it is a link data group containing key-only data to support a many-to-many relationship. For example, entity C could be a detail entity to master entities A and B, with the key of entity C being ab. Another trick is that if the key-only relation initially appears in second or third normal form then the relation should be tested for inter-key dependence.

Analysis of our key-only relation shows that the data items of the key of ab are in fact dependent on each other in some way. One therefore needs to apply the first basic normalisation test of "for a given value of the key(s) is there just one possible value of the attribute". The second basic test of "is the data directly dependent on the key(s)" is clearly irrelevant. There is no data. By applying the first test it is established that a repeats for a given value of b. A is therefore extracted as a new relation dependent on b.

After the second and third normal form steps are undertaken all dependent non-key data items will have been removed as part key dependences. Therefore the remaining dependences are always between part-keys. Thus an alternative description sometimes heard is that fourth normal form is about removing inter part-key dependences from a key-only relation.

The basic tenet of 5NF is that existence of certain table rows implies the existence of some others. For example, there is a teacher entity and a course entity. The course is a multi-day course on information technology covering database, data communications, expert systems and object oriented design. There are a number of teachers who each have a number of relevant skills. Teacher 1 has database and expert system skills, teacher 2 has database and data communication skills and teacher 3 has object oriented design and data communication skills. The teachers have no applicability to the entire course because none of them by themselves have all the skills appropriate to giving the course by themselves. It is only when the three teachers are combined to form a team to take the course that a new table (teaching team) and table row (team occurrence—composed of teachers 1 to 3) is created. Where there are "overlapping" dependences (the teacher skills overlap as regards the information technology course) an "extra" relation is required to indicate the overlap explicitly.

The difference of 5NF from all other normalisation steps is that it works at table row level, with the key value of each row having to be individually assessed as to the significance in business terms of the key value in relation to other key values prior to applying the rule. In the above example it is specifically teachers 1, 2 and 3, not any teacher, with their combination of skills that is significant. Contrast this with, say, the rule of 1NF—for a given value of the key can there be more than one possible value of the attributes? There is no reference to specific key values like teachers 1, 2 and 3—just any repeating value of the non-key attributes for a value, not a specific value, of the prime key. The significance is the repetition of values, not the values themselves.

5NF is not a spurious notion nor is it that difficult to understand. What is difficult and requires a practised eye to accomplish is to identify where the step is required.

Sixth normal form (6NF) is concerned with entity sub-types. The rule is to test whether certain data attributes in a relation are dependent on the prime key based on certain conditions, conditions set by the application. A simple example illustrates the point. A vehicle relation is composed of vehicle number, wing span, sail area and weight. This relation is in third normal form because it satisfies the two tests for normalisation. However, it is obvious that wing span relates to vehicle of type aircraft and sail area relates to vehicle of type sailing boat. The relation needs to be split into two sub-types within vehicle—of aircraft with the attributes of vehicle number and wing span and of sailing boat with the attributes of vehicle number and sail area. The vehicle relation would contain the attributes of vehicle number and weight and be the super-type relation to the aircraft and sailing boat entity sub-types.

A complete hierarchy of entity sub-types can be obtained. Consider customer in figure 3.31. It contains generic attributes relevant to all customers, such as name and address. It also contains attributes relevant only to non-creditworthy customers—debt amount, litigation mechanism—and attributes relevant only to creditworthy customers—amount loaned, repayment period, loan type. The entity sub-types creditworthy customer could simultaneously be an entity super-type as well as an entity sub-type. Creditworthy loans could be secured—description of security, security value—and unsecured—amount taken unsecured. There is no potential limit to the degree of entity sub-type decomposition.

An entity sub-type can relate to more than one super-type. The loan type attribute in the entity sub-type credit customer acts as a foreign key to the entity super-type loan type. Thus the entity sub-type credit customer has two super-types: customer and loan type. A complete hierarchical and network data structure of entity sub-types and super-types can therefore exist.

The final area of relational data decomposition, given current thinking, is the Boyce/Codd normal form (BCNF) rule. It is now recognised that 3NF suffers from a number of deficiencies. It does not deal adequately with relations that have multiple candidate keys, relations with compound keys and where the candidate keys overlap (candidate keys ab and bc). The objective of BCNF is to provide a tighter definition for these circumstances.

Dr Codd's colleague and business partner Chris Date in his book *An Introduction to Database Systems*, Volume 1, defined BCNF as "a relation is in BCNF if and only if every determinant is a candidate key". A determinant is any attribute on which some other attribute is functionally dependent. Date is talking in terms of candidate keys, not just the primary key. A candidate key is an attribute in a relation that could be used, either by itself as a simple primary key or with other attributes as a compound/composite key, as the prime key to the relation. A relation can have any number of candidate keys. To be in BCNF the prime key of a relation must be the only determinant in the relation. Relations with attributes that can also function as primary keys need to be decomposed further.

Consider the situation of a relation composed of pupil, subject and teacher, the first two attributes forming a compound key. Pupil and teacher could equally effectively form an alternative compound key to the relation. There is

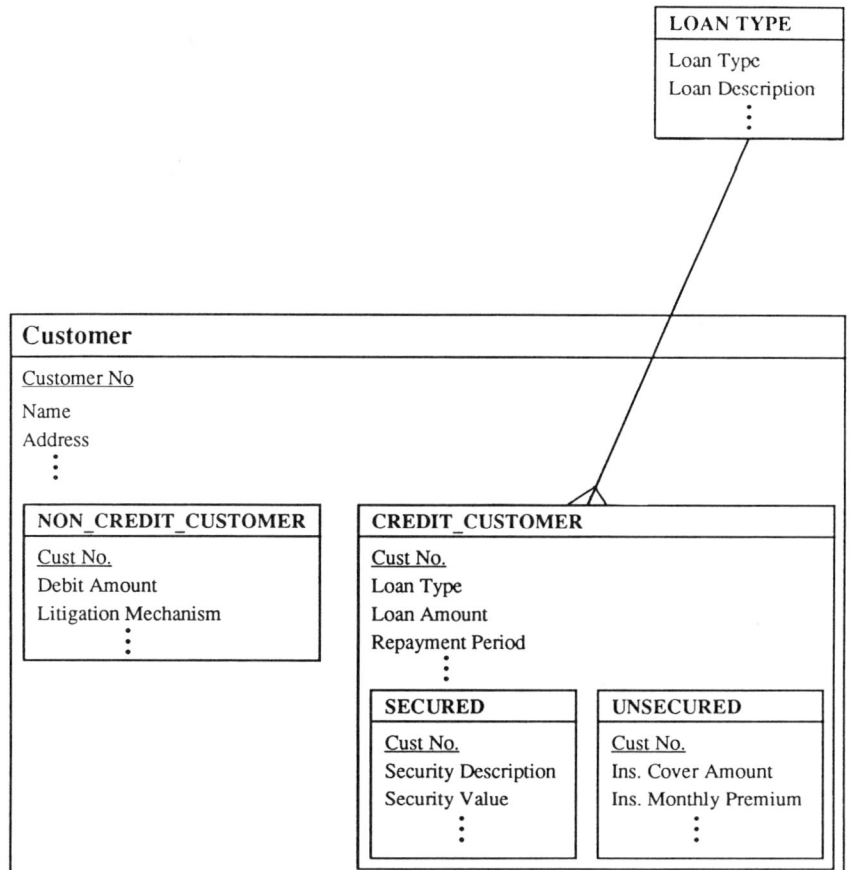

Figure 3.31 Sixth normal form/entity sub-types

thus the situation where alternative candidate part-keys functionally overlap. However, close analysis of the attributes show that each pupil is taught by only one teacher for a given subject, each pupil can learn only one subject, each teacher teaches only one subject and there can be several teachers per subject. The matching data structure is illustrated in figure 3.32a. Therefore neither of the compound key relations is in third normal form because it contains a dependence—if you know the value of teacher you know the value of the subject. There would therefore inevitably be some update anomalies. With the candidate compound key of pupil and subject if you delete the pupil Smith who is studying geography you would also delete the subject geography. With the candidate compound keys of pupil and teacher if you delete pupil Smith who is studying with teacher Moriarty you would also delete teacher

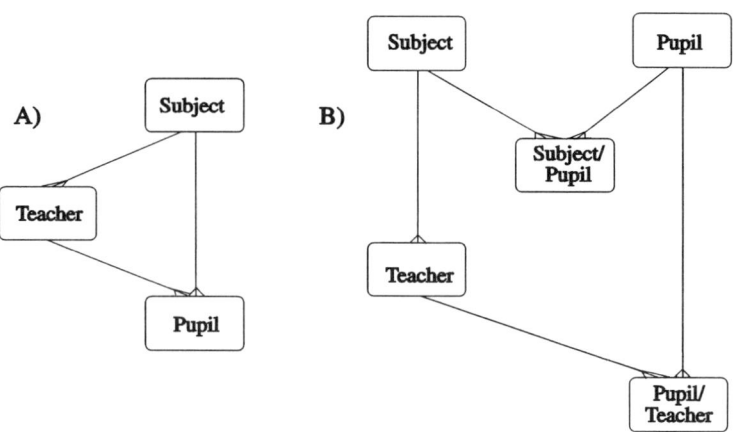

Figure 3.32 Boyce/Codd normalisation

Moriarty. The difficulties are that the attribute pupil is a determinant, not a candidate key. The difficulty is resolved by creating a set of many-to-many relations between the entities pupil and teacher and pupil and subject as illustrated in figure 3.32b.

To be frank, the author has found 4 NF, 5NF and Boyce/Codd normal to be of more academic than practical interest. He has never applied them in practice.

A little trick the author uses. Never apply relational data analysis to the output data flows from an application. They are nothing more than the input data flows plus a process. No data is lost and the amount of data to be analysed can be reduced by up to half.

3.2.4 Combining entities

There are occasions when there is a one-to one relationship between entities. The suggestion has been made by SSADM that one solution to this is to combine the entities, but no examples are given.

Assume a one-to-one relationship between the entities Member of Parliament (MP) and Constituency. They could be combined into a single entity. Whenever this is possible a number of questions need to be asked:

- What should the name of the combined entity be? Whichever name is chosen from the two source entities the name not selected will be lost as will the role of the previously named entity. It should be borne in mind that the two entities play different roles and that a MP is not a Constituency.

Assume that the key of the MP entity is MP Code and the key of the Constituency entity is Constituency No. Which of these two keys should be used for the combined entity? Certain file handlers, such as IMS, use the facility of prime and secondary keys.

- Has flexibility been lost by combining the entities? The answer is invariably yes. What if one subsequently requires to keep a history of the MPs in a constituency? The implemented model would not enable this to be supported. The real world situation, of course, is that over time there is a many-to-many relationship between these two entities.

The principle followed by the author is never to combine one-to-one relationship entities. Where a one-to-one relationship is initially established it is always the case that only the present state of affairs is required to be recorded. At any point in time there is only one MP to one Constituency. The author always asks "What if you require to record the history about the entities?" Invariably one entity is more stable than the other, the MPs being more frequently changed than the Constituency, so that a Constituency can have many MPs. The question, however, is not enough. Having established that one entity is more dynamic than the other, it is also necessary to establish whether the entity occurrences can change in relation to each other over time. The standard question to establish this is, using the above example, "Can an MP relate to more than one Constituency over time and can one Constituency have more than one MP over time?" The answer is, of course, yes to both questions and a many-to-many relationship is established.

3.2.5 Additional points of detail

This section addresses each of the SSADM design techniques and identifies points of improvement that can be made.

3.2.5.1 Entity relationship modelling

The entity model should only be built on the data retrieval business requirements. The reasons for this are discussed in section 2.2.2.4 and are based on the concept of design on data retrieval business requirements.

3.2.5.2 Entity model set cardinalities

It is essential to record four cardinalities between a master and its detail entities—minimum, maximum, average and working average. The differences between these cardinalities and how to ascertain them are detailed in section 2.4.2.3/3.

3.2.5.3 Entity Life Histories (ELHs)

(1) SSADM now supports entity sub-types. Currently relational file handlers do not support entity sub-types fully in that there is no property inheritance of the sub-type(s) to the super-type. Entity sub-types are treated as separate tables and linked explicitly in the SQL data *manipulation* language constructs using table joins. What is needed is for the entities to be implicitly linked in the schema *definition* language in the dictionary as defined in the SQL data definition language. Dr Codd has proposed an extended relational RM/T model. RM/T includes, *inter alia*, support for entity sub-types, although still with no recognition of property inheritance. When describing the RM/T model in volume 2 of his book referred to earlier, Date proposed an enhanced SQL data definition language to support entity sub-types and super-type relationships.

An example of an enhanced SQL data definition language syntax as proposed by Date to support figure 3.31 is:

Create Table Customer
Create Table Non_Credit_Customer Sub-Type Customer
Create Table Credit_Customer Sub_Type Customer, Loan_Type
Create Table Secured Sub_Type Credit_Customer

Thus "at the physical level" table sub-types are still stored as separate tables of data, but will eventually be supported implicitly by schema definition rather than by explicit data access calls.

The important point to remember is that entity sub-types and their relationships to entity super-types can still be treated as normal standalone entities. As far as the author is concerned the technique does not require modification.

It will seen in chapter 6 that the semantic net/class object facility of expert systems/object oriented systems is perfectly suited to supporting entity sub-types with additional facilities to those provided in the schema definition language, facilities such as class and property inheritance.

(2) The SSADM manual provides examples as to how ELHs can be used to support referential integrity. To be frank, the author has never used ELHs in this regard. For example, on the deletion of a customer entity the author has never drawn a customer ELH to ensure that the customer's orders have all been deleted previously. Standard referential integrity constraints for inserts (always check the foreign key value in the detail entity is matched by a master entity prime key value), updates (...) and deletes (...) are specified as business rules in physical process design and can be defined as rules in the database schema definition if the file handler supports this facility, otherwise in the application program. ELH state indicators do not enrich the referential integrity specification mechanism.

Systematic Tricks

Enhancements to the SQL data definition language in version 2 have been made to support referential integrity constraints through database triggers. The formal version 2 syntax has not yet been finalised by the ANSI committee, but several database vendors, such as Cincom for SUPRA and Computer Corporation of America for DATACOM/DB, have already created their own syntax.

For example, when an order is inserted or a foreign key is modified on an existing order a referential integrity constraint is that a customer table row appropriate to the foreign key value of customer number in the order table being inserted/modified must also exist. The constraint will be defined in DATACOM/DB as:

Create Table Order
Column
Foreign Key (Customer_No) Reference Customer.Customer_No

Equally, when deleting a customer a constraint check should be made that no related order exists.

The SQL syntax is:

Create Table Customer
Column
References Order on Delete Check (search condition)

Given the ease with which referential integrity constraints can be specified logically in the Process Models and physically in the schema data definition language the "hassle" of drawing ELHs to reflect the potentially large number of referential integrity constraints possible (how do you handle the n exclusive relationship constraints that can occur on a broad network data structure as in figure 2.6) is not worth it.

3.2.5.4 Transaction access path analysis

SSADM includes access path analysis in the form of the EAP and the ECD techniques, but as has been said many times the techniques do not count the accesses. This is a crucial weakness, and means that EAPs and ECDs cannot be used to test the LDM for efficiency. That is one of the prime purposes of the technique of transaction access path analysis. The technique has already been discussed in part when reviewing "tricks of the trade" to rectify the conceptual weaknesses of SSADM. However, the technique has not been described. That is the purpose of this section. The readers familiar with earlier versions of SSADM will recognise that the contents of the access path forms are from the Process Outlines.

The examples given below show the different access paths that a single business requirement can have and how to ascertain which one to adopt for the logical design specification.

Consider the LDM in figure 3.33. For a sample business requirement "For a specified customer list all unpaid invoices for a specified product" there are three possible entry points to consider—customer, product and status. It is necessary to find which entry point provides the optimum logical access path. This is the same requirement as for testing the LDM and transaction access paths for efficiency. New techniques to do this are described in section 3.1.

Representative access paths for the three entry point possibilities are detailed in figures 3.34–3.36, 3.34 entering on product, 3.35 on customer and 3.36 on invoice status. The logical access paths would be drawn pictographically in the EAPs and the ECDs. Users of version 3 of SSADM will recognise the form. It contains information that ought to be detailed in the EAPs and ECDs.

The details required to be recorded against a transaction path are:

- Transaction reference/business requirement identification.

- Name/number of the entity being accessed.

- Type of access to the entity—read, write, update, delete and whether relationship links to entities are being established or broken. Increasingly the relational set constructs of select, union, difference, intersect and very occasionally divide require to be specified when accessing/joining multiple entities.

- Data item(s) used for selecting the entity occurrences.

- Access path—to identify whether the entity is being accessed from a master or detail entity or directly as an entry point.

- Access via master or detail—to identify from which master or detail entity previously accessed the current entity is being accessed from. This is only necessary if there are multiple master and detail entity relationships to access via.

- Number of accesses to the entity for each execution of this transaction.

- The data item(s) in the entity to be accessed. It may be that no data items are to be accessed as the entity is only being accessed as a "bridge" to another entity. In this case identify in the comments column that the entity is for "access only".

- Probability of execution of the access operation. The author records this information in the comments column.

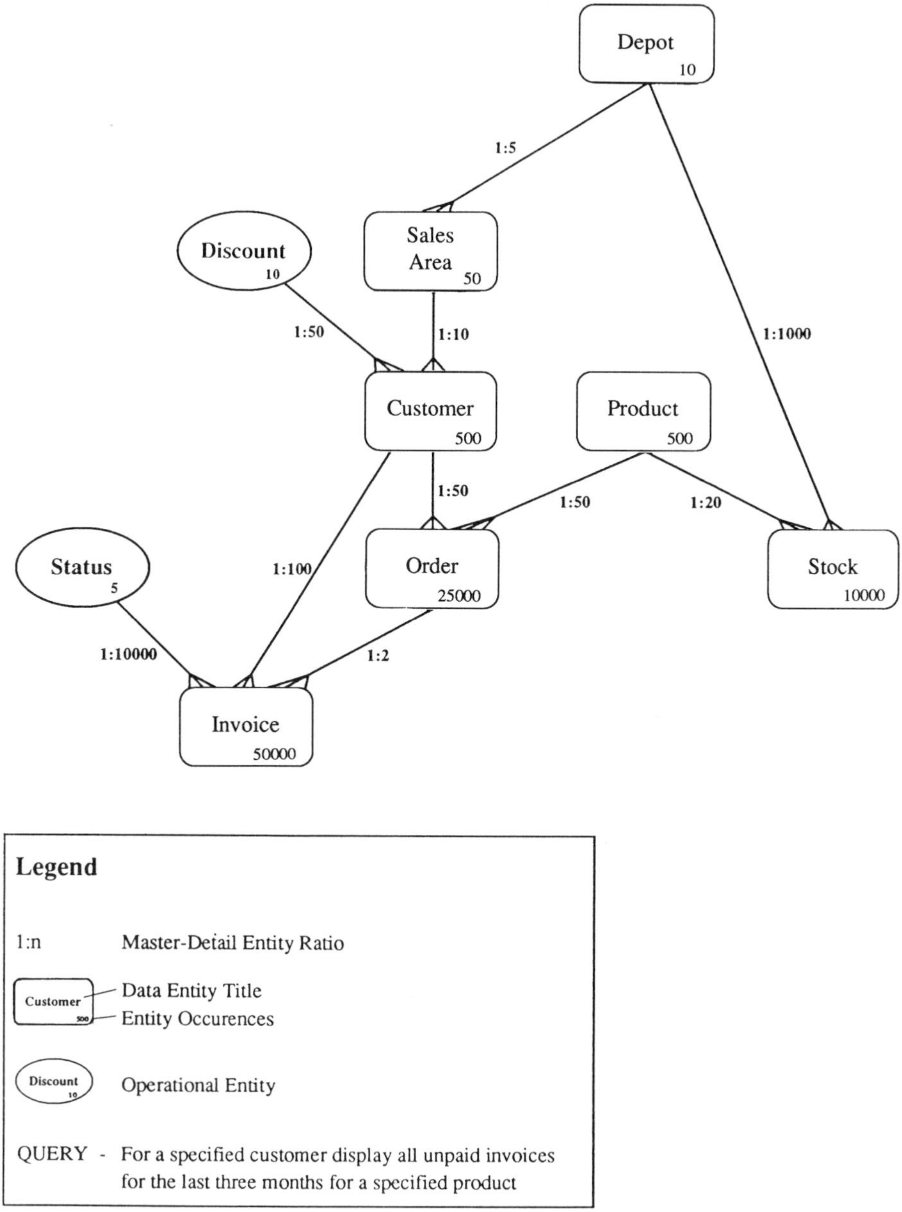

Systematic Tricks

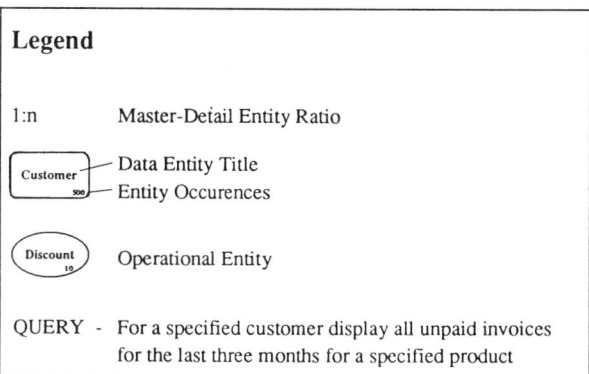

Figure 3.33 CLDD data model

SYSTEM: EXAMPLE 1						
AUTHOR: J. HARES DATE: 1990			PAGE: 1		CONT. PAGE NO: 1	
FUNCTION NO:		FUNCTION NAME			TRANS ID	
ENTITY NAME	ACC TYPE	READ PATH	ACCESS VIA	NO. ACC.	DATA ITEMS	
Customer	R	DIR	—	1	⋮	
Invoice	R	MAST	Customer	100	⋮	Read entire set. Retrieve those with invoice status value of unpaid. Assume 10% hit ratio - therefore 10 invoices retrieved.
Order	R	DIR	—	10	⋮	For access only. Assume 1 invoice matches desired product.
Product	R	DIR	—	1	⋮	
				112		

ACCESS TYPE
I - INSERT   L+ - ADD TO LINK PATH
M - MODIFY   L- - REMOVE
R - READ
D - DELETE

READ PATH
DIR   - DIRECT
SEQ   - SEQUENTIAL
DET   - VIA DETAIL
MAST - VIA MASTER

Figure 3.34 Logical access map—1

- Data items used for ordering the entity occurrences accessed. The author records this information in the comments column.

- The assumed storage sequence in which the data in the logical data model is held. If it is assumed that an entity is stored in a particular sequence, for example customer is in the sequence of customer number, if the business requirement requires to retrieve customers in customer number order, then access to the customer can be logical sequential and not sorted. This would indicate that indexed access to customer for a relational file handler would be a valid database solution for this transaction. Clearly the transaction is but part of a total summary access path map picture, but it is indicative, particularly if the transaction is high volume and other transactions also require similar access, of the need for a particular storage sequence.

Systematic Tricks

SYSTEM: EXAMPLE 2						
AUTHOR: J. HARES DATE: 1990 PAGE: 1 CONT. PAGE NO: 1						
FUNCTION NO: FUNCTION NAME TRANS ID						
ENTITY NAME	ACC TYPE	READ PATH	ACCESS VIA	NO. ACC.	DATA ITEMS	
Product	R	D	—	1	⋮	
Order	R	MAST	Product	50		Read entire set. For access only.
Invoice	R	MAST	Order	100		Read entire set. Retrieve those with invoice status value of unpaid and Customer No. specified in Query. Assume 1 invoice matches unpaid, Cust. No. and Product Code specified in Query.
Customer	R	DIR	—	1	⋮	
				152		

ACCESS TYPE
I - INSERT   L+ - ADD TO LINK PATH
M - MODIFY   L- - REMOVE
R - READ
D - DELETE

READ PATH
DIR   - DIRECT
SEQ - SEQUENTIAL
DET - VIA DETAIL
MAST - VIA MASTER

Figure 3.35 Logical access map—2

- Once the assumed storage sequence of the entities is known then the forward or backward scrolling of a set of occurrences of an entity can be ascertained. This information is particularly important if the implemented design uses pointer chain technology to link record types together.
- Comments. This column should be well annotated with a full explanation of the access logic as detailed in the above columns. For example, the document may record that 50 orders are to be accessed for one access to a customer entity. This does not explain that the 50 accesses represent accessing the entire set of orders for each customer accessed. In fact what is meant is that the transaction reads the entire customer/order set until no more orders are obtained. It so happens that the set contains on average 50 orders.

SYSTEM: EXAMPLE 3						
AUTHOR: J. HARES		DATE: 1990		PAGE: 1		CONT. PAGE NO: 1
FUNCTION NO:		FUNCTION NAME				TRANS ID
ENTITY NAME	ACC TYPE	READ PATH	ACCESS VIA	NO. ACC.	DATA ITEMS	
Invoice Status	R	DIR	—	1	⋮	Status of unpaid. Assume 70% of invoices are unpaid.
Invoices	R	MAST	Invoice Status	8750	⋮	Read set reverse and retrieve on Customer No. specified in Query and date. Assume invoices in time sequence and stored one year. Therefore read 1/4 set. Assume 0.2% hit rate.
Order	R	DIR	—	18		Search on Product No. specified in Query. Assume 1 invoice relates to Product and Customer.
Product	R	DIR	—	1	⋮	
Customer	R	DIR	—	1	⋮	
					35073	

ACCESS TYPE
I - INSERT   L+ - ADD TO LINK PATH
M - MODIFY   L- - REMOVE
R - READ
D - DELETE

READ PATH
DIR   - DIRECT
SEQ   - SEQUENTIAL
DET   - VIA DETAIL
MAST - VIA MASTER

Figure 3.36 Logical access map—3

The comments column can also be used as a "dump" for those items of information that are less frequently filled in, such as sort order and probability of execution. A transaction access path document with all the above columns would be very large.

Additional points of detail that need to be considered are:

- whether the entry point type is direct, logical sequential (when the sequence of key is important) or physical sequential (when the sequence of key is not important);
- whether access to the table is purely for access only or as part of an access path to another table. Too many such "for access only" accesses indicate an inefficient access path, such that either the LDM or the transaction access path needs modification;
- whether the entire set requires to be accessed or whether it is to be searched on a combination of keys in either a forward or backward direction;
- the recording of referential integrity constraints. Full use of L+ and L— link calls inserting or deleting relationships between tables must be made. The fact that certain file handler types implicitly support some referential integrity constraints (IMS automatically ensures integrity on inserts to one master entity) should not be used as an excuse not to define referential integrity logically. The overheads of referential integrity are still there and must be measured.

A set of example transaction access path maps is now described. They illustrate the degree of access detail the author has found necessary to record. The examples also show how the transaction access path maps can be interpreted as to their impact on subsequent database design.

- Consider figure 3.35. When accessing order from the product master entity the entire set of orders for the product needs to be specified as being accessed. To say that 50 orders are being accessed without putting any description of the access would indicate to the uninformed application programmer to issue 50 read order commands and then stop, whereas the given product which is accessed may actually contain n orders from 0 to infinity. An additional point to make is that the access to the orders is for access purposes only, to provide an access path to invoices. The data within the entity is actually not required.
- The description relating to invoices illustrates another point. The logical data model indicates that the database cardinality of orders to invoices is on average two. In fact for this transaction, the transaction cardinality is assumed to be one. There is therefore the concept of transaction cardinalities. SSADM does not recognise transaction cardinality.
- Figure 3.34 illustrates, when accessing invoices from customer, the concept of hit rate—a variation of transaction cardinalities. It is assumed that only one in ten invoices accessed are relevant to the query. The Navigator version of the Information Engineering method identifies this as probability of execution. SSADM does not address the issue.

- Figure 3.36 illustrates yet another point. The query is to retrieve invoices received during the last three months and which are unpaid. Invoices are assumed to be stored in date sequence and are stored for one year in the database. The access path chosen is therefore to read the invoice set from invoice status in reverse time sequence. This means that the file handler needs to be able to support reading a table of information in a reverse sequence according to the relationship to its master entity, in this case invoice status. Only part of the set needs to be accessed. A relational file handler would have some difficulty with this type of access by not providing pointer chain technology to link table rows within a table together. If the query had been to retrieve the last unpaid invoice then a relational file handler would have even more difficulty because the last invoice is not known. With the earlier and much verbally abused pointer chain technology type file handlers this kind of access would, of course, be easy to achieve.

Full and precise specification of the access requirements to each entity type must be specified in the EAPs and ECDs and in the Process Models where the access paths are converted into access calls. Only by such attention to detail can the mechanisms which the file handler requires to use to support the logical data accesses be ascertained. Problem areas can be identified to the database designer.

If the file handler has data access limitations these have to be overcome in the application programs. Processing solutions to file handler problems can be included in the Process Models to forewarn application programmers.

3.2.5.5 First-cut physical

The author has already mentioned in section 2.4.2.3/12 that the first-cut database rules do not take data access into account when producing the initial database design. The rules only consider the data structure component of database design. 50% percent of the input to database design is ignored. The database design will work (the generic rules are legal after all), but all too often the performance of the first-cut database is "like a winter's day—short, dark and dirty". Bear in mind that the first-cut rules, which ignore the 50% that is most measurable, expensive and tuneable, are applied to a LDM that itself has not been tested for efficiency. The result is potential inefficiency on inefficiency! It is not surprising that much performance assessment is often required—and unnecessarily required.

The enhancements needed for this technique are described in section 3.1.4. The port authority application illustrates how the enhancements were supported. The transaction access paths were recorded individually into the

Systematic Tricks 157

CASE tool dictionary. Unfortunately the tool did not summate the logical accesses to produce a set of summary access path maps as described in section 3.1.2. A simple routine was therefore written. It read all the accesses from the dictionary and multiplied for each transaction the number of accesses to the entry point entities and along the relationships between the entities by access type by the frequency of the transactions per day (a set of daily summary access paths were decided upon). A manual exercise was then undertaken to record the summarised access totals onto the already optimised LDM (the LDM had already been tested for logical efficiency using the techniques described in section 3.1.1).

The summary access paths were then interpreted as described in section 3.1.3 and the appropriate SQL data definition language code written. The data definition language code generation capabilities of the tool were not used because they, like SSADM, only take data structure into account. Writing the data definition language schema code is not a major task and should only occur once per database design. Little extra effort was expended on coding the handcrafted database design approach that takes total data structure and total data access into account when producing a balanced database design as compared with the automated code generation approach that takes only data structure into account. The performance of the database, as anticipated, caused no problems. Major effort savings in performance assessment were achieved at little expenditure in hand-writing schema code generation.

3.2.5.6 Performance assessment

The SSADM manuals describe only the features that a database administrator should consider when reviewing the performance of the database and the transactions. Unlike many other leading structured design methods a database optimisation document and supporting explanatory description is provided.

SSADM identifies most but not all of the elements that should go into transaction resource utilisation timings. SSADM correctly identifies the two main elements as processor and disk I/O time. The processor time elements the method identifies are application program overhead, file handler access calls (these are the logical I/O data accesses to the buffer pool), teleprocessing monitor/operating system overheads (these are both interrupt handlers and with some database vendors are in fact combined as a single facility) and disk I/O. The disk I/O are physical accesses to the database data, the indexes and the journal file. The processor elements that the SSADM manuals do not identify are the overheads that are incurred in the paging of control software (operating system and teleprocessing monitor) and application programs, if a multi-processor is being used, if the processors are split into client/server architectures and the disk I/O overheads that will be incurred through referential integrity support. Information Engineering uses a general 20% overhead for the operating system/teleprocessing monitor. Neither does the method show how the processor loadings can be calculated using the transaction resource utilisation timings. Each element omitted is considered below.

It is unfortunate that SSADM does not identify the difference between transaction resource utilisation time and transaction response time. Transaction resource utilisation time is the combined processor and disk I/O time that will be required to perform the logic that must be executed for the transaction to satisfy a business requirement. Transaction response time is the transaction resource utilisation response time plus any telecommunications overheads and the wait times that will occur between disk I/Os. The file handler and operating system/teleprocessing monitor issue calls to get data or software from disk in order to continue processing. Given that disk I/O is slow (remember it works at mechanical speed whereas the processor works at electronic speed) the processor will put the transaction into a "wait" condition.

It is essential to include the processing and disk I/O overheads incurred in maintaining referential integrity. Consider figure 2.6. When a rate is inserted disk I/O is incurred in checking that all the related master entities/table rows exist. Rate was clustered on disk via City Pair. It was assumed that some half of the rates for each City Pair were in the same page as the related City Pair, so that 50% of the time there was only logical I/O to the buffer pool for the referential integrity access to City Pair. Disk I/O was incurred to all the other tables to which rates was related. The disk I/O incurred to support referential integrity was greater than all the disk I/Os incurred in the rest of the application and was the reason why the performance of this application caused the company unsatisfactory response times when the application went live.

It should be noted that the transaction timing described above requires to know processor size in terms of processor main memory. How much main memory is available for the operating system, teleprocessing monitor, application programs, buffer pools, the file handlers, query languages etc. affects the amount of disk I/O for data and software. Clearly if there is sufficient main memory for a large buffer pool then the volume of data disk I/O will be reduced. Equally, if a large area of main memory can be set aside for control and application software the volume of disk I/O incurred through virtual paging is also reduced.

Paper timing exercises to calculate transaction resource utilisation and response times as advocated by SSADM have the overwhelming advantage of being a proactive database design philosophy that enables the designer not only to identify those transactions which do not perform satisfactorily but also to identify where a performance problem within a transaction is to be found.

If the performance problem exists there are three *and only three areas* where optimisation can take place. These areas are:

- the processor;
- the transaction;
- the database.

Some guidelines as to where a solution can be found are:

- if the bulk of the transactions have a poor performance and no design errors in the transactions (e.g. poor access paths within the database because no entry point is available) are found then the processor is likely to be of insufficient power;
- if a transaction has a poor performance on a "standalone" basis then it is likely that the transaction requires re-design;
- if the bulk of the transactions consistently have a poor performance on a particular record type/table then the database design is likely to be flawed in that area.

Once the design is optimised the processor loadings can be calculated. By taking the processor overhead of each transaction, multiplying it by the number of times the transaction runs during a processor time slot, for example an 8 hour processing day, summating the processor overheads of all the transactions and dividing the summated figure by the processor time slot, an average processor loading is calculated. Clearly the loading figures can be optimised to reflect the peaks and troughs of transaction throughputs. The author has used this technique many times to advise clients as to the likely processor speed requirements.

These transaction and processor performance measurements are crucially dependent on accurate path length instruction figures for each file handler function from the database vendors.

3.3 TRICKS SPECIFIC TO VERSION 4 OF SSADM

Experience of the application of version 4 has shown that some consistent problems are arising. The problems are consistent as they have each occurred on a number of projects.

The solutions are based on a set of worked examples of two business requirements that show how the techniques and deliverables fit together. Some of the examples are generic to the solutions, that is, the example is relevant to more than one of the problems. The examples are based on the logical data model in figure 3.37 and the two business requirements,

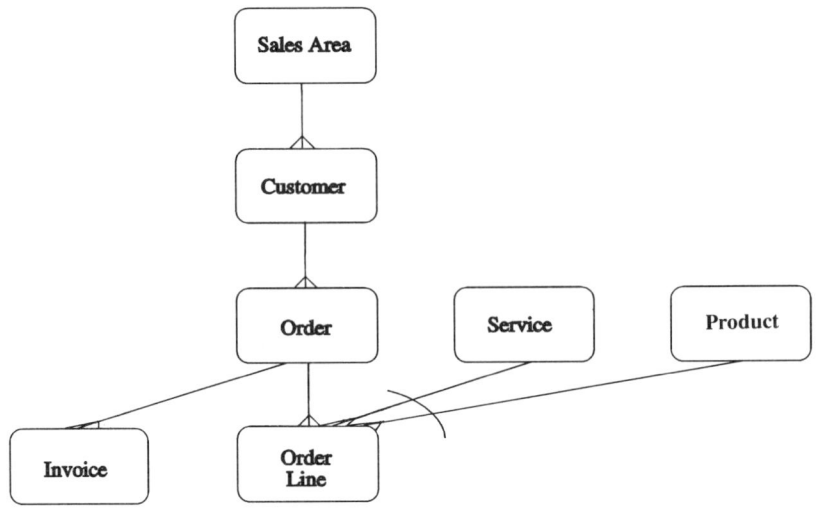

Figure 3.37 Worked example logical data model

one enquiry for Display Customer Information ("For a specified Customer display Sales Area and all Order and Order Lines for Products and Services that costs more than £10 and the Invoices that have been paid in the last six months") and one update for Do This ("Insert an Order and its Order Lines and update the total value on order of the Customers, Products and Services as appropriate").

3.3.1 Client-server architecture

Client-server architecture has been designed to cater for the increasing use of intelligent workstations front-ending the main back-end processor. This has the benefit that processing is undertaken where it is most appropriate. The man/machine interface processing typical of the user, such as word processing, spreadsheets, input data editing and screen handling, is undertaken on the front-end "client" processor. The back-end "server" processor, by contrast, is concerned with supporting the main application system processing against the database and the production of bulk batch reports. The man/machine interface and database processing are thus kept separate. Any communication between the client and the server processors is by explicit calls requesting a service, of the client type "Can you send me some data?" or "Can you process me some data?" and providing server responses of the type "Here is the data" or "Here is the result".

The server to one set of clients can in turn be a client to another server—they can therefore have a many-to-many relationship.

Tricks Specific to Version 4 of SSADM 161

SSADM does not support client-server architecture, in that operations are only added to the Process Models that cater for database processing and not to the Dialogue Structure Diagrams that cater for the man/machine interface.

The techniques for the front-end client processing are the I/O Structure and the Dialogue and Mean Menu Structure Diagrams. To these need to be added the operations for the logic of the front-end processing type. The operations are the full array *except for the database access commands.*

The techniques for the back-end server processing are the Entity Life Histories and the enquiry and update Process Models. The EAPs and the ECDs are not concerned with the specification of operations and are therefore not part of the client-server problem. All the operations for the processing of the database data can be specified against the Process Models.

3.3.2 Transaction access path analysis

The author re-introduced the technique of transaction access path analysis on the projects using version 4. The technique is an extension of that used in version 3 and transaction access path examples are described in section 3.2.5.4 with the notation added to the EAPs and ECDs.

The technique is needed to get round the deficiency of the EAP and ECD techniques. As shown in the beginning of this chapter, transaction access path analysis is the only technique that can answer the two questions "Which entry point and access path should be used?" and "Is the logical data model efficient?" *SSADM should bring the technique back.*

3.3.3 The one-to-one lassoing of entities

This affects two techniques—Process Modelling and the I/O Structures. Both require similar modification.

The problem occurs where there is a one-to-one access between the entities in the transaction access path maps and therefore in the EAPs and the ECDs. Such entities are lassoed and bundled together into one process box in the subsequent Enquiry and Update Process Models (EPMs and UPMs). Extremely strange results were obtained when the lasso facility was applied to many of the EAPs and ECDs to produce the Enquiry and Update Process Models.

Some of the business requirements had long access paths against the logical data model, with one access to each of the entities. There were also a number of business requirements that accessed an entity at the bottom of the LDM and from there accessed up to the top of the model, each access up being one access to another entity. There were yet other business requirements to insert/access one occurrence of an entity and then insert/access one occurrence of another entity and so on.

With such access paths the laasoing produces the ridiculous result of one process box in the Process Models. There is no structure to the Process Model and all the operations to n entities are grouped together. The sequence of access to the entities is lost, and great blocks of less manageable code are potentially produced. The solution was to get rid of the lasso.

Given that the lasso was got rid of in the Process Models it was necessary to do the same to the I/O Structures for the retrieval business requirements as they are merged with the EAP Enquiry Output data structures to produce the EPMs. There is no such merging of the update business requirements, as the output information is typically nothing more than an error message or successful update indicator, and thus has a trivial data structure.

The lasso as a diagrammatic tool is not used in the I/O Structure where there is a one-to-one relationship of attributes from different entities, but the effect is the same. What happens in the standard technique is that where there is a one-to-one relationship in the entity attribute values being input or output then the entities to which the attributes relate are combined into a single element box. A lassoing effect is achieved although not shown diagrammatically.

The I/O Structure Diagrams in their Enquiry Output data structure form for the retrieval business requirements are merged with the EAP Enquiry Input data structures to produce the EPMs. If the combining of the element boxes is not stopped then the I/O Structure Diagram output data structures will be at variance with the EAP input data structures, and there would be a structure clash in the merging process. It was therefore also necessary to get rid of the rule that where the values of the attributes do not repeat in the entities that are input/output in sequence in an I/O Structure the attributes are merged into a single element box. The I/O Structure Diagrams will then match the EAPs.

There was no risk in what was done. To the Process Models was restored the concept of the sequence of access that was in SSADM version 3, with the added benefit that the process logic for the massaging of the data was matched with the access logic in the transactions access paths. For the I/O Structures the element boxes are later grouped by the LGDE lassoes to produce the Dialogue Structure Diagrams, so that the diagrams are no different from standard SSADM. The same data attributes will be combined in an LGDE as previously, the only difference being that there may be more element boxes (now based on the entities) in the LGDE lasso.

3.3.3.1 The Enquiry Access Paths (EAPs)

The EAP for the "Display Customer Information" business requirement with the application of the lassoes is shown in figure 3.38 and the resultant Enquiry Input data structure in figure 3.39. This is the standard version 4 approach.

Tricks Specific to Version 4 of SSADM

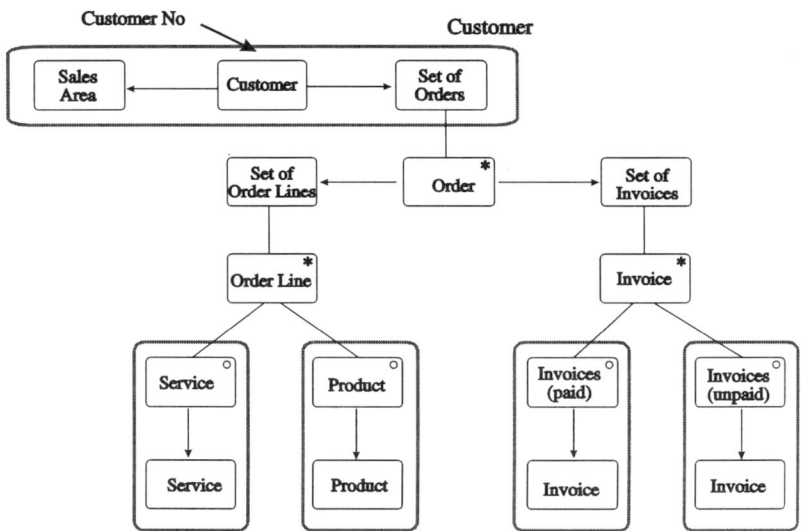

Figure 3.38 Group accesses on the enquiry access path

The lasso that is of significance is the one that groups more than one entity because of one-to-one access, in the example that being the entities Customer and Sales Area, with one Customer being related to one Sales Area. The other lassoes for the optional accesses to Services, Products and Invoices and the Set of Orders, Set of Order Lines and Set of Invoices are not entities, even though they have an entity access arrow into them. They are in the EAP for Jackson structuring purposes.

Given the access path to the logical data model there is a sequence of one access to the Customer and from there one access to the Sales Area, one Customer to one Sales Area. The entities are therefore lassoed in the EAP. The problem with the lasso is that it gets rid of the sequenced access. In the Enquiry Input Data Structure resulting from the lassoing it can be seen that the Sales Area box has disappeared and is "swallowed up" by the Customer box, which is the name of the lasso in the EAP.

The solution is not to lasso the one-to-one entity accesses. The revised Enquiry Input data structure is shown in figure 3.40, showing a sequence of access on Customer and Sales Area. The structure has been drawn to conform to the Jackson standards—the iterated Order is extracted to show its sequence of access before the Order Lines and paid and unpaid Invoices, and the top "root" box to the structure is the event rather then the entry point entity. The Customer entry point entity is now the first access in the entity access sequence of the structure. The remainder of the structure is unchanged.

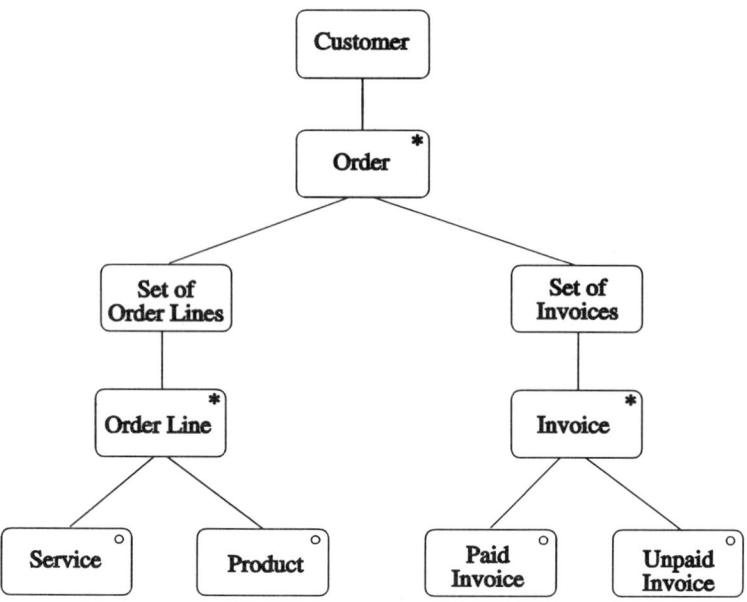

Figure 3.39 Enquiry input data structure

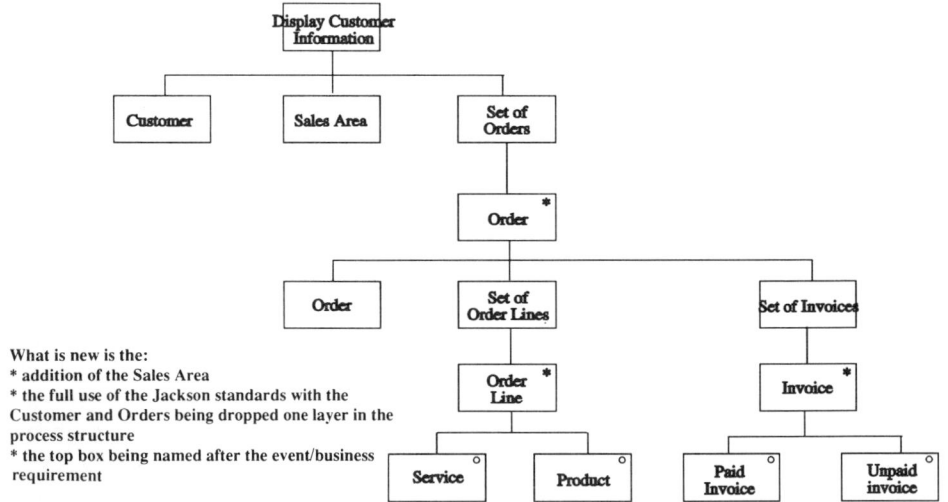

Figure 3.40 Enquiry input data structure for Display Customer information

3.3.3.2 The Effect Correspondence Diagrams (ECDs)

The ECD access path for the "Do This" access path in the ECD is shown in figure 3.41, with the lasso around the one-to-one access of Order to Customer and to Set of Order Lines. The revised Process Model from the absence of lassoing is shown in figure 3.42, showing that the Customer entity is now separate, such that the sequence of processing the Order and then the Customer can be seen. With the lassoing facility this sequence would be lost.

The Process Model has been modelled to be compliant with Jackson standards, with the Process Order line box being shown as the first access in the iterated sequence of accessing Order Line, followed by the access to the Product or Service. The "root" box to the Process Model structure is also named after the event and not the entry point entity.

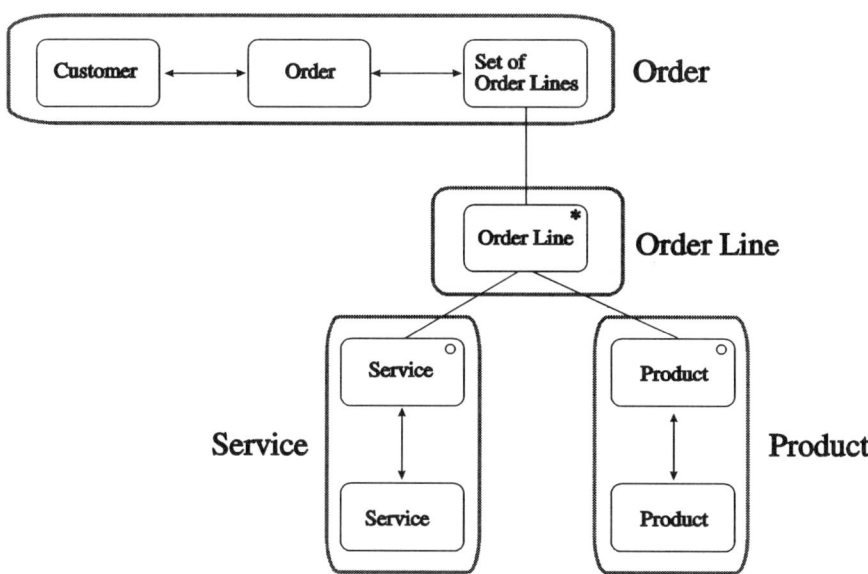

Figure 3.41 "Do this" effect correspondence diagram

3.3.3.3 The I/O Structure

There are several modifications that need to be made to this technique:

- The I/O Structures need to be "unlassoed", particularly the data retrieval business requirement. I/O Structures, as converted into the Enquiry Output data structures, are merged with the EAP Input Data structures to

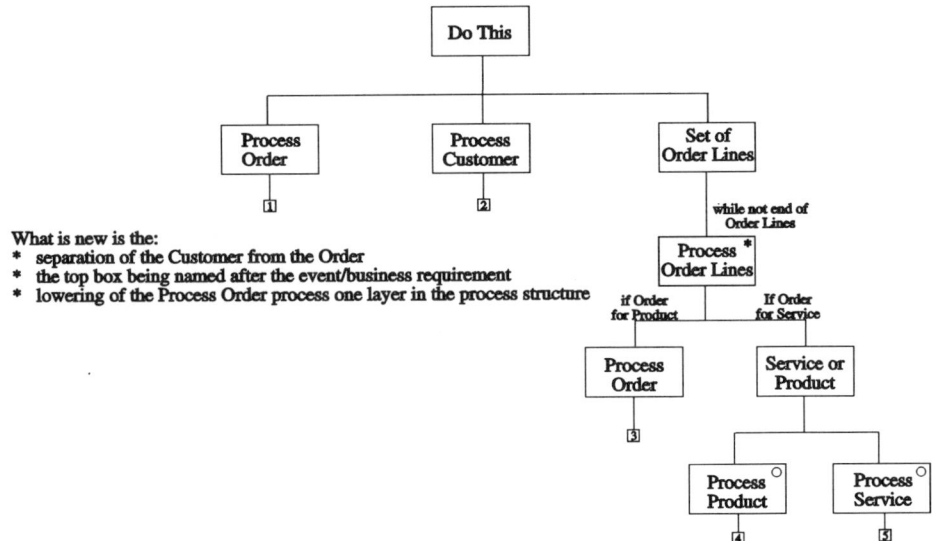

What is new is the:
* separation of the Customer from the Order
* the top box being named after the event/business requirement
* lowering of the Process Order process one layer in the process structure

Figure 3.42 "Do this" update process model

form the EPMs. Consider the I/O Structure in figure 3.43. This shows the structure for the Display Customer Information business requirement with the Sales Area information incorporated with the Customer in the element box B, because there is one Sales Area for the Customer. The sequence that the Customer is accessed before the Sales Area is lost, just as in the EAP.

- The revised I/O Structure Diagram is shown in figure 3.44 with the Sales Area "extracted" from the Customer and shown in element box C in sequence after access to the Customer. The supporting I/O Structure Description and Dialogue Element Description documents would be similarly modified. The I/O Structure diagram is now synchronised with the Enquiry Input Data Structure diagram to which it must be merged to produce the EPM.

- The fact that there is no risk to the overall man/machine interface design in what we are doing can be seen by comparing figures 3.45 and 3.46. These show the grouping of the element boxes in the I/O Structure Diagram by the Logical Groupings of the Dialogue Elements (the LGDEs), that is, the groupings of the element boxes as convenient components of a screen format. A LGDE can be a full screen or merely part of the screen. The LGDE 2 grouping in the revised "unlassoed" Dialogue Structure Diagram (figure 3.45) has the same data attributes in it as the previous unrevised

Tricks Specific to Version 4 of SSADM

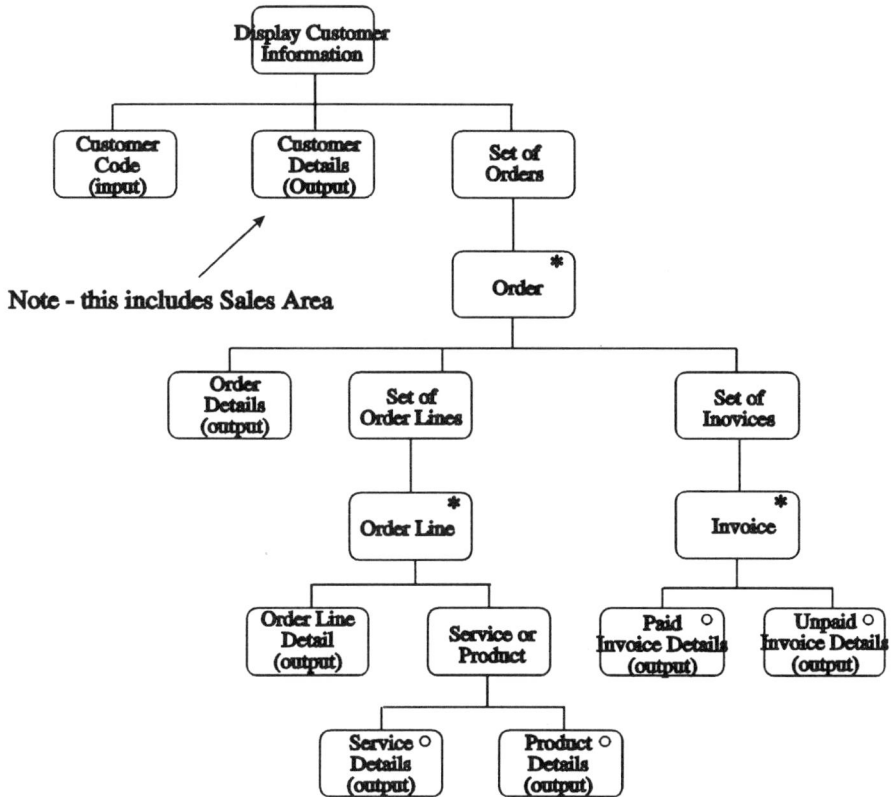

Figure 3.43 I/O structure for Display Customer information

Dialogue Structure Diagram (figure 3.46), the only difference being that the entity Sales Area is an element box in LGDE 2. The results as regards the man/machine interface are unchanged.

3.3.4 Dialogue design

These diagrams need to have the screen processing operations added to support the client "front-end" processing. The operations and constraints are the same as those that can be used in the Process Modelling techniques minus those that are concerned with data access. The "front-end" processing is of two main types:

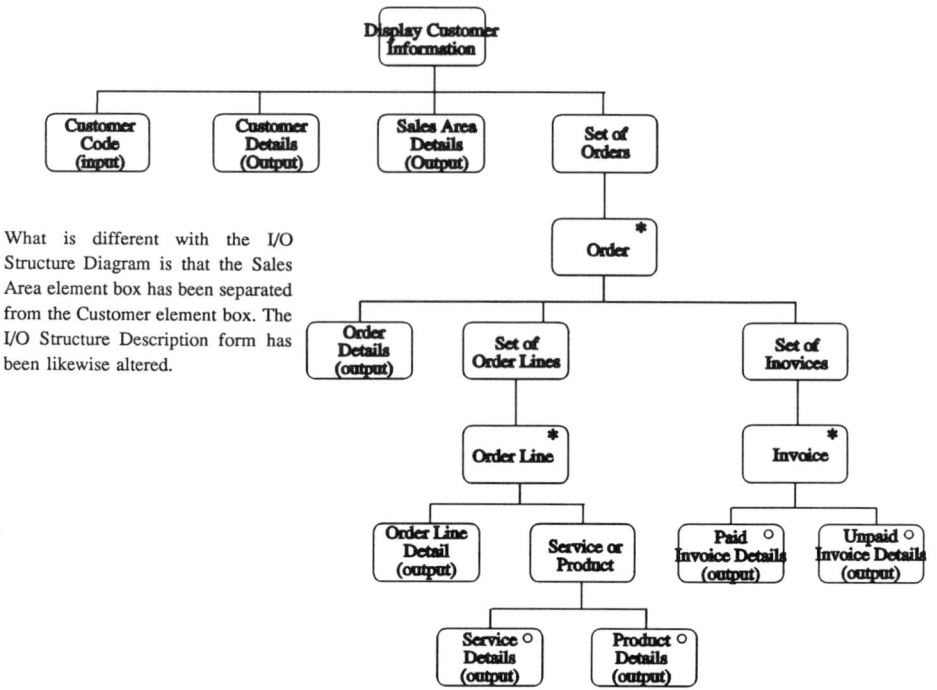

What is different with the I/O Structure Diagram is that the Sales Area element box has been separated from the Customer element box. The I/O Structure Description form has been likewise altered.

Figure 3.44 I/O structure for Display Customer information

- Simple editing of the data item, typically with a built-in function such as numeric editing. There can also be cross validation of the attributes on the input screen. However, where it is necessary to access the database for cross validation of entities and attributes not on the screen then the editing is done in the Process Models.
- Input and output screen formatting. Given the facilities of a modern database/4GL software, then it is likely that this will consist of merely naming a panel for display. If it is necessary to change the screen attributes for any of the variables or literals or modify the screen in some other way for the business requirement then operations for this need to be specified.

3.3.5 The Enquiry Process Models

The resultant unlassoed Enquiry Process Model for the enquiry Display Customer Information is shown in figure 3.47. It can be seen that the Sales Area entity continues to be modelled and the "root" box of the structure is named after the business requirement.

Tricks Specific to Version 4 of SSADM

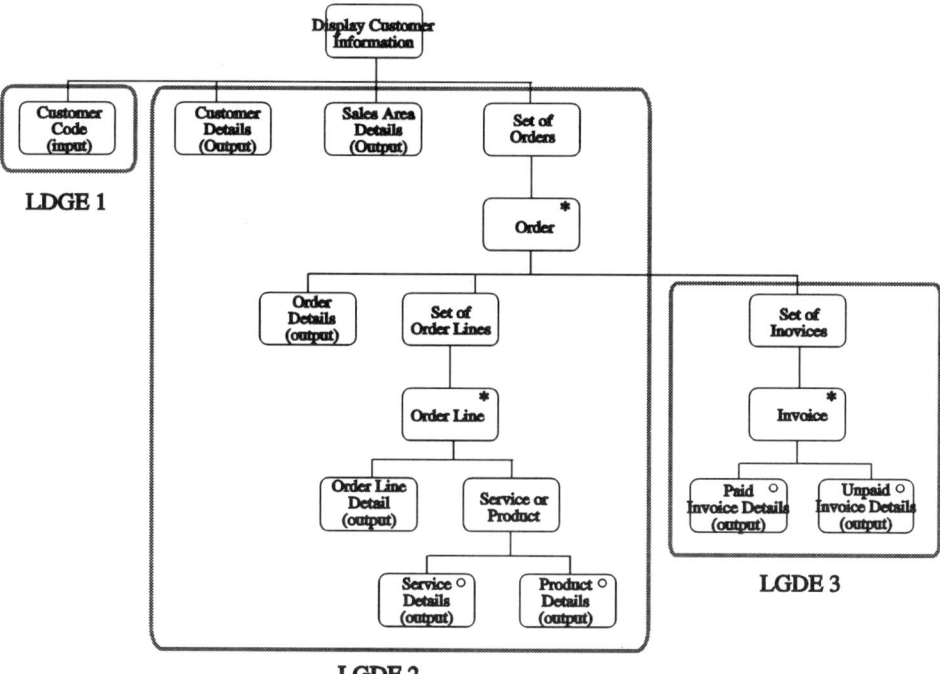

Figure 3.45 Revised I/O structure for Display Customer information

3.3.6 The need to support full Jackson standards

The problem being faced is that SSADM only produces Jackson-like diagrams. These "like" standards are used in the I/O Structure Diagrams and the EPMs and UPMs. Compounding the problem further is that the standard is applied differently in the techniques. For example, the top box in the I/O Structure is the event name, whereas in the Process Model, also at the event level, it is the name of an entity. We shall see that there are other examples of inconsistency.

The policy that has been followed is that the Enquiry Input and Enquiry Output data structures input into the Process Models remain unchanged, but that their conversion into their respective Process Models be modified to reflect Jackson standards.

There are a number of problems to solve:

- The version 4 standard is that the box at the top of the Process Model is the name of the first entity accessed by the event. This needs to be altered to that of the event name in the Function Catalogue (it is suggested that the numeric reference to the event be included as well), and that the first

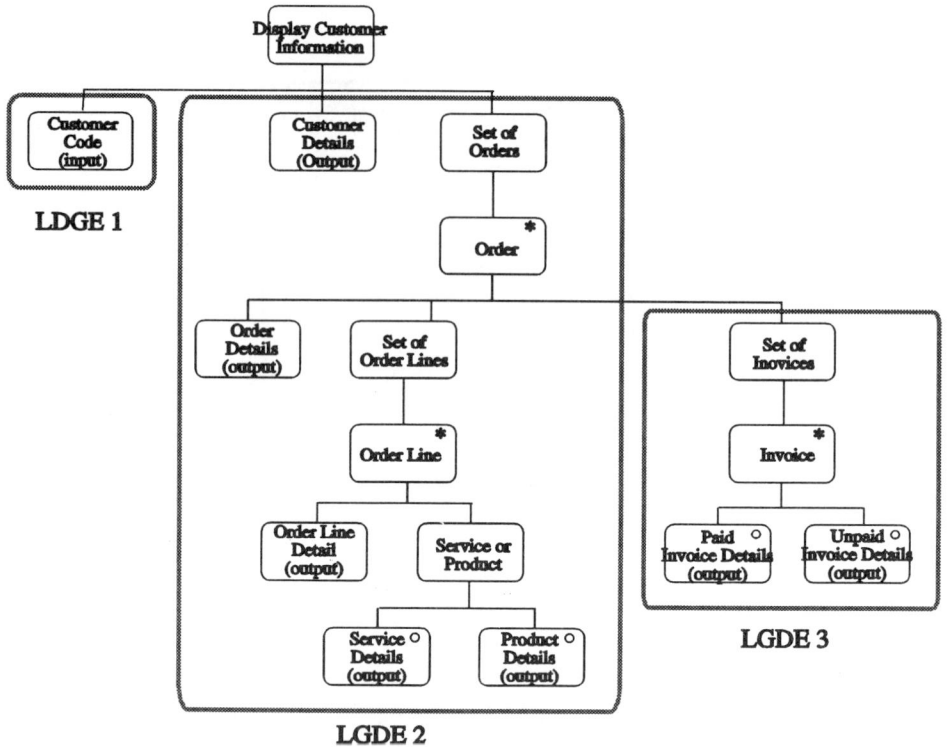

Figure 3.46 Standard I/O structure for Display Customer information

entity in the access sequence in the Process Model is the entity that was the top box. This can be seen by comparing figures 3.47 and 3.39, figure 3.39 representing the revised Jackson compatible input structure to the EPM for the Display Customer Information business requirement. The EPM is shown in figure 3.47.

- Where the entity being accessed is an iteration or selection, is not the bottom of the process model tree and is followed by a sequence of further accesses to other entities, then the iteration is to be treated as a sequence of access to the iterated entity and to the other entities. For example, in figure 3.47 the iteration of Orders and from there to access the Order Lines and the paid and unpaid Invoices is drawn with the Order as the first part of the iterated sequence of Orders, Order Lines and the paid and unpaid Invoices. The Order has been created as a leaf entity in what will become a Process Model. Compare figure 3.47 against figure 3.39 to see the difference.

Tricks Specific to Version 4 of SSADM 171

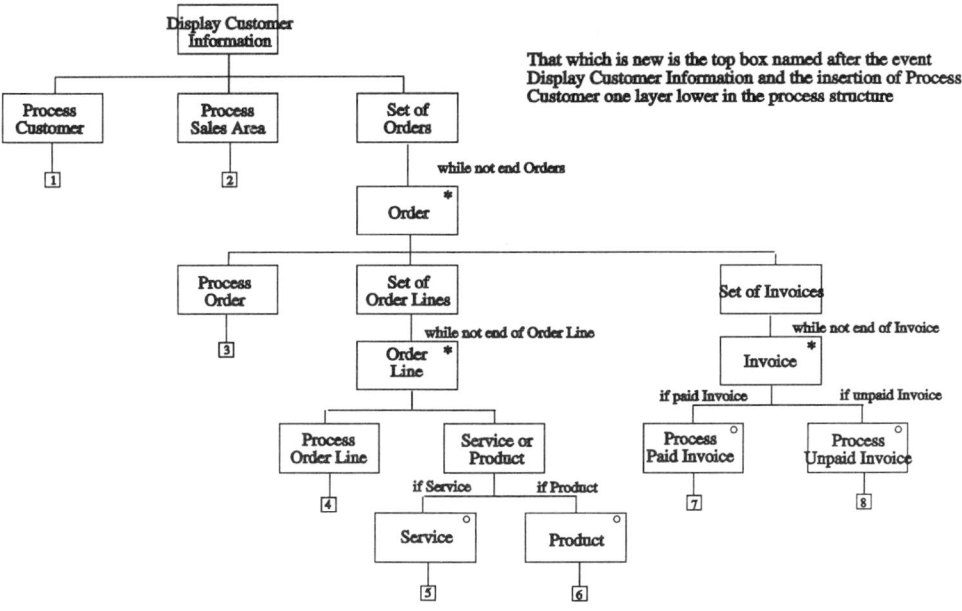

Figure 3.47 Enquiry process model for Display Customer information

3.3.7 Structure clashes

It was mentioned earlier that there is often difficulty in recognising structure clashes and the types of structure clash. Structure clashes can only occur for the Process Models that are composed of merged input and output data structures, that is, the Enquiry Process Models.

The clash illustrated in figure 2.11 is an order clash. The sequence of the data components (the accessed entities) of the input data structure are in a different order from the data components (the comparable output data groupings) of the output data structure. A simple output sort routine would solve this problem.

A boundary clash cannot be easily seen in the different structures of the EAP and I/O Structures because they do not show the attributes in the entities. It may be that the attributes in the output data groupings do not match those of the input entities. Assume that the input entities from the EAP are A, B and C with attributes 1 to 10 in entity A, 11 to 20 in entity B and 21 to 30 in entity C. The output data groupings are also about entities A, B and C but the attributes are not grouped by entity but by a different arrangement of the attributes on the screen. On the first display line on the screen are grouped under a general heading of some kind the attributes 1, 2, 7,15, 17 19, 25 and 28; on the second display line under another heading are displayed the attributes 3, 4, 9, 11,

13 and 22 and so forth. The physical display groupings of the attributes is different from the logical groupings of the entities. There obviously needs to be an output attribute arrangement routine of some kind.

The interleaving clash is where the both the input entities and the output data groupings are not in third normal form. This is usually the case when the database design has been "bent" from the logical data model for performance reasons and "bits of data" are stored in entities to which they do not "belong". The solution to this is to design a process that normalises the "untidiness" of the input data so as to reduce the "seriousness" of the interleaving clash to a boundary clash. The process to do this is similar in design to that of the Process Data Interface in Physical Design.

3.3.8 The Function Component Implementation Map (FCIM)

The SSADM manuals do not give an example of an FCIM. It is possible therefore that the example presented here will not be in conformance to what was intended by the developers of the method.

The manuals state that the FCIM serves as a system network diagram, showing the relationships of all the logical components of all the functions to the physical. It is also necessary to define which physical facility is to be used to implement each component, each "fragment". The logical components include:

- the Process Models (for the server/database part of the processing);
- the Dialogue Models (for the client/human computer interface part of the processing);
- common processes;
- business rules for attribute editing, checking the exclusivity of the logical data model relationships and referential integrity);
- success units at the appropriate level, such as the state indicators from the ELHs at the entity level and at the end of processing for the event level.

The physical components include:

- a procedural programming language, typically to be used for the coding of the operations on the Process and Dialogue Models;
- a query language, to be used where the enquiry is merely to retrieve data;
- a screen painter/report writer for the formatting of the output data;
- a menu generator for the menu screens;
- schema definition of the business rules, if the file handler provides this facility. If it does not then the rules must be coded procedurally;
- a database access language, such as SQL.

It has proved impossible in practice to create the three-dimensional FCIM map that the above would require—one dimension is the list of the logical components, another is the relationship of the logical components to the physical components and the third is the relationship of the physical components to the implementation facilities. The solution is to create two FCIMs, one a system network diagram and one for each function (in reality the processing for the event—the Process and the Dialogue Models are at the event/business requirement level).

The network diagram would be all the logical components to their named physical components, usually a one-to-one pairing for the procedural coding, the database access calls to a physical call, the business rules to a named rule in the schema definition, the common processing to a named process and the success units as added code to the procedural component.

The FCIM for the function was a simple case of defining which physical facility was to be used for each of the logical components of the function, for example the BEGIN and COMMIT facility for transaction success units. It was this diagram that was used by the project management to plan which skills were required for which part of the system to be developed, by whom and by how much effort.

4

Additional Data Processing Environments for SSADM—Distributed and Realtime

4.1 Distributed Systems

The transmission of data between locations is becoming increasingly important, for commercial, technical, competitive and performance reasons. Two types of companies are emerging, the megabig, multi-site international companies offering a comprehensive range of products and services, and the small, usually single site, focus/niche companies concentrating on a single product or service, often in only a single sector of the market. An example of the former type is the major international airlines, with numerous offices in multiple sites and countries, offering passengers an integrated set of products (flights) and ancillary services (hotel bookings, car hire and holiday arrangements). Each site has its own data requirements. Typical focus/niche companies are those specialising in information technology training, and in a particular structured design and development method, such as SSADM.

There is an increasing speed and complexity of trade. Given the efficiency of modern communications, the length of time to conduct business over long distances is continuously reducing. Customers are able to place a product order at site 1, have the stock allocated against the order from sites 2 and 3 and have the product packaged at site 3 as a single continuous function, for a total delivery time of, say, half a day. The Kobler Unit of Imperial College, London, found that of the six top rankings of the benefits of information technology the most important two were concerned with speed of service—shorten the time taken from order to delivery and accelerate response to new customer queries. If the company is located at many sites then a dynamically linked distributed system of some kind would be a prerequisite to achieve these top two requirements.

The megasized companies also need to control their multifarious activities. High speed and great complexity demand synchronised control. The wide geographical multi-site spread of these large companies means that the synchronised control of a company's data across sites can often only be provided by a distributed system of some kind.

The dynamic and synchronised nature of distributed systems also enables these large companies to be competitive with the focus/niche companies. The small specialised companies can often offer a fast and personalised service to customers and thereby can gain a competitive edge over their big rivals. By dynamically linking their often widespread activities in a distributed database the large companies can at least offer integrated and fast if not personalised service.

The removal of trade barriers in 1992 within the European Community is forcing these large companies to reposition themselves *vis-à-vis* each other. This is being achieved by a shifting balance of business alliances, takeovers and mergers. The computer systems of previously distinct companies need to be integrated, preferably dynamically—hence distributed systems.

A number of technical advancements are also encouraging the moves towards distributed systems. There is changing cost profile for processor hardware. For example, IBM is moving towards three different processor architectures—enterprise systems (ES) based on the S370 type processors, application systems (AS) based on the new AS400 type processors and the personal systems (PS) based on the PS/2 type processors. The cost/ratio factor of these processors is currently on average some 10, 2 and 0.2. This illustrates the tremendous reduction in costs per unit of processing power with smaller processors.

Using the cost/ratio as a basis, many smaller distributed but interlinked processors could be a cheaper architecture than a single centralised mainframe type processor. The widespread adoption of intelligent workstation/PS client-server processors reflects this cost/ratio. Some relational database products are now supporting co-operative client–server processing, with front-end application development software running on a number of low cost workstations, and back-end application support software running on a reduced size host processor.

A distinction needs to be made between distributed database and other forms of distributed system. *Distributed database* is designed to provide all the facilities of a distributed system. Of all the other forms of distributed systems technology *distributed database is the only technology that can support the two concepts underlying distributed systems. The other forms of distributed technology are "poor man's" distributed database with a more limited array of facilities.*

There is little doubt that distributed database is not yet widely adopted, partly because it is currently a financially costly strategy, partly because the technology is recent and not fully developed and partly because there are cheaper and proven low risk alternative technical solutions. The slow speed and high cost of data transmission on wide area networks has also proved a barrier. Distributed database has therefore not achieved "critical mass" as regards sales.

Two trends are reducing some of these impediments. The leading relational vendors have developed distributed database software and, as their technology is developed, will eventually "anoint" the subject. IBM have a particular incentive to develop distributed database software. Much of their software is not portable between their three processor architectures. For example, their DB2 relational database and MVS operating system run on the ES type processors but not on the AS and PS type processors. IBM are thus not able to move the software to the data—they must therefore move the data to the software. The full solution to this problem is distributed database.

The other trend is the popular adoption of much faster and cheaper local area networks for distributed systems that are not so geographically dispersed, typically within a building or between buildings a few miles apart.

The final technical reason for adopting this technology is the demand for high performance through distributed and co-operative processing. This latter technology is becoming ever more popular and will probably win the day. Distributed processing is being added to co-operative processing to further optimise distributed performance. The different types of distributed system are addressed in section 4.1.2.

4.1.1 Distributed database concepts

The two unique features of a distributed system—location and update and recovery synchronisation—are the basis of the two concepts that distinguish distributed systems from traditional database processing at a single site—location transparency and multi-site synhcronisation.

Many applications function across multiple locations and transmit data between the locations as required. However, the data transmission has been simplistic—initially through offline transfer, possibly by tape, or through bulk transmission down a fixed topology telecommunication line, mostly at fixed and predefined times and with the location to which the data is to be transmitted known in advance. The data processing at each location has been localised and not dynamically synchronised with other locations. Such offline distributed processing is the hallmark of decentralised systems. The concepts of location and update synchronisation are therefore not relevant.

Distributed systems are different. A distributed system is one where data and programs used by an application are stored on more than one processor in different locations *and* the processors are dynamically linked in a synchronised manner and the data and programs function together as a synchronised whole. Furthermore, the data transmission between sites[1] is *ad hoc*, the messages of data between locations are unformatted and the location of data storage and program execution is not known in advance. The concepts of distribution are therefore highly pertinent.

There is one further aspect to consider. The concepts of location transrency and multi-site synchronisation require to be hidden from the users of the system. The users should not be aware that the data may be distributed—the overall concept of a "single image" system.

4.1.2 Types of distributed system

There are a number of ways in which the technology for distributed systems can be viewed.

4.1.2.1 Distributed database

Distributed database is concerned with the transmission of data between multiple file handlers of the same type at locations that are geographically widely dispersed. A system could have some of its data in Frankfurt, some of its data in Paris and some of its data in London, all using, let's say, the ORACLE relational file handler. Distributed database therefore requires to support the location of data, but does not worry about data access between incompatible file handlers between sites.

Distributed database incorporates the technology that supports the two concepts underlying distributed systems. It is the "all singing, all dancing " technology for data distribution. All the other approaches that are used to support the distribution of data and processing require to "do it yourself" to support the concepts. However, the author is not aware of any of the vendors claiming to support distributed database software providing all the facilities that are required.

The structure of distributed database software is illustrated in figure 4.1. There are multiple databases at multiple locations, the location databases together forming the corporate database. The two features that distinguish distributed database come into play. Firstly location. The data is potentially spread like "grass seed" across locations, yet the locations need to be hidden from the users of the distributed system. The distributed database software

---

[1] Note: the terms site, location and node are used interchangeably and in the context of the usual terminology when discussing the different components of distrubuted database. For example, people talk about location transparency and multi-site synchronisation.

Distributed Systems

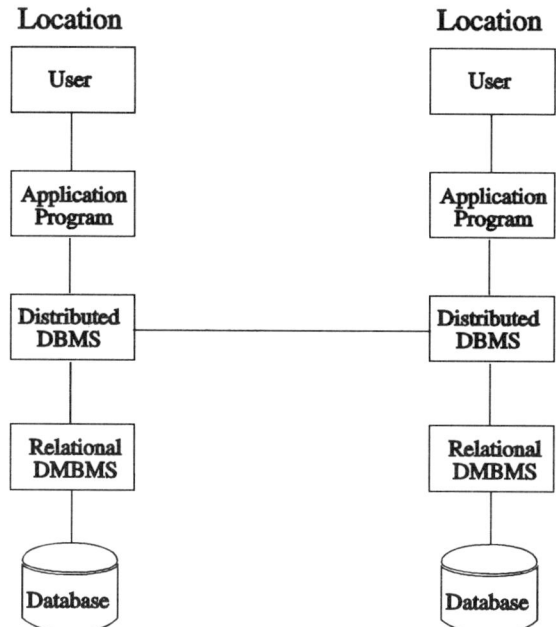

Figure 4.1 Distributed database

must therefore not only support the location of the data but also have facilities for finding the data automatically. Secondly synchronisation. The location databases are linked dynamically, such that multi-site updates and software and hardware recovery must be synchronised. The figure shows that there is a distributed database manager at each location between the applications programs and the local file handler, and controlling data searching and update synchronisation across locations, with the local file handler then being invoked to access the "now ascertained to be local" data. *The important point is that the distribution and synchronisation of data is handled by file handler technology.* It is not a case of "do it yourself". The distributed file handler intercepts the SQL DML calls of the application program and searches the global dictionary for the location of the data or uses a multi-site searching routine if the global dictionary is not adequate. The SQL calls are routed to the appropriate remote distributed file handler, from which a response is ultimately received.

4.1.2.2 *Client-server*

Figure 4.2 shows client-server processing. Another term often used is co-operative processing. Client-server architecture has been designed to cater

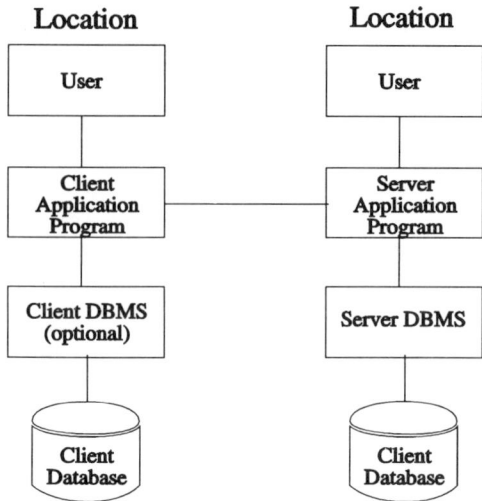

Figure 4.2 Client-server database

for the increasing use of intelligent workstations (IWS) front-ending the main back-end processor. This has the benefit that processing is undertaken where it is most appropriate. A client is a requester of a service and the server is the provider of a service. Any communication between the client and the server processes is by explicit calls requesting a service, of the client type "Can you send me some data?" or "Can you process me some data?" and providing server responses of the type "Here is the data" or "Here is the result". There can be a hierarchy of such facilities, so that a client processor can also simultaneously be a server processor.

There are three major components to the client-server processing software:

- The presentation man/machine interface component typically provides screen formatting, data editing built-in functions, spreadsheets, word processing and help facilities and is undertaken solely on the front-end "client" processor.

- The IWS client application component would typically include application bespoke data editing facilities and utility functions, often with package software, such as spreadsheets and word processing. The application component on the server process would typically include database handling, bulk data messaging and management reports.

- The access component to the corporate database on the server would be

Distributed Systems

via a file handler, usually relational with SQL on the host processor, and is run on the "back-end "processor. The back-end "server" processor is concerned with supporting the main application system processing against the database and the production of bulk batch reports. The file handler would be a local file handler without any capability to support distributed database, that is, being able to find the data and synchronise updates across locations. Any communication between the front-end and the back-end processors is explicitly invoked and controlled in the application program by the application programmer. The man/machine interface and database processing are thus kept separate.

Client-server processing is very similar to distributed processing. Both send messages to application processes on different processors, as can be seen by comparing figures 4.2 and 4.4. The reason it is separated is that there is file handler technology specifically developed to support it, with specialised calls between named client and the server processors, whereas distributed processing has long been used before client-server architecture was available and required procedure calls between application programs or parts of application programs that happen to be resident on different processors.

Client-server processing does not support the two concepts underlying distributed systems. The author regards client-server processing as the "poor man's" distributed database, yet of the different technology classifications for distributed systems in this section it is the one that is becoming most widely used. Several of the relational database vendors have rewritten the file handler software to support client-server facilities, one such being Ingres Inc. (now owned by ASK Inc.) and Oracle Corporation. Software AG are also rewriting ADABAS to be client-server. Sybase produced their SYBASE database product from the outset as being client-server.

Note that there is usually only one corporate database in client-server processing. The IWS may contain personal databases, but they are not synchronised with the corporate database. Equally, there are only two location types—the IWS and the host. If the data is not on the IWS it must be on the host processor and vice versa. The location of the data is therefore implicit. Client-server processing therefore does not require to support either location transparency or update synchronization. It facilitates distributed data access rather than distributed data. There is therefore no technical overlap between client-server processing and distributed database.

There is one other difference between client-server processing and distributed database. With distributed database all sites are regarded as equal, each fulfilling the same role—there is no concept of a master site. Such is not the case with client-server processing. In figure 4.2 the host processor has different functions to fulfil from the IWS.

4.1.2.3 Remote database access

Remote database access (RDBA) is an example of distributed data but without the distributed database manager software being used. Consider figure 4.3. It shows an application program issuing a request to a local file handler and/or to a named remote file handler supporting a database containing the remaining information required. Apart from the Connect call the access to the remote file handler is the same as to the local file handler. The DML commands are the same. The significant points here are:

- The application program is responsible for issuing a Connect statement to the named remote file handler, so there is no location transparency or automatic update synchronisation with the two-phase commit facility as far as the application programmer is concerned. This must be taken care of by the application programmer.

- The data communications will be high. The remote file handler returns the raw data requested, that is, all the orders won. There is no local processing of the remote data at the remote location.

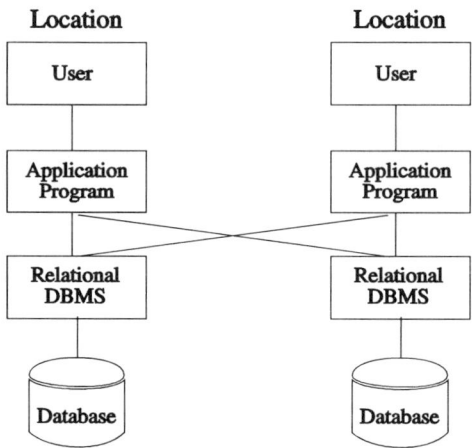

Figure 4.3 Remote database access

4.1.2.4 Distributed processing

Distributed database is only concerned with the distribution of data. However, the transmission of data between locations is costly, with the processing overheads being up to 100 times the cost of accessing data from the local disk. If there is much data to be transmitted between locations in support of a business requirement then it may be more efficient to process the data at the remote location(s) and, instead of transmitting the raw data back to the requesting location, to transmit merely the result, the "reply". There would then be correlation processing of the returned result(s) at the requesting location before output presentation. This approach is distributed processing, often referred to as request or transaction shipping. The structure of this software is illustrated in figure 4.4.

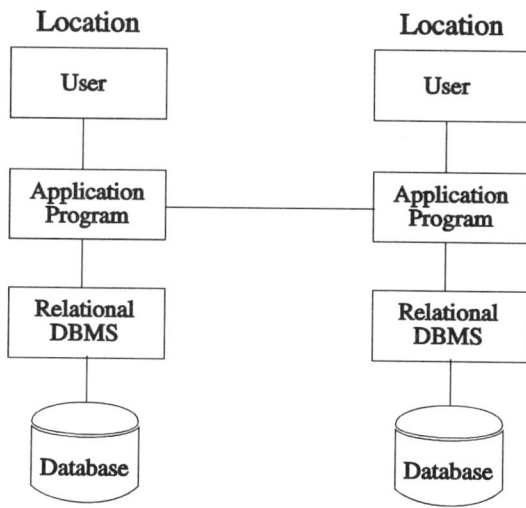

Figure 4.4   Distributed processing (1)

There is no support for the two concepts of distributed systems with distributed processing. The local process can facilitate the searching for the location of the remote data by accessing the global dictionary that indicates the location of the remote tables of data, but that is the only readily available "help" that is obtained from distributed database if it is installed.

With distributed processing it is therefore "do it yourself" in the application program software regarding the two features of distributed database. The communication between the processes is via a normal transaction code (with any associated parameters and data). That is why the figure shows that communication between the sites is via the application component, not the file handler. As will be seen later in this chapter, distributed processing can provide higher performance than distributed database, such that the cost and effort of distributed "do it yourself" may be justified.

Distributed processing occurs where program-to-program communication (PTOPC) is used. Consider figure 4.5.

The process at the triggering client site is the master process and sends a

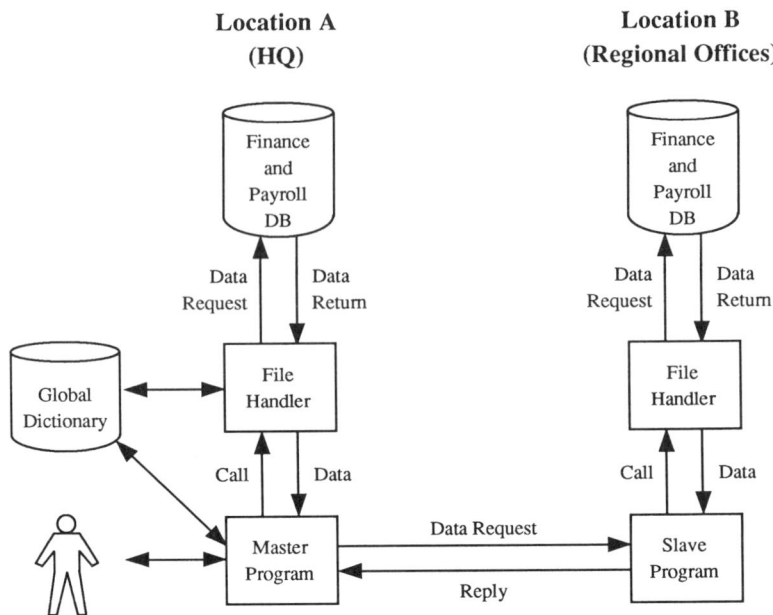

Figure 4.5 Distributed processing (2)

data request as a normal teleprocessing type transaction message (that is, a transaction code and call parameters) to the remote teleprocessing monitor, which in turn invokes the appropriate slave process, not a file handler, at the remote site(s). The master and slave processes invoke their local file handlers to retrieve the appropriate local data. Each process then processes its local data. The remote slave process(es) transmit a processed reply, not the raw data as in distributed database and remote database access, back to the master program at the triggering location.

The example in figure 4.5 assumes the business requirement is "For all salesmen calculate their sales bonus for the last six months and add it to their current month's payroll cheque". The master process at the head office triggering site invokes the file handler at the head office to access the salesman's pay information. At the same time the master process also sends data requests to a slave process at the regional office(s), which in turn invokes the local file handler to access the salesman's order information. Each slave process then processes all the orders and calculates the bonuses. The bonus and not the raw data in the orders is transmitted as the reply to the master process. When the master process has received the responses from all the slave processes the individual bonuses can be added to the salesman's basic pay, the appropriate tax and other deductions can be computed and the pay roll cheque issued.

The advantage of PTOPC processing is that:

- data communication is reduced, with processed rather than raw data being transmitted;
- because it is the transaction that is shipped, the slave process(es) are not restricted to the same file handler as the triggering site. The client and the server file handlers can be different.

The disadvantage is that the programmer requires to understand the concept of location, either by accessing the global dictionary to find the location of the data or through some multi-site searching technique as described earlier, *and* the requirement to synchronise the responses from the remote slave processes. If the processing is to be distributed in order to reduce data communication overheads then the distribution and synchronisation of the data has to be supported in the process rather than in the database software.

4.1.2.5 Heterogeneous systems (gateways)

The author has found that there is considerable confusion between a distributed database and a heterogeneous system. Both after all are concerned with sending data between multiple file handlers.

Heterogeneous systems are concerned with the transmission of data between multiple file handlers of different types. For example, some of the data could be stored and accessed using IBM's IMS file handler, some of the data could be stored and accessed using DEC's Codasyl DBMS-32 file handler and some of the data could be stored and accessed using Ingres Inc.'s relational file handler INGRES. These file handler types each have different data storage and access mechanisms and languages. Gateway software mapping the different database data structure and data access facilities of the file handler types is required before data communications between the file handlers can be achieved. It is the responsibility of the gateway software to make the data in the various file handler types of compatible format.

A particular concern not often appreciated is that, with multiple file handlers involved, it is, like distributed database, necessary to ensure update synchronisation across the file handler types. The facility for this is what is called a two-phase commit procedure. Even though both of the file handler types may have their own two-phase commit facilities it is unlikely that they are compatible. The gateway software is therefore also responsible for update synchronisation.

The rest of this chapter is primarily concerned with distributed database. Once the facilities of distributed database are understood then the other forms of distributed systems, being less capable, can also be easily understood.

4.1.3 SSADM and distributed systems

SSADM version 4 does not support the concept of location. The concept is not included in any of the design techniques or the supporting documentation. The method therefore does not support distributed systems. Planned updates to the method are underway, with work done by Ed Dee of Edinburgh University (figures 4.1–4.4 reflect this work) and others, and a Interface Guide for Distributed Systems as indicated in the version 4 manuals will be produced.

Fortunately, of all the data processing environments which SSADM does not support, the distributed environment is the one to which the method can be most easily upgraded. The reason for this happy state of affairs is because of the policy followed by all the suppliers of distributed database software. The policy has been to preserve *without change* the existing database technology for processing in a centralised environment and to "add on" extra software to support data distribution.

The policy results in a number of significant benefits. *All investment in existing database technology and structured design techniques is wholly preserved—one need unlearn nothing.* Database technology and the techniques used in structured design and development methods, such as SSADM, require enhancement by addition rather than by modification. A further result of this policy is that distributed database is a partnership between equal and independent but co-operating centralised systems using current technology, with the triggering site responsible, as far as possible, for finding the location of the distributed data and synchronising multi-site processing.

The enhancements to SSADM to support distributed systems are considered from two aspects—the physical technology and the logical and physical design and development techniques. The physical technology is considered first. Although this may appear to be "putting the cart before the horse", it is necessary to understand the technology in order to appreciate the requirements for and implications of data distribution in the techniques and their deliverables. The implementation technology has always preceded the design techniques for the simple fact that there is no point in having the design techniques if there is no means of implementing the logical design specification the techniques produce.

The enhancements to the SSADM logical and physical design techniques to support location transparency and multi-site update synchronisation are described, along with worked examples. The examples are based on two major distributed applications the author has been involved with.

4.1.4 Distributed database technology

All the facilities, bar one, for distributed database are to be found in database software for batch and online processing at a site. All the facilities relate to

Distributed Systems

data handling. *Distributed database is concerned with the distribution of data, not logic.* As indicated earlier, the other forms of distributed systems are nothing more than centralised processing with "do it yourself" enhancements to handle the searching of data across multiple locations plus multi-site database update and recovery synchronisation. The facilities are:

- set level processing;
- concurrency control;
- dynamic transaction rollback;
- system rollforward;
- transaction utilisation monitor;
- query optimisation;
- directory maintenance;
- location transparency.

It is location that is the driving feature of distributed systems. If there were no multiple locations then location transparency and multi-site synchronisation would not be required.

4.1.4.1 Set level processing

Set level processing is a *sine qua non* for distributed database software. The basic additional overhead in distributed systems is remote I/O, the sending of messages between sites. The area of greatest processing overhead in any computer system generally is I/O. In centralised processing there are two kinds of I/O—disk I/O and terminal I/O. Distributed database adds a third type of I/O—remote I/O.

In an IBM mainframe processor environment a typical disk I/O to read a table row is some 3500 path length instructions (PLI) additional to the 1500 PLI for the logical I/O accessing the buffer pool. A typical terminal I/O under the CICS teleprocessing monitor software (assuming that the CICS software is in the processor main memory) is some 50 000 PLI for an average message pair (assumed to be 100 characters input and 500 characters output). Advice the author has received is that a remote I/O message on a wide area network can cost up to 500 000 PL/I, depending on the telecommunication network design. A remote I/O on a local area network is apparently some one third the processor overhead.

It can be seen that remote I/O is extremely expensive in processor resources—up to 100 times greater than local disk I/O—and therefore requires to be kept to an absolute minimum. (Note: the support for remote I/O is largely offloaded from the main processor with front-end processors dedicated to handling data communications.) It is therefore essential to support set level processing as found in a relational file handler, where data is sent as a single remote I/O stream of information rather than n record-at-a-time remote I/Os to be found in pointer chain technology file handlers, such as in IDMS and IMS.

The advantage of set level processing is that, irrespective of whether one is retrieving a specific table row (display customer number 12345) or multiple rows (display all customers with red hair), the data is transmitted between sites as a single remote I/O data message. With record-at-a-time processing each table row will be transmitted as separate remote I/O messages.

Thus if the retrieved set incurs 1000 rows (i.e. there are 1000 customers with red hair) then the overhead would be $1000 \times 500\,000$ PLI record-at-a-time messages rather than $1 \times 500\,000$ PLI set messages. It is not surprising that earlier attempts at distributed database with record-at-a-time file handlers have not proved successful.

4.1.4.2 Concurrency control

Concurrency control of database data in a multi-user environment is as necessary in distributed as it is for centralised processing. Distributed concurrency control is, however, only necessary where data is distributed horizontally. Horizontal data distribution is discussed in section 4.1.4.8.

There are two basic distributed concurrency control mechanisms—multi-phase locking and timestamping. The mechanisms are not mutually exclusive and n variations of distributed concurrency control are possible by combining certain features of each.

Locking places a "hold" on the table row to be accessed, the type of hold being appropriate to the type of access. Locking can lead to a deadlock situation, that is, a table row is locked by a process at one site and is required by a process at another site and vice versa. Timestamping does not incur a deadlock possibility, as all transactions are executed serially only in timestamp order. Timestamping is a crude mechanism as transactions are executed in the sequence they are triggered, rather than in transaction priority order. Both mechanisms require a degree of message passing between sites, the locking mechanism for maintaining the wait-for-graphs and the timestamping mechanism for cross-checking and maintaining the timestamp of the last access to the database records.

In centralised processing locking rather than timestamping is universally used. Given the policy followed by the distributed database vendors, most products appear to have adopted distributed locking for distributed concurrency control. The author is not aware of any of the leading non-research/commercially available distributed database products using timestamping. Much research is underway regarding distributed timestamp mechanisms. The debate of the relative merits of these two concurrency control mechanisms in a distributed database is still open.

There are two main kinds of lock levels—the "S" lock and the "X" lock. An "S" share lock prevents an update to the data object (usually table row but can be at escalated page or table level) from occurring until the lock is released, but allows other read transactions to access the data object. The "X" exclusive lock prevents any other transactions accessing the data object occurrences until the "X" lock is released, either through an explicit application program COMMIT call or through an implicit lock release at the end of processing. Transactions wishing to access an "X" locked data object are simply held in abeyance until the "X" lock is released. Such a procedure as described above will always be correct. This approach is often referred to as request or transaction shipping.[1]

In a distributed system where a transaction requires to read a data object an "S" share lock is required on at least *one occurrence* of the object in the system. Where a transaction requires to update a data object an "X" update lock is required on *every single occurrence copy* of the data object, even when the data object occurrences are stored at multiple sites. An "X" lock requires to take into account data distribution, an "S" lock does not. Within a distributed system these locks are maintained in a "wait-for-graph" at each site. The graphs show "who is waiting for whom". Each graph is a matrix relating the type of lock with the transaction ID, table ID and row ID.

In a distributed system multiple file handlers are used, one at each site. Should a data object occurrence be stored by more than one file handler, it is essential that, where locking is used, the locks of the data object occurrences are synchronised and occur against all occurrences of the data object at all sites or not at all. The locks therefore require to be shipped around the distributed network to the site wait-for-graphs. The wait-for-graphs are stored and accessed as standard data tables. It is therefore necessary to synchronise updates with a multi-phase concurrency control procedure similar to that later described for the two-phase commit procedure.

Where such "X" locking is used deadlock can occur. In centralised processing this deadlock is called "deadly embrace", and occurs where two application programs require to access each other's "X" locked data object in order to continue processing. In such a situation a deadlock has been achieved. In a centralised environment, the procedure is to backout temporarily one of the transactions and to continue normal processing. The backed out transaction is then restarted. In a distributed environment the deadlocked programs may be residing in different processors at different sites. Such a situation occurs where transaction T1 holds B at site 1 and

---

[1] There is a third kind of lock, a "U" lock that prepares an "S" lock for upgrading to an "X" lock. The "U" ltype lock is not found in relational technology. It is appropriate to the earlier pointer chain technology file handler type.

requests A at site 2, while transaction T2 holds A at site 2 and requests B at site 1. A global deadlock resolution procedure taking into account site distribution is therefore required.

The multi-phase locking overhead is particularly severe as up to $5n$ (where n is the number of sites) messages between sites can occur. In reality only $4n$ messages occur as the fifth message is concatenated with the fourth. We shall see later that this concatenation is, in the author's opinion, a dangerous feature as the multi-phase commit is not sufficiently robust. This all involves remote I/O, which, as we have seen, is expensive. When a global deadlock is detected the wait-for-graphs usually have to be merged in order to ascertain the total distributed deadlock situation. This merging involves yet further remote I/O.

It has been argued that multi-phase locking is an additional two-phase commit procedure for multi-site locking prior to the two-phase commit procedure for multi-site database updating. The author has strong doubts on this. The more he delves into this subject, the more he believes that the locking for concurrency control is actually integrated in some way as part of the database two-phase commit procedure—see later in section 4.1.4.3—possibly with the wait-for-graphs lock being set on the first broadcast message from the co-ordinating site and being released on the second part of the two-phase procedure when the databases have been updated. Putting it another way, it is probable that the vendors of distributed database do not support distributed concurrency control, being prepared to live with the virtually non-existent chance of distributed deadlock.

Enquiries against a number of vendors of distributed database software has failed to identify whether the two-phase distributed concurrency procedure is incorporated as part of or is distinct from the two-phase commit procedure for distributed database update, or indeed is even bothered with. The clinching evidence has not yet been obtained. This book assumes that the two procedures are integrated.

A problem that multi-phase locking must resolve is the recognition when a message response to a message call is not being received, either because of a hardware/software failure or poor response performance at the remote locations. This must be handled by a timeout mechanism, which recognises that if a response message is not received after a specified period of time then a deadlock must be assumed and the locking procedure must be reversed and restarted.

If, as the author believes, multi-phase locking is integrated with the two-phase database commit procedure timeout can be a standalone mechanism by which a global deadlock across multiple sites can be resolved. The principle is simple—if the transaction timeout is exceeded assume a deadlock has occurred and begin the normal distributed backout procedure. Two transactions are in deadlock and hence waiting potentially indefinitely for each other to release the lock. The timeout time will eventually be exceeded.

Each site recognises this and rolls back and restarts its own transaction. The expense of maintaining and merging the distributed wait-for-graphs can be avoided. Concurrency control remains a local function. The SUPRA product from CINCOM has adopted this approach.

The danger of this simplistic approach to concurrency control is that a deadlock may not have occurred, the delay merely being part of a transaction performing badly. More rollbacks than are justified is the potential adverse result. Many distributed database vendors consider the saving on distributed communications worth the small risk of unnecessary rollback.

Timestamping has an advantage in that no distributed "S" and "X" locks are set and the remote I/O of locking and deadlock detection are avoided. Deadlock cannot occur. Shipping wait-for information around the network and testing for timeouts is therefore not required. Transaction backouts and restarts are also not required. Although local "X" locks are still required to ensure correct local concurrency control of the database, the global deadlock resolution software does not use them. Instead every transaction is assigned a globally unique timestamp.

Because transactions can occur simultaneously at multiple sites, a timestamp composed solely of a system time clock is inadequate. In order to make timestamping globally unique, the location ID also requires to be added to the clock time as a minor key. Each data object in the database requires to store a unique timestamp for the last read and update transaction that accessed the data object. When any transactions are in conflict, the younger transaction with a later timestamp value is backed out. Where the timestamps are of equal value the choice of a transaction to backout is usually made on transaction or location priority.

There are many degrees of sophistication in the various timestamping mechanisms. One such mechanism is conservative timestamping.

Conservative timestamping eliminates conflicts by not performing a transaction where a transaction conflict could occur. A transaction waits until all older transactions (i.e. those transactions with a lower timestamp value) have completed processing. Where a conflict could occur timestamping requires to send messages for each transaction to all the remote participant sites to ascertain the "age" of other processing transactions. This is clearly an expensive approach in processing resources because of extensive and expensive remote I/O messaging, and inevitably concurrency is also reduced. It also means that *all* transactions originating at a site, and not just those in a conflict, must execute COMMIT statements in timestamp order, irrespective of their priority, and even though transaction T2 may reach a COMMIT point before transaction T1, which started processing first.

Unless this expensive approach is adopted an update request for a transaction with a "younger" timestamp could be received and processed by a remote site before an update request for a transaction with an "older" timestamp. The older request would have to be rejected and restarted later.

This heavy inter-site communication of multi-phase locking and timestamping needs to be reduced. Two optimisation mechanisms are available—transaction classes and conflict graph analysis. Transaction classes recognise those transactions where conflicts cannot occur and hence applies the multi-phase locking or timestamping concurrency control procedures only where conflicts can occur.

In order to identify the possibility of conflict, each transaction requires to be allocated to a transaction class. As part of system design each transaction access path requires a specification of the tables, rows and columns that it reads and changes. Transactions are then assigned to one or more transaction classes, which contain a definition of what can be accessed as a readset or writeset. The transaction class is rather like the user view facility of a relational file handler.

A representative specification of transaction classes is given in figure 4.6, where it can be seen that the readsets and writesets of transaction classes TC1 and TC2 do not conflict, but that the TC3 readset is in conflict with the TC2 and TC4 writeset and the TC4 writeset is in conflict with the TC2 writeset and TC3 readset. Only where a writeset of one transaction class intersects with a readset or writeset of another transaction class can conflict occur.

In order to minimise the possibility of conflict it is important that the distributed database designer define the transaction classes as tightly as possible. The benefit is that inter-site communication for synchronising locking or timestamping in transaction classes where there is no conflict or the conflict is readset to readset is not required. The net effect is to reduce inter-site messaging and increase concurrency.

The ability to pre-define transaction classes is fine where the update business/transactions are known in advance and specified in the logical design. But what about *ad hoc* updates, which by their nature cannot be predefined? In order to provide the ultimate safety net for this type of situation a global class also needs to be defined. The global class needs to be defined as being able to access the entire database on both the readset and writeset.

The global class mechanism, of course, nullifies the very advantage of transaction classes, because all other transactions will be in a conflict of some kind. *Ad hoc* unpredefined updates need to be strongly discouraged in a distributed system. The database administrator of a distributed system needs to be able to define very precise security mechanisms in order to prevent *ad hoc* unauthorised updates to a distributed database.

Conflict graph analysis is a further refinement to transaction classes, such that concurrency is further increased. The graphs identify possible readset and writeset conflicts across transaction classes and identify if it is possible to have transactions interleaved rather than synchronised. A conflict graph is drawn in figure 4.7. The conflicts are the horizontal and diagonal lines. These indicate that synchronisation *may* be required. If there are no horizontal

TC1	Readset:	Select * from Person Where name = "Smith"
TC1	Writeset:	Select * from Person Where name = "Smith"
TC2	Readset:	Empty
TC2	Writeset:	Select * from Person Where name = "Morris"
TC3	Readset:	Select * from Person Where name between "Harris" and "Thomas"
TC3	Writeset:	Empty
TC4	Readset:	Empty
TC4	Writeset:	Select * from Person Where name between "King" and "Roberts"

Figure 4.6 Transaction classes

or diagonal lines connecting classes then no transaction synchronisation is required.

The objective of conflict graph analysis is to decide for every pair of transaction classes where conflict occurs what level of synchronisation is required. If the conflict is on two "X" locks then a different concurrency control is required than if the conflict is between an "S" and an "X" lock. With the former transaction synchronisation is achieved by backing out one transaction and restarting it. With the latter, the two transactions can be interleaved. Both transactions run more slowly but no transaction backout occurs.

Transaction interleaving in order to increase performance is not a feature unique to distributed database. It is to be found in normal centralised processing and is described in Chris Date's excellent book *An Introduction*

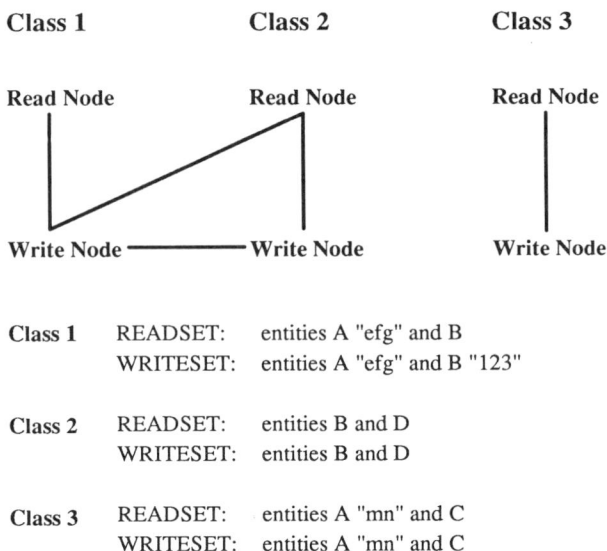

Figure 4.7 Conflict graph analysis

to *Database Systems*, Volume 2. The point that must be understood is that conflicting transactions triggered at different sites can still be interleaved rather than backed out.

4.1.4.3 Distributed transaction rollback

Transaction rollback is the recovery of a failed running application program. In a centralised system transaction rollback is achieved through reading the before image log in reverse time order to the last commit/checkpoint record for the transaction. The problem in a distributed environment is that the transaction may wish to update the same data record at multiple sites. In such a situation the transaction must execute successfully at all the sites or not at all. This requires a mechanism whereby the transaction at the triggering site checks all the remote participant sites that it is OK to update before executing the updates. Obviously a routine of message passing between the triggering and participant sites is therefore required.

The policy followed in distributed database is that the triggering site controls the execution of the message passing. A site in a distributed system is a "master" site *for the duration of a transaction only,* thus preserving as much as possible the concept that all sites in a distributed system are equal. Each

Distributed Systems

message pair (a "call" from the triggering site to the remote participant sites and an "echo" message in response) is called a phase. Distributed transaction rollback is achieved through a "two-phase commit" procedure, as two sets of message pairs between the triggering and participant sites are required.

A single-phase commit procedure is inadequate for distributed transaction rollback. As illustrated in figure 4.8, a triggering transaction at site 2 could fail as it is receiving "OK to update" echo messages from the participant sites involved in the multi-site synchronised updating. The figure illustrates that the triggering site has recorded that some of the participant sites, in this case site 1, have committed their database updates because successful echo messages have been received, but not have recorded that all the participant sites, site 3, have committed because of a failure during the message receiving phase. Triggering site 2 has received an OK message from site 1 but not from site 3, because of its own or telecommunications failure. The distributed database is therefore not synchronised. All three sites may have committed but site 2 does not yet know that site 3 has achieved a successful commit, in this case because of failed communications between sites 2 and 3. A multi-phase commit procedure is required.

A two-phase procedure has been universally adopted and is illustrated in figure 4.9. In this procedure the triggering site broadcasts a "prepare to update" call message to the participant sites, which "agree" to commit a table row, record the fact in their log file with all relevant information about the database table row, and then reply by issuing an "OK to update" echo message back to the triggering site. When all the remote sites messages have been received the triggering and participating sites are synchronised at the point of commit. Any failure at this point will be sensed by the triggering site and a rollback command immediately issued to the participant sites for local recovery from the log file.

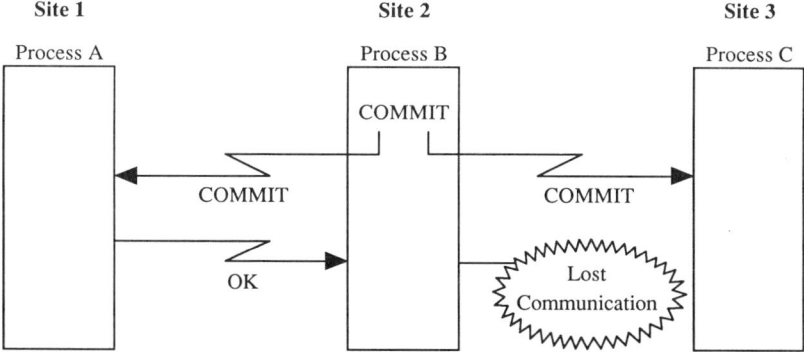

Figure 4.8 Distributed transaction rollback (single phase commit)

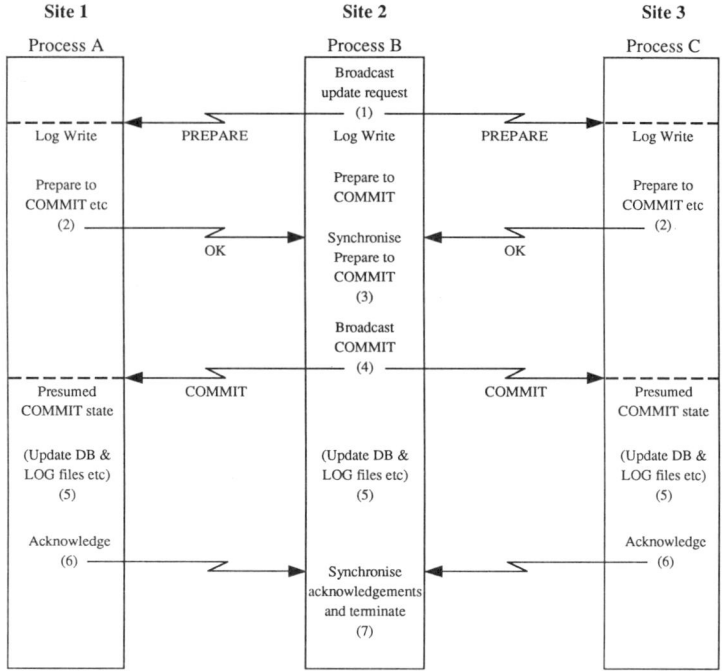

Figure 4.9 Distributed transaction rollback (two phase commit)

Assuming all is well the triggering site synchronises all the "OK to update "messages. Clearly the commit procedure can only progress at the speed of the slowest site. When the triggering site has received a message from all the participant sites it broadcasts a "commit" call message to the participant sites. All the sites update their logs and databases in that order and issue commit calls to release the database X lock/timestamp previously established by the concurrency control mechanisms described earlier. One site could fail, such that the commits are not executed at all the sites. On completion of a commit each participant site then transmits a commit "acknowledgement" echo message back to the triggering site, which synchronises all the acknowledgments and then terminates. Again the co-ordinating site is paced by the slowest of the remote sites.

A two-phase commit procedure is inherently more stable, because should the triggering site fail after issuing the commit call message at the beginning of the second phase of the commit procedure then the log records are available at each site for the normal dynamic transaction rollback procedures found in centralised processing. The distributed software will send messages to all the participant sites to indicate a failure has occurred and normal local dynamic transaction rollback is then undertaken and the local locks/timestamps on the table rows are released.

In spite of the fact that the two-phase commit procedure has been universally adopted by the relational file handlers supporting distributed database, *the author is not convinced that the two-phase commit procedure is totally robust*. Consider the situation in figure 4.9. Site 1 has updated the database and log files at point in time 5 and issued an acknowledgement message back to site 2, the co-ordinating site (point in time 6). Normal processing at site 1 continues and many updates to the local database are made. Site 3 meantime cannot issue an acknowledgment message to site 2 because there has been a communications failure after point in time 4 and before point in time 5. The timeout time is exceeded and the co-ordinating site issues a broadcast message to the participating sites to rollback. Too late—there have been n updates to the database at site 1, the transactions have committed and the users who triggered those transactions have made their decisions on the information processed and have taken the appropriate action.

The only solution to this situation is for there to be a two-and-a-half-phase commit procedure, with the co-ordinating site issuing a fifth and final message to the participating sites when the acknowledgement messages have been received from all the participating sites. There will thus be a final committing of the updates to the database only when all the participating sites have successfully completed the passing of messages to the co-ordinating site that the updates have been successful.

The downside of this extra rigour is that, even when there is a failure of communications between the participating and co-ordinating site for whatever reason, the other participating sites must wait for recovery. *All the sites in a distributed system can thus only function at the speed of the slowest site—* and this must occur twice, once when the co-ordinating site is synchronising first the prepare to commit message from the participating sites and second the acknowledgement messages.

Distributed systems are not totally robust in other areas as well. There is a period of risk between the time the co-ordinating transaction at the triggering site issues the prepare to update and commit messages and the participant sites respond with their OK to update and acknowledgement messages and vice versa. This period of risk cannot be totally eliminated, for the simple fact that the processors executing the transactions are faster than the telecommunication lines transmitting the messages. It can only be reduced through issuing even more levels of commit procedure, such as three-phase commits and four-phase commits. This inevitably increases the remote I/O processing overheads. The system could then be spending more processor time servicing the message passing than supporting the business requirement.

Any distributed locking procedure requires a timeout mechanism to test for the possibility of failure or excessively poor distributed performance.

A triggering co-ordinating site sending an update message to a participant site waits in expectation of a response. The participant site(s) can fail at any time without the co-ordinating site being aware. The co-ordinating site continues to wait for a potentially indefinite period of time. Clearly, such a situation cannot be tolerated for long, for obvious response time reasons. A timeout needs to be specified. When the timeout is exceeded, the co-ordinating site re-transmits the call messages as many times as necessary, at a frequency to be specified, to ensure that the messages eventually reach their intended participant sites, again potentially for an indefinite period of time. After a further period of time elapses, when the timeout end point has been reached, the co-ordinating site assumes a failure and conducts a local recovery procedure, as well as instructing the participating sites to rollback.

In addition, if a site failure occurs at any point during the two-phase commit procedure (either at the co-ordinating or participant sites) the recovery procedure at the failing site must communicate with the other sites (what message status are you in?) to ascertain what must be done to ensure that the commit procedure is completed successfully. When the communication is sent every site examines its log to see if it has any record of the transaction in question. If the message is that they have received messages from the co-ordinating site, then the failed site continues to issue messages to the co-ordinating site on the assumption that it, the co-ordinating site, is continuing to function. If the message is that they have not received messages from the co-ordinating site then the failed site must assume that the co-ordinating site has also failed and conducts a local rollback. The other sites will come to the same conclusion when the timeout end point has been reached.

4.1.4.4 Distributed transaction rollforward

System rollforward is required to recover lost data. The data can be lost either through an unscheduled processor shutdown or through a disk crash. With centralised processing the rollforward is achieved by copying the last database dump copy and reading the after image log records accumulated since the last dump in a forward time order direction. An enhancement to this simple procedure is required for distributed systems.

Most vendors of distributed database software provide for lost data by dynamically duplicating the database. When data is lost in the master database, processing is automatically switched to the database copy until the master database is recovered. This is a perfectly acceptable solution, but has the obvious overhead of doubling local database processing overheads.

A finessed approach is illustrated in figure 4.10, whereby during the period of time that a site is unavailable the remaining running sites update a local pending file of all distributed update transactions for the failed site during its unavailability. The entire distributed database transaction is restricted to the

Distributed Systems

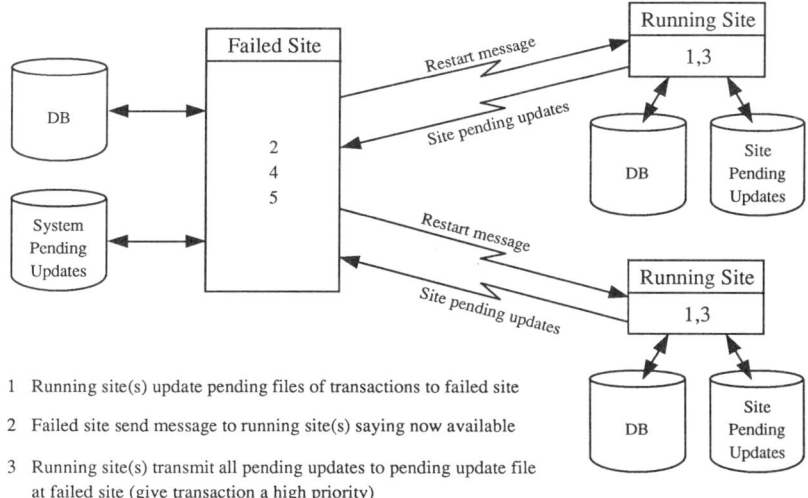

1. Running site(s) update pending files of transactions to failed site
2. Failed site send message to running site(s) saying now available
3. Running site(s) transmit all pending updates to pending update file at failed site (give transaction a high priority)
4. Failed site timestamp co-ordinates pending file(s)
5. Failed site update database as a standard process and keep trail of remote updates during restarts and applies

Figure 4.10 Distributed system rollforward

pending file. All sorts of co-ordination difficulties arise if the local updates are allowed to proceed, along with the updates to the other running remote sites, without waiting for the failed site to recover. Continued application processing at the running sites with n further updates would mean the failed site getting further and further out of step with the remainder of the system. Synchronised recovery would be impossible.

When the failed site becomes available it sends a restart message to the running sites, which flush the failed site pending updates to the now recovered site. The recovered site becomes responsible for the rollforward recovery. It timestamp co-ordinates the pending file updates from the running sites and updates the dump/rollforward recovered database and the various remote databases of the running sites which created the pending files with the pending updates, using the normal two-phase commit procedure to ensure distributed integrity of the updates. A temporary pending file is maintained at the recovered site during the recovery process for dynamic updates to the database that might occur during the pending update process. The temporary pending file is then progressively updated to the database when the failed site pending files updates have been completed. The author is not aware of any database product providing this facility, although developments are underway.

4.1.4.5 Distributed query optimisation

Relational file handlers have query optimisers that dynamically ascertain the optimum access path to the database data. The degree of sophistication varies considerably from relational product to relational product. The best optimisers work on the basis of database cardinality lists, which are statistics files that record the set cardinality of related tables based on a common key. For example, the table customer has a set cardinality with the table orders, in that a customer can relate to n orders. Some products, such as INGRES, ADABAS and DB2, keep a count of the cardinality between each table row of customer and the related table rows of order based on the common key of customer code. Other products, such as Tandem's NonStop SQL, keep an average count of the cardinality between the tables. Most products run an offline statistics utility program that counts the set cardinality between the tables on any specified key(s). ADABAS keeps a dynamic count in the ISN lists. By keeping these set cardinalities the optimiser can rapidly ascertain the number of table rows in each table to be accessed for a given set of search keys.

The sending of data between sites incurs expensive remote I/O overhead. Furthermore, given the slowness of the telecommunication line data transmission speeds, particularly wide area telecommunication lines, the transmission of large data volumes needs to be minimised. It is therefore a sound policy to send minimum data to maximum data. This can only be achieved if the local cardinality lists are accessed to ascertain the number of table rows requiring access in a distributed transaction.

Currently there are four basic strategies, of varying degrees of sophistication, that can be used to send data between sites. Beginning with the simplest they are:

- Ship remote accessed table data unmerged to the triggering site.
- Merge the remote accessed data and ship to the triggering site.
- Ship minimum remote merged table data to maximum remote merged data in ascending size of merged tables, before shipping to the triggering site.
- The triggering site to calculate the total "picture" of alternative remote access and data transmission strategies between sites before instructing execution.

Only the second and third distributed query optimiser strategies have been adopted by the current database products claiming to support distributed database. The second strategy is the most widely used approach. The third strategy is more in line with the policy of shipping minimum data to maximum data. It may involve shipping data around the network before

Distributed Systems

shipping it to the triggering site. The first two strategies do not require to use the database set cardinality lists.

Note that the only criterion of measurement is the number of table rows to be transmitted. Other issues not currently considered are the length of the data stream to be transmitted (i.e. number of table rows × row data length), telecommunication line speed between sites, telecommunication network layout, whether the local data should be included in the distributed query optimisation algorithm and the number of remote I/Os.

A worked example of each strategy shows the different performance implications. Assume the scenario in figure 4.11. The distributed request is triggered at site 3 and requires to access all information concerning certain values of tables A and B at sites 1, 2 and 4. The number of table rows to be accessed in each table and the merged results are indicated against the hatched portion of each table.

The first strategy would involve the triggering site issuing access requests to the remote sites, which return 270 table rows, 40 from site 1, 80 from site 2 and 150 from site 4. Six remote I/Os are incurred, 3 to request the data at the remote sites and 3 to send it back. The second strategy merges the remote accessed tables at their local sites. Thus tables A and B at site 1 produce a merged result of, say, 5 rows, a similar merging at site 2 a merged result of 15

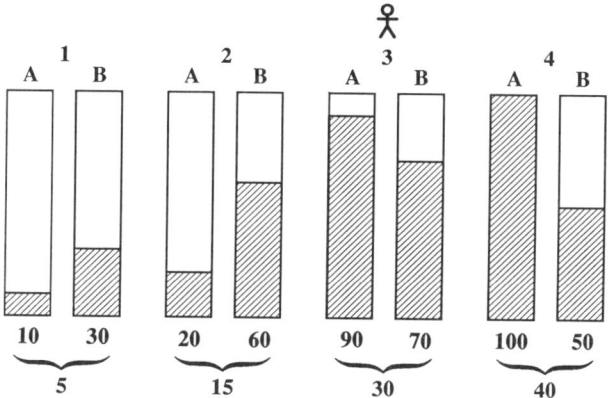

1. Ship 270 rows to Site 3 + 6 remote I/Os

2. Ship 60 rows to Site 3 + 6 remote I/Os

3. Ship 5 rows from Site 1 to Site 2
   Ship 4 rows (say) from Site 2 to Site 4 } 39 rows + 12 remote I/Os
   Ship 30 rows (say) from Site 4 to Site 3

4. Less than 39 rows + 12 remote I/Os

Figure 4.11 Distributed query optimiser strategies

rows and at site 4 a merged result of 40 rows. The merged data of 60 rows is sent to site 3. There would be still be 6 remote I/Os, 3 to request the data and 3 to send it back.

The third strategy would be for the triggering site to read the remote set cardinality lists first and ascertain the amount of data to be accessed and send the smallest merged table to the next largest merged table and so on until sending the final result to the triggering site for final merging. In the example in figure 4.10 this would entail shipping the merged tables of site 1 (5 rows) to site 2, sending the combined merged result (say 4 rows) to site 4 and sending the combined merged result (say 30 rows) to site 3. This access strategy would involve sending 39 rows between the sites and 12 remote I/Os, three to request the remote cardinality data, three to send the cardinality statistics back to the triggering site, three to send transmission instructions to the remote sites and three to transmit the data from the remote sites to the triggering site.

It is an astonishing fact that the only distributed database product that undertakes this level of optimisation is INGRES. The trouble with this strategy is that the size of the merged table sent from site 2 to site 4 and eventually from site 4 to site 3 is not calculated in advance at the triggering site. With hindsight (which can be calculated with the fourth strategy by combining the separate location cardinality statistics to calculate the size of the transmitted merged tables), it may be that sending the merged table of 40 rows would produce a better, i.e. smaller, merged result than sending it to site 2 with its merged table of 15 rows. The fourth strategy would thus transmit less than 39 rows between the sites, but still incur 12 remote I/Os.

There is thus a balance shift—the greater the sophistication of the distributed query optimiser, the fewer the table rows to be transmitted between sites and, once cardinality lists are used, a doubling of the number of remote I/Os. So long as the telecommunication line speeds are slow the emphasis must be on reducing the length of the data stream transmitted between locations. This clearly favours the more sophisticated optimisers. If the line speed increases dramatically (as perhaps on local area networks) without a commensurate decrease in the remote I/O processing overheads then the less sophisticated optimisers come into their own. Distributed database designers must ascertain the distributed query optimisation mechanism, as it can significantly affect processor and telecommunication overheads and thereby transaction resource utilisation and response times.

4.1.4.6 Distributed transaction utilisation nonitor

Relational file handlers have many advantages, one of which is easy data access specification through command and menu driven query languages. SQL is a command driven query language and is the industry standard for relational file handlers. Each relational database vendor has also developed their own menu driven query language. Users are therefore able to pose *ad hoc* questions. Many of the *ad hoc* access requirements may be to retrieve data in bulk. Without realising it, users may easily issue requests for a large volume of data and thereby "clog up" the processor and degrade performance. Given the overheads of remote I/O and slow data transmission speeds, such a situation in a distributed system could well be catastrophic.

It is necessary to provide a "governor" facility that can dynamically prevent users unwittingly issuing excessively resource-hungry requests for data. Such a governor should measure the processor, disk and terminal I/O resource utilisation overheads of a request. Currently the governor facilities provided, as for example by INGRES and DB2, only measure processor time. If a system is distributed it may well be that the data the user is requesting is spread across the locations like "grass seed". The governor should therefore also measure the remote I/O and telecommunication overheads incurred. IBM provide a distributed governor, which is presumably an extension of their Resource Limit Facility product local governor for DB2. The author has not been able to obtain any details. One would assume that the distributed governor reads the remote cardinality list of the local query optimisers, ascertains the number of remote table rows required to be accessed and thereby calculates, at a minimum, remote I/O and, hopefully, telecommunication overheads.

4.1.4.7 Distributed dictionary

Each site in a distributed system is equal to all the other sites. No site should be considered a master site. However, a distributed transaction must be co-ordinated. Co-ordination requires control. The policy that has been followed is that the site that triggers the transaction acts as the controlling site, *but only for the duration of the transaction*. Co-ordination during live running of the system is therefore at the transaction and not at the system level.

Given this principle of equal sites, there should be no master dictionary recording the distribution of data across sites. What is required is that each dictionary at each site records fully the distribution of data, screen formats, report formats, application programs etc. for all sites. Thus each local dictionary is also a global dictionary. Currently most products define local information in local dictionaries and contain only a reference to the tables of data in other locations in a distributed network. There are major adverse performance implications of this approach, for example the dynamic parsing of every *ad hoc* SQL query requires access to remote dictionaries if remote data access is required to parse the query specification for correctness. This increases transaction overheads and lowers performance.

Clearly where the local dictionaries are also global there is a need to synchronise the updates of the dictionaries across multiple sites. Given that a dictionary is nothing more than standard data stored in standard tables, the normal mechanism of the two-phase commit procedure is appropriate and sufficient for online updates.

As an insurance of global distributed dictionary integrity and consistency distributed software should ensure dictionary synchronisation is part of the start-up procedures. For this to occur one site is designated a master site. This is the only occasion that the concept of a master site is used at the system level (the recovered site controlling hardware recovery has not yet been developed, see section 4.1.4.4).

4.1.4.8 Location transparency

Users should not be aware that the system they are using is distributed. They should not have to know the distribution of data between sites. The location of the data should be transparent.

The mechanisms required to support location transparency are crucially dependent on how the data is distributed across the locations. Data can be distributed either vertically or horizontally, as illustrated in figure 4.12.

With vertical distribution there is a one-for-one match between an entity and its location. A location can store many entities (physically stored in the form of tables); an entity can be stored at one location only. The great advantage of vertical distribution is that the location of an entity can be identified as an attribute of the entity in the schema description within the global dictionaries at each location. The software merely has to access the global dictionary and the location of the entity and its occurrences can be ascertained. This means that complicated software to search for the location

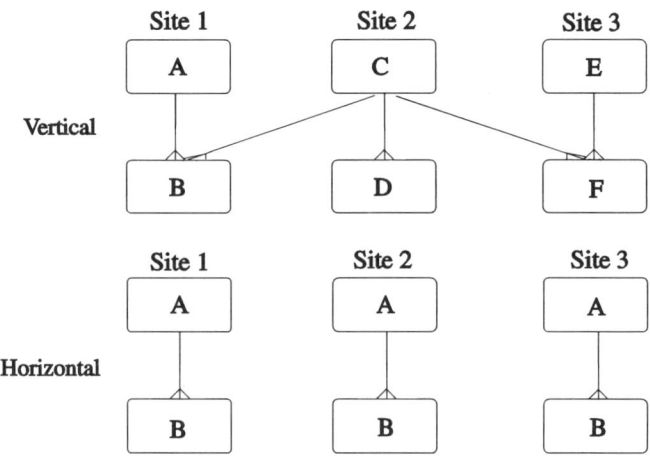

Figure 4.12 Data distribution

of an entity and its occurrences is not required. If the retrieval request is " For a specified occurrence of C display all related Bs and Ds" then the local global dictionary would indicate that tables C and D are to be found at location 2 and table B at location 1. The recording of the location of an entity in the global dictionaries is simple and provides highly efficient support for finding the location of data. Vertically distributed systems have this efficiency and simplicity.

Unfortunately, the bulk of distributed applications distribute data in a horizontal manner. With horizontal data distribution there is no longer a one-for-one match between a location and an entity type. The problem is the entity occurrences. If it had been possible to store the location of each entity occurrence in the global dictionaries, the disk storage requirements and, above all, the maintenance of the entity occurrence location identifier would be prohibitive. Dictionaries only define entities at the entity type, not the entity occurrence level. The global dictionaries cannot therefore be used to identify the location of horizontally distributed data.

An alternative mechanism for locating horizontally distributed data is required. The mechanisms are:

- Fragmentation within entities.
  The entity is fragmented such that specified table columns (vertical fragmentation) or table rows based on a specified value range of the key data item(s) (horizontal fragmentation) relate to specified locations. Such fragmentation vertically structures in a physical design an entity that is horizontally distributed in a logical design. Fragmentation can be specified in the global dictionaries with all the advantages thereby provided.

  The trouble is that such fragmentation is of little practical value. The real world does not usually work in an entity fragmented manner. For example, vertical fragmentation of data columns makes little sense. It is not a real world situation to store the columns containing the order delivery information in Tokyo, the columns containing delivery contact person information in New York and columns containing the order product information in London. The only relevance of vertical fragmentation identified by the author was in a manufacturing environment, where certain columns of a product table were appropriate to drilling, certain columns appropriate to lathe machining and certain columns appropriate to robotic processing. The columns of data would be distributed to the processors built into the drill, lathe and robot.

  Horizontal fragmentation is also of equally limited practicality in normal commercial environments. It has relevance if one is accessing the table on the key column which is used to provide the value for horizontal fragmentation, for example orders number 1 to 99 to be stored in Tokyo, orders 100 to 199 to be stored in New York and orders 200 to 299 to be stored in London. This is fine so long as one wants to access a specific order on order number. However, the real world is that one usually requires to access an order on other than order number. One usually wishes to access orders on a specific customer or product.[1] One cannot then use the local global dictionary as the basis of locating the data.

- Partitioning between entities
  Where a logical data model indicates that data is distributed horizontally, a solution can be to "bend" the physical design to be at variance to the logical design by stipulating that an entity, which logically is found at multiple locations, is to be stored physically at a single location. One is converting the logically horizontal distribution of data to be physically distributed vertically, again with all the benefits of global dictionary access.

- Multi-site searching
  Multi-site searching is the only mechanism that fully provides real world support for location transparency where data is distributed horizontally. Any product claiming to support location transparency in all environments (vertical and horizontal data distribution) must provide a minimum of one multi-site searching mechanism. There are four such mechanisms—broadcast; "round robin"; replicated index; replicated database:

  — With broadcast searching a data access request is shipped from the triggering site to all the sites. The advantage of broadcast searching is that the data request response times can be uniform, in that the request is shipped simultaneously to all sites and, if each site processes the transaction immediately, the response times will be made more or less uniform. This, of course, is an ideal situation, as the database and processor circumstances at each of the remote sites will not be identical. A disadvantage is that the distributed data retrieval access overheads are high as all sites are accessed, even though they may contain no data relevant to the request.

  — "Round robin" searching involves a progressive walk round each site in turn until the requisite data is found. An advantage of round robin is that only half the sites within the distributed system require to be accessed on average, such that the distributed data retrieval access overheads are only half those of the broadcast mechanism. A disadvantage is that the data access response time is not uniform and varies depending on the number of sites accessed.

  — Replicated index replicates all or selected indexes of each site at all the sites. A data access request therefore accesses the local site indexes,

---

[1] There is a little "trick of the design trade" to get round this. Put the secondary index on customer number and product code on top of the primary key on order number. IBM's VSAM keyed sequential data set and Tandem's NonStop SQL file handlers provide this facility. One can then access on the Product Code and Customer Number secondary keys, which then point to the prime key of Order Number.

all of which except the indexes of the triggering site are replications, to ascertain which site index points to the requisite data. The site pointed to is accessed. An advantage of replicated index is that multi-site searching is not required on data retrieval transactions. A disadvantage is that the replicated indexes require expensive multi-site updating.
— Replicated database replicates all or selected user data tables and indexes at all the sites. The advantage is that all data retrieval requests to the replicated data are local. The obvious disadvantage is that any data or index changes require to be replicated at all the sites.

- There is clearly a balance shift in the data retrieval to data maintenance overheads from broadcast to replicated databases, with the degree of balance shift depending on the number of sites within the distributed system. With broadcast the data retrieval overheads are greatest, as all sites require to be accessed. Because no track of where the data is to be stored is maintained, the data maintenance overheads are the lowest. At the other extreme is replicated database, where data retrieval access is minimal as it is entirely local to the replicated data, while the data maintenance overheads are maximised because of the replication of user data and indexes. It should be borne in mind that updates in a distributed system require, at a minimum, two message pairs between the co-ordinating and participating sites for each set of table rows and supporting log. The number of databases and log files in the set equals the number of sites in the distributed system.

4.1.5 Enhancements to SSADM

One of the two unique features of distributed systems is controlled by the distributed database software. Synchronised multi-site databases updating and hardware/software recovery is handled by the distributed concurrency control, the two-phase commit and pending files procedures. SSADM does not therefore require to be concerned with the synchronisation aspect of data distribution during logical design, only during physical design.

The policy followed by the vendors of distributed database has been to "add on" the requirements of data distribution to the existing facilities for centralised database processing, the overriding necessity being the preservation of existing investments (financial, technical and human) in information technology. The great benefit of this policy is that what has been used and learned previously is preserved unchanged. What is true of the technology is true for the techniques. *SSADM is wholly preserved for distributed systems*. The only enhancements required are that the:

- concept of location needs to be "added on" to the logical design techniques and supporting documentation;

- concepts of synchronisation *and* location need to be "added on" to the physical design and development techniques and supporting documentation.

The enhancements to SSADM will be described using two distributed applications the author was involved with. In both cases SSADM based design and development techniques were used. The enhancements about to be described were applied and proved successful.

The first was a government system that required to issue public documents on demand. Any government requires to support this function and the locations used in this case study are purely illustrative and chosen at random. Assume there are five regional offices: London, Derby, Edinburgh, Manchester and Cambridge. London also had head office functions, such as accounting. Manchester, Edinburgh and Cambridge each had the same volume of business, with Derby being substantially less. Once the documents had been issued to the public a central reference repository was maintained at another location.

There were thus six locations but only three location types.[1] There were four regional offices of the location type regional office and one location each for the location types head office and central repository. The important thing to understand about location types is that the business at each location type is the same, only the data and business requirement volumes varying. The degree of business difference between location types can be from small to totally different. A location can belong to only one location type; a location type can have $1 - n$ locations. Identifying the location types substantially reduces the amount of logical design work required. In this system it was effectively limited to three logical designs rather than six.

The second distributed system was an island port authority, which handled general as well as passenger traffic. The case study was only concerned with monitoring, in realtime, the planning and control of general cargo between the four islands, the mainland of Britain and the continent of Europe.

There were differences in the way the business of the ports were run. Only

---

[1] The concept of location type is important to distrubuted systems. One of the major problems of analysing distributed systems is that if the business varies across locations then there is a need to undertake a full logical analysis and produce a full logical design for each of the locations. The design effort is substantially increased, and the author's experience is that it is invariably underestimated. Given the business is the same within a location type the scope for reduced analysis and design effort is considerable, particularly if there are many locations per location type. The only information that will vary is the volumes and cardinality ratios of the entities and the volumes of the business requirements.

What can often occur is that there is not a clear dividing line between a location and a location type. It can be that some of the business requirements are the same across locations and that some of the business requirements are diffrerent across location types. This was the case of the main cargo port and the smaller ports, where the only significant difference in the business was that there were no lorries at the small ports. They were of the same location types but were not totally homogeneous. This has to be watched out for.

the main port on the largest island contained a warehouse for the storage of the general cargo. All the small ports loaded and unloaded the cargo direct from and onto the lorries, so there was no need for fork lift trucks and the more sophisticated business of planning the loading and movement of the cargo pallets. The cargo movement management function was to be found only at the main port.

The head office was located on the main island and, as far as the case study was concerned, monitored the general cargo movements on a summarised daily basis. The port authority had five locations, with three location types.

The logical design techniques   The logical design techniques to be reviewed are those which SSADM uses and include enhancements suggested as necessary in chapter three. It is assumed the reader is familiar with the suggestions.

4.1.5.1 Dataflow diagrams.

The dataflow diagrams are built on the data maintenance business requirements. The first thing that requires to be ascertained therefore is the location at which the data maintenance business requirements are triggered. Given that the main design principle of distributed systems is to store the data at the location where it is most frequently accessed, so as to keep remote I/O to a minimum, the location at which the transactions are triggered is clearly essential information. Once the data maintenance business requirements have been identified and catalogued by location within location type the design and construction of the dataflow diagrams can be undertaken as normal, with a set of dataflow diagrams per location type.

Do not expect the dataflow diagrams to be always uniform across location types. Variations within the "common" location type model can occur. For the government system all the regional offices conducted their document support business in an identical manner. For this location type the dataflow diagrams were identical and did not require n dataflow diagrams for n locations. The head office included an accounting function so that additional external entities, dataflows, datastores and processes were included in the "common" dataflow diagrams for the regional offices. The reference centre had a totally different dataflow diagram.

The port company dataflow diagrams varied between the main port and the smaller ports in that the medieval ports were not concerned with fork lift trucks. A much simplified diagram of the event level processes for the general cargo function for both the main and the small ports is illustrated in figure 4.13. The small port is missing the processes relating to fork lift trucks. Variations in the dataflow diagrams continued to the next level of decomposition, the problems-to-solve. The main port contained many bays

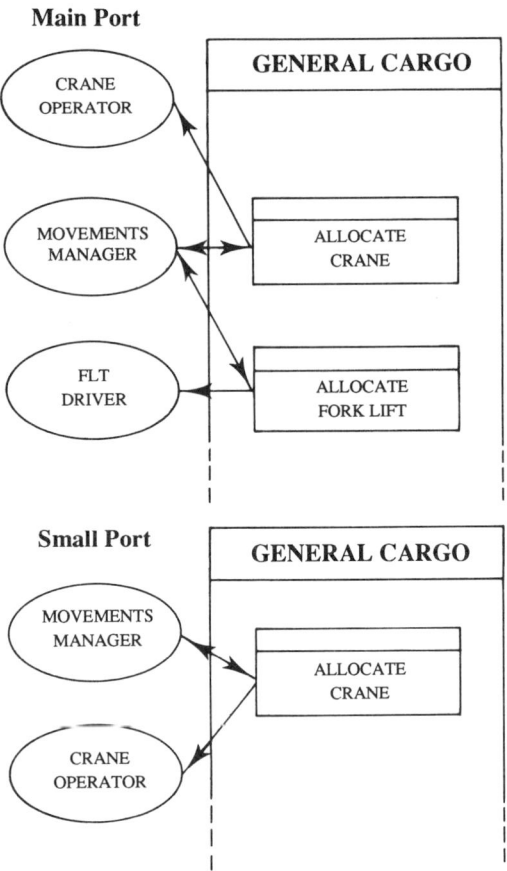

Figure 4.13 Distributed DFDs: 1. Note: the processes are at the event level

at which vessels could be moored for loading and unloading. The small ports contained only one bay. Therefore the problem-to-solve of ascertaining a bay for a vessel to moor at was not necessary. The different dataflow diagrams can be seen in figure 4.14.

4.1.5.2 Logical Data Model (LDM)

As explained in section 2.2.2.4 the physical database and its logical precursor the LDM should be built from the data retrieval business requirements. As with dataflow diagrams the first thing that requires to be ascertained is the location at which the business requirements are triggered. Once the requirements have been identified and catalogued by location within location

Distributed Systems

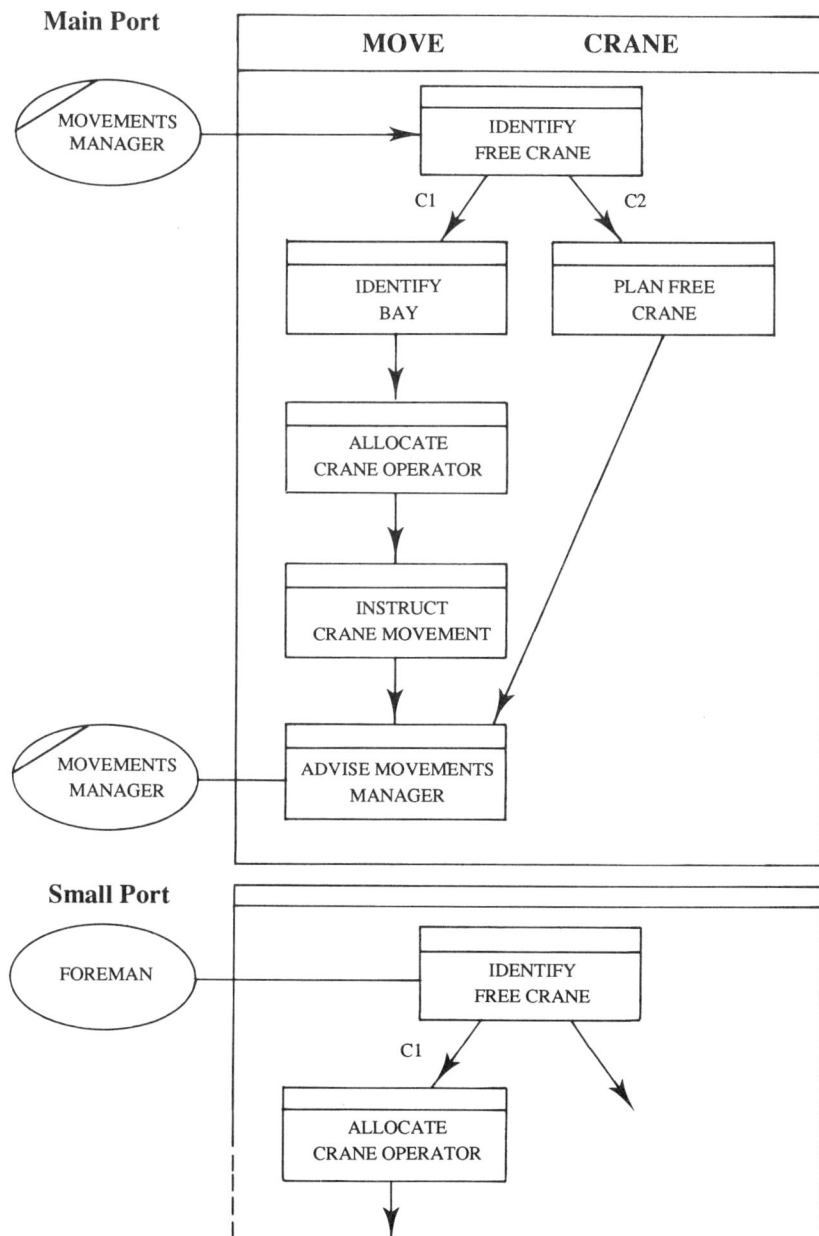

Figure 4.14 Distributed DFDs: 2

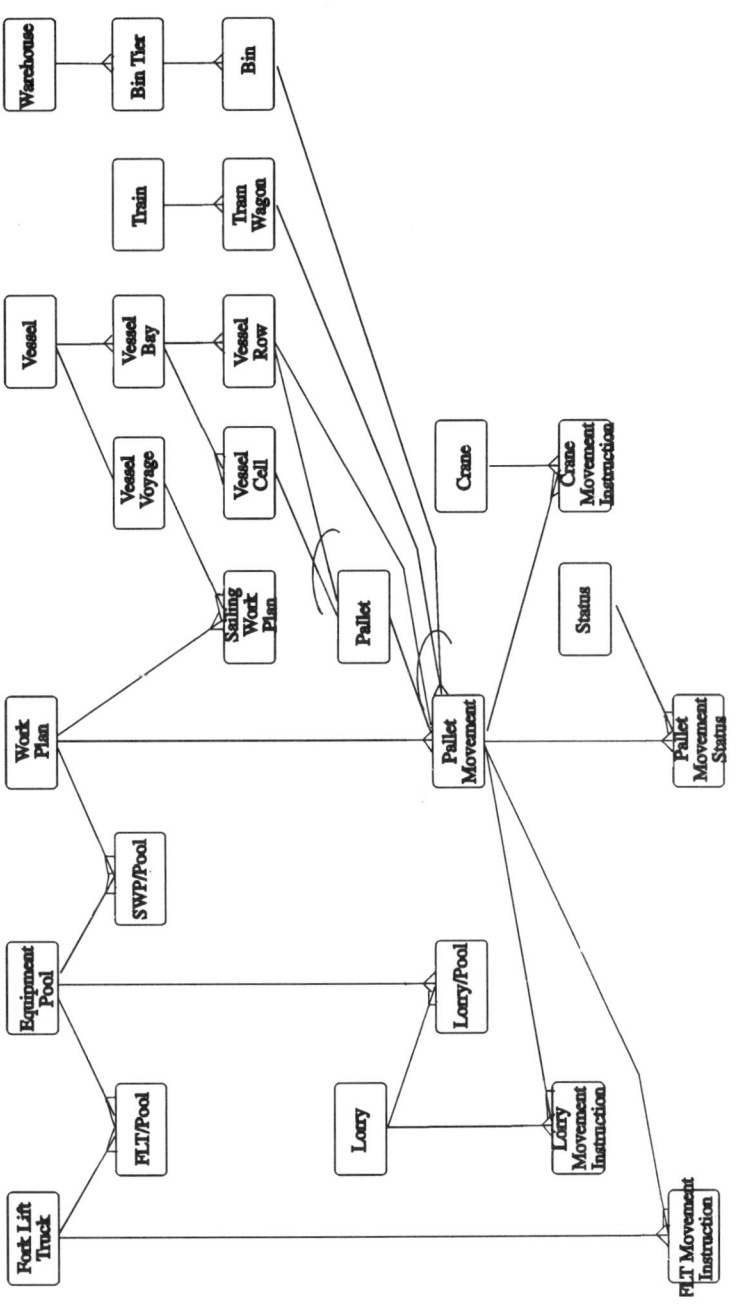

Figure 4.15 Main port logical data model

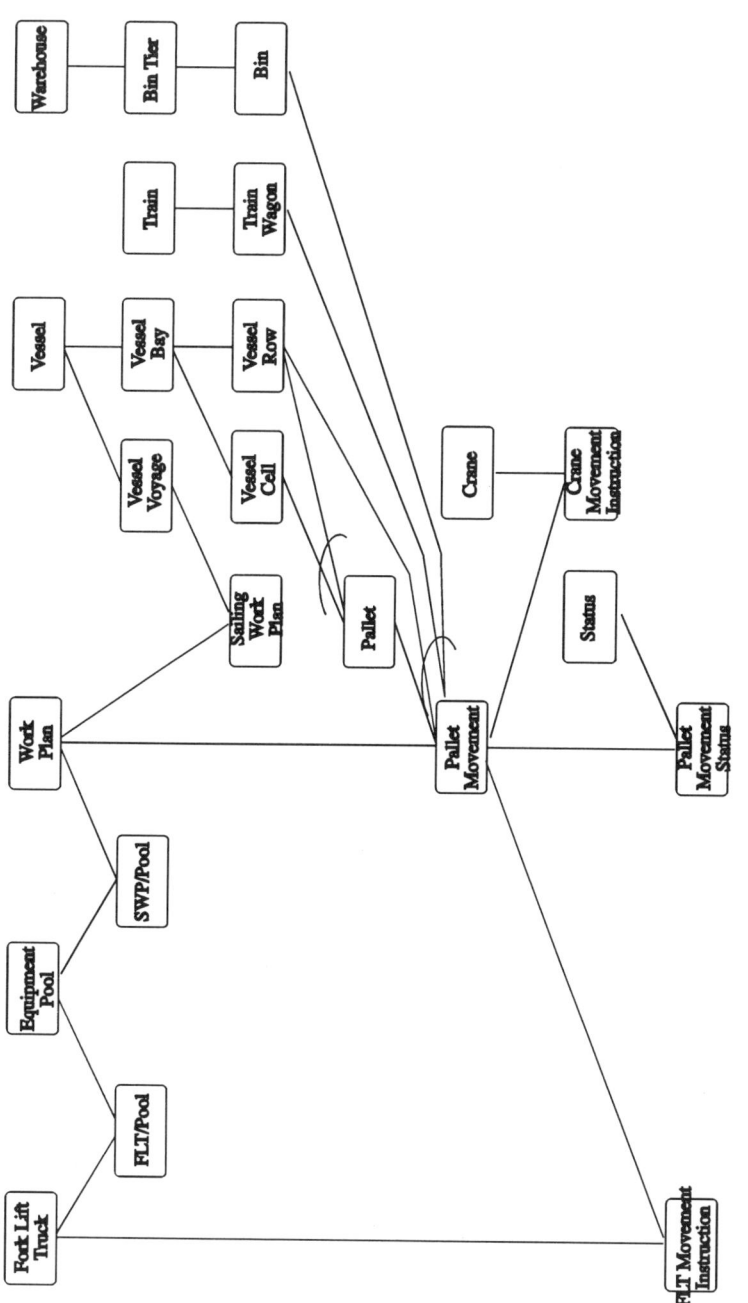

Figure 4.16 Small port logical data model

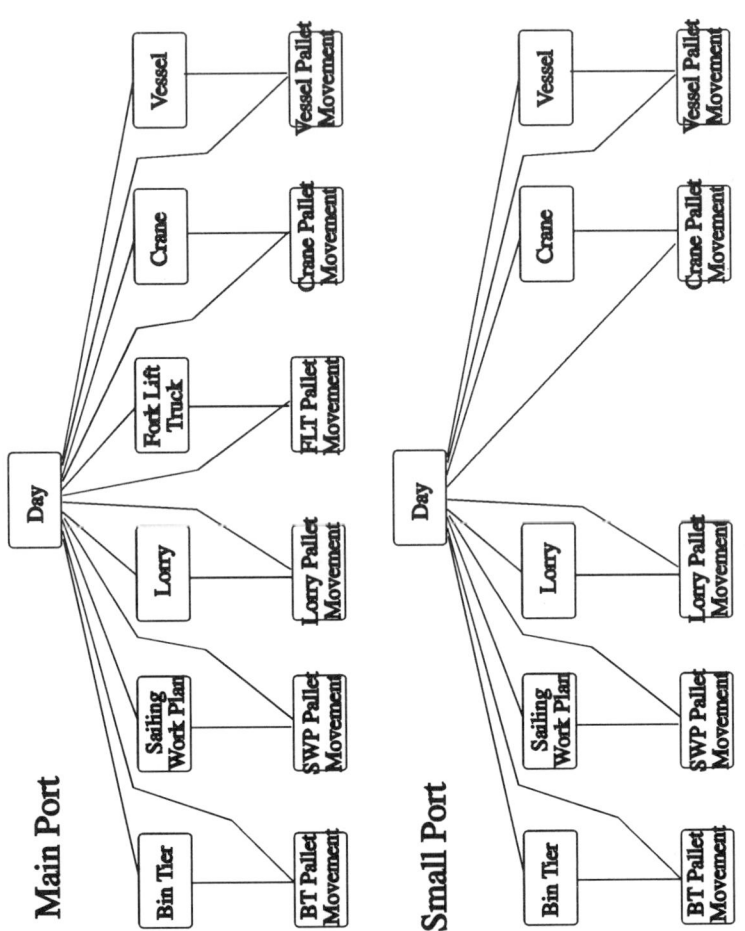

Figure 4.17 Head office logical data models

Distributed Systems 215

type the design and construction of the LDM by location type can proceed as normal.

The LDM varied by location type for the two case study applications in exactly the same way as for the dataflow diagrams. The port authority data models are illustrated in figures 4.15 and 4.16. The data models for the main and small ports are identical except that the small port model contains no entities relating to lorries. The head office model is entirely different. The part that is relevant to this book relates to the fact that the head office required to monitor the daily movements of cargo pallets from a number of viewpoints and not by individual pallet, as illustrated in figure 4.17. The head office data model was therefore radically different from the data models for the other ports. Furthermore, the head office had two data models, one for the main port and one for the small ports, and again the model for the main port excluded the fork lift trucks. This has been the only occasion the author has required to produce two data models for one location type.

The head office LDMs could have been combined as a single LDM with another entity of Port Type at the top providing the link. While perfectly valid it does not hide the fact that there are, in fact, two data models reflecting two business applications hidden within one LDM.

What can, and usually does, vary by location type and even location, are the by-products of the LDM—the data items in the entities (usually little variation) and the entity volumes, lives and set cardinalities (almost always vary). Consider figures 4.18–4.20. The examples are based on the

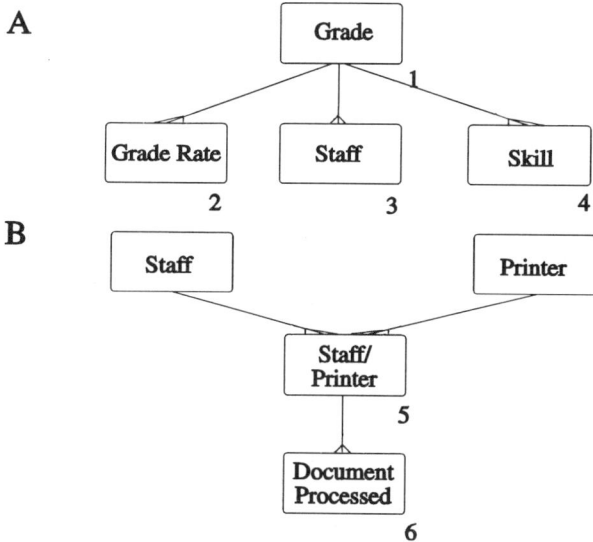

Figure 4.18 Distributed volumes

Figure 4.19

SYSTEM:			DATA GROUP NO: 1		NAME: GRADE				
ATTRIBUTES					**OCCURRENCES**				
CODE NAME	PK(/) FK(*)	USER NAME	TYPE/ SIZE	COMMENTS/EXAMPLES	ACTIVE	ADD P/M	GTH P/M	LIFE (M)	DEL/ARC P/M
	/	Grade code Description ⋮	X2 X20		15			As req'd	

	DEPENDENT DATA GROUP RATIOS					
	GROUP NO.	MIN	MAX	MEAN	WORK AVE	RATIO COMMENTS
STAFF:	3(LDN)	0	60	6	7	10% have zero
	3(DY)	0	8	1	4	70% have zero
	3(OTH)	0	60	6	9	10% have zero
GRADE RATE:	2	0	5	2	2	
SKILL:	4	1	15	4	3	

Figure 4.19 Data group contents/volumes/set ratios (1)

Figure 4.20

SYSTEM:			DATA GROUP NO: 5		NAME: STAFF/PRINTER				
ATTRIBUTES					**OCCURRENCES**				
CODE NAME	PK(/) FK(*)	USER NAME	TYPE/ SIZE	COMMENTS/EXAMPLES	ACTIVE	ADD P/M	GTH P/M	LIFE (M)	DEL/ARC P/M
	/ / / / /	Printer No. Staff reference no Task code Work date Start time End time ⋮	N2 X10 X? N6 N4 N4	 YYMMDD YYMMDD HHMM log on time HHMM log on time	LDN-25 DER-10 OTH-35			1 day	

DEPENDENT DATA GROUP RATIOS					
GROUP NO.	MIN	MAX	MEAN	WORK AVE	RATIO COMMENTS
(LDN) 6	0	400	70	70	0% have zero
(DER) 6	0	400	6	7	0% have zero
(OTH) 6	0	400	85	85	0% have zero

Figure 4.20 Data group contents/volumes/set ratios (2)

government system. Figure 4.18 represents part of the LDM and figures 4.19 and 4.20 the supporting documentation. In figure 4.18A it was ascertained that at the regional offices the volume of grade did not vary by location type, with 18 grades per location, that the grade rate (a figure for the calculation of salaries etc.) and skill set cardinalities also did not vary by location type, whereas the set cardinality of grade to staff did vary by location type. This can be seen in the numbers in figure 4.19. It was ascertained that there were two grade rates per grade at all the location types and three skills per grade at all the location types. It was found that the head office at London had some

Distributed Systems

seven persons per grade on average, the office at Derby had four per grade and the other regional offices throughout the United Kingdom had nine staff per grade. Here the variation was at the specific location, not location type.

Figures 4.18A with 4.20 illustrates some additional points. They show that a member of staff would be allocated to a printer for the printing of the public documents. While the member of staff was at the printer he/she would process a number of documents. Unlike the previous example, the volumes of the entity, in this case staff/printer, do vary by location with London having some 25, Derby some 10 and the remaining offices some 35. The set cardinality of the documents processed per staff/printer also varied as illustrated.

4.1.5.3 Relational data analysis

The technique is unchanged. The practitioner must be aware that the same data may not be found at all locations and also check that the meaning of the data items is the same across location types. If it is not then the relations will vary by location type.

4.1.5.4 Entity Life Histories (ELHs)

The technique is unchanged but must be applied to all the entities for each logical data model at all the locations. Again the practitioner must be aware that the events that update the entities can vary by location type. Clearly where they do not vary the ELH for an entity is valid for all location types. The potentially mammoth task of applying the technique on n data models for n locations can thereby be significantly reduced.

The operations that are added to the ELHs are mostly those to do with the insertion and deletions of the entities and the maintenance of the relationships between the entities. If an event occurs at a location it is very unlikely that these operations will vary between locations. They are, after all, only concerned with the basics of the maintenance of the database and not with the business processing of the updates—that is, database housekeeping. Certainly the author has not found this to be the case.

For the government ministry application the ELHs did not vary by location type. Each entity had exactly the same life at all the locations. There were, however, subtle variations in the life of some of the entities in the port authority application. Simplified but representative ELHs are shown in figure 4.21. As with the DFDs, it illustrates that the cargo at the small ports was not moved by a fork lift truck. However, the life of the cargo pallet at the head office had a radically different life—see figure 4.22.

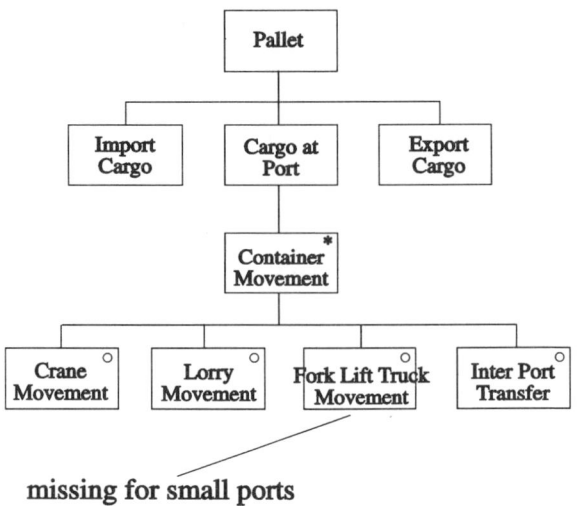

Figure 4.21 Pallet life at main and small ports

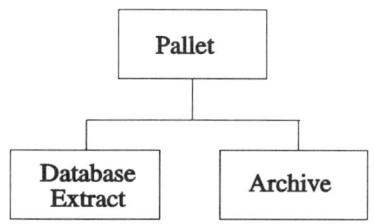

Figure 4.22 Pallet life at head office

4.1.5.5 Transaction access path analysis

The technique as described in sections 3.1.1 and 3.2.5.4 is unchanged, but once again the practitioner must be aware of the possibility of variations by location type, not only in the access path to the entities (different entry points and different access paths between entities) but also in the access volumes.

It was found at the main port that the entry points to the data model for a given business requirement can vary by location. Consider figure 4.23. The business requirement was "For a specified pallet display all pallet movements for a specified sailing work plan". The entry point possibilities were pallet and sailing work plan. At the main port access via pallet would yield a better access path than via sailing work plan, solely because the

Distributed Systems

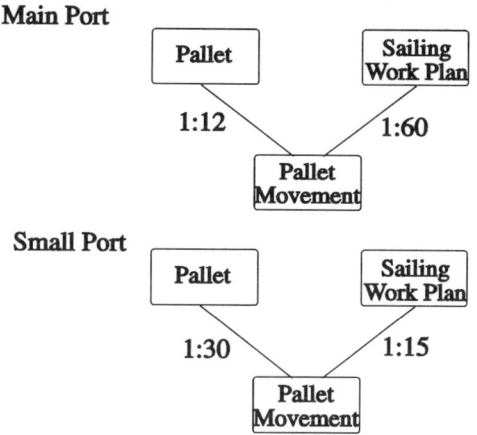

Figure 4.23 Different entry points by location

set cardinality of pallet-to-pallet movement was less than via sailing work plan. The reason for this ratio was that at the main port, which had more space, better efficiency could be obtained. A sailing work plan could be a much larger set of tasks for moving pallets and a pallet would require much fewer movements between entering and leaving the port.

At the small ports, with the constraints of geography and space, the sailing work plan was a limited set of tasks for moving pallets, the pallets would have a slower turnaround time and require many movements and "shuffling" around the port before being ready to be loaded onto a vessel. The cardinalities of sailing work plan and pallet-to-pallet movement were the reverse of the main port, such that the optimum entry point in this case was via sailing work plan.

The access paths illustrated in figures 4.24–4.27 have been pasteurised in that the data group names have been converted to numbers. These numbers are not significant as regards the point being made. The access paths are based on the government system. The business requirement Cash Reconciliation has the same access path across all locations, but the access volumes for the entities varied considerably. If all the transactions in the application had the same variations of access overhead the processing power at the London head office would need to be some six times the power of the Derby office.

However, the access volumes do not have to vary for a given transaction. The business requirement Receive Application was exactly the same at all the offices concerned with processing the public documents. An application is an application is an application!

FUNCTION NO:			FUNCTION NAME: DERBY CASH RECONCILIATION			
DATA GROUP	ACC TYPE	READ PATH	ACCESS VIA	NO. ACC.	DATA ITEMS	CONDITIONS & COMMENTS
1	R	D	–	2		
2	R	P	1	8		
3	R	P	2	440		
4	R	P	3	465		
				915		

ACCESS TYPE
I - insert   L+ - add to optional link path
M - modify   L- - remove from optional link path
R - read
D - delete

READ PATH
D - direct
PS - physical sequential
LS - logical sequential
C - via child
P - via parent

Figure 4.24 Distributed transaction access path—1

FUNCTION NO:			FUNCTION NAME: UK LOCATIONS (excl. London) CASH REC'N			
DATA GROUP	ACC TYPE	READ PATH	ACCESS VIA	NO. ACC.	DATA ITEMS	CONDITIONS & COMMENTS
1	R	D	–	2		
2	R	P	1	33		
3	R	P	2	1815		
4	R	P	3	1905		
				3755		

ACCESS TYPE
I - insert   L+ - add to optional link path
M - modify   L- - remove from optional link path
R - read
D - delete

READ PATH
D - direct
PS - physical sequential
LS - logical sequential
C - via child
P - via parent

Figure 4.25 Distributed transaction access path—2

Distributed Systems

FUNCTION NO:			FUNCTION NAME: LONDON CASH RECONCILIATION			
DATA GROUP	ACC TYPE	READ PATH	ACCESS VIA	NO. ACC.	DATA ITEMS	CONDITIONS & COMMENTS
1	R	D	–	2		
2	R	P	1	55		
3	R	P	2	2915		
4	R	P	3	3060		
				6032		

ACCESS TYPE
I - insert   L+ - add to optional link path
M - modify   L- - remove from optional link path
R - read
D - delete

READ PATH
D - direct
PS - physical sequential
LS - logical sequential
C - via child
P - via parent

Figure 4.26 Distributed transaction access path—3

Location common access path

FUNCTION NO:			FUNCTION NAME: RECEIVE APPLICATION			
DATA GROUP	ACC TYPE	READ PATH	ACCESS VIA	NO. ACC.	DATA ITEMS	CONDITIONS & COMMENTS
22	I	D	–	1		
14	L+	C	22	1		Create link
16	I	P	22	1		
22	L+	C	16	1		Create link
10	R	D	–	1		For access only
6	R	P	10	1		Read reverse. Assume last. For access only.
53	R	P	6	1		Read reverse. Assume last.
7	M	P	53	1		
4	I	P	7	1		

ACCESS TYPE
I - insert   L+ - add to optional link path
M - modify   L- - remove from optional link path
R - read
D - delete

READ PATH
D - direct
PS - physical sequential
LS - logical sequential
C - via child
P - via parent

Figure 4.27 Distributed transaction access path—4

There is no rule as to whether the transaction access volumes vary by location or not. It has to be ascertained on a transaction basis.

It was also found that the volume of transactions varied by location type. This is illustrated in figure 4.28. This document, along with the individual transaction access path maps, was crucial to producing summary access path maps per location type.

Type: Online

Function No.	Function Name	Function Description	Frequency		Location
1	Print document	Print a document of a specified application	Peak (x day) 2250 450 6200	Off Peak 440 45 1450	London Derby Manchester
2	Close application	Close specified applications	Peak (x day) 25 12 75	Off Peak 4 - 14	London Derby Manchester
3	Withdraw application	Withdraw specified applications	Peak (x day) 29 7 16	Off Peak 4 - 17	London Derby Manchester

Figure 4.28 Distributed update function catalogue

4.1.5.6 *Summary access path maps*

Summary access path maps are crucial to the successful design of a distributed database, particularly where the data is distributed horizontally across locations. Without the summary access path maps the design of the distributed database will be left to the old fashioned "seat of the pants" approach of design practised before the adoption of structured methods. Remember, where the data is distributed horizontally the database dictionary cannot be used to ascertain the location of data and an expensive multi-site searching mechanism across locations for the data is required. The greater the number of sites, the greater the expense of searching.

The purpose of the summary maps is to enable the designer to ascertain objectively whether:

- it is possible to convert the horizontally distributed logical design into a vertically distributed physical design and thereby take advantage of the benefits of access to the global dictionary to ascertain the location of data;

Distributed Systems

- the balance of data access, particularly remote access, is data retrieval or data maintenance. If the balance is data retrieval then the "broadcast" multi-site searching mechanism would be the most expensive. If the balance is data maintenance then the replicated database mechanism would be the most expensive.

The summary access path maps crucial to distributed database design are the data maintenance and data retrieval summary maps by location or location type as appropriate. Summary maps by location are needed where the data access requirements are unique to a particular location; summary maps by location type are needed where the data access requirements are common across a number of locations. Figures 4.29 and 4.31 illustrate how

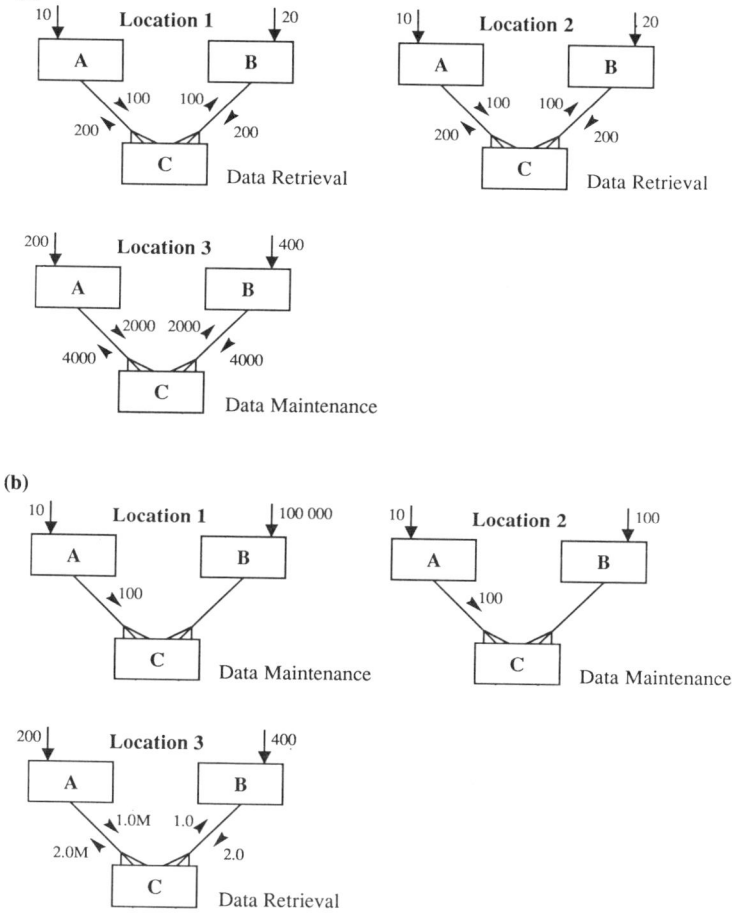

Figure 4.29 Distributed summary access path maps—1

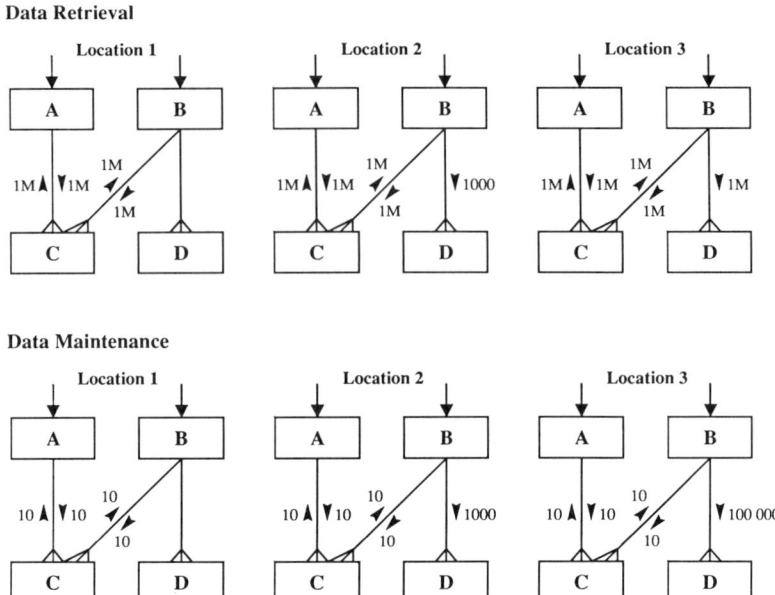

Figure 4.30 Distributed summary access path maps—2

these summary maps can be used to provide answers to both of the above issues.

The figures are not case study based, are somewhat unrealistic and are for teaching purposes only to show how the maps can be used. The maps assume that all the data accesses are distributed. Figure 4.29a has all data maintenance accesses at one location and all the data retrieval accesses at the other locations. The figure illustrates an extreme situation of moving all of a location's data to another location—in this case moving the data of locations 1 and 2 to location 3. The reason for this is that 90% of all data access is at location 3. The overhead of this strategy would be remote I/O from locations 1 and 2 to location 3 to access "their" data, but the overhead would be small as the data retrieval access costs are only some 10% of the data maintenance access costs.

Figure 4.29(b) shows that distributed database design can be at the table level—that an individual table can be moved to another location. There are high data maintenance accesses on table B at location 1, very little data maintenance accesses at location 2 and very high data retrieval accesses on tables A and C at location 3. The message here is to put tables A and C at location 3 because 99% of all data access is there, put table B at location 1 for the same reason and to store no data at location 2.

Figure 4.30 is a further enhanced and more realistic example showing, *inter alia*, which tables and indexes to replicate and at what locations. It shows:

- Do not store table D at location 1, as it never occurs there.
- Replicate tables A, B and C at locations 1, 2 and 3, because there are low data maintenance and high data retrieval accesses.
- Replicate indexes on table D at location 2 and hold table D at location 3. This is because table D is accessed only a little at location 2 and a large amount at location 3.

What is being achieved in these three examples is the "bending" of the horizontally distributed logical data design to make it vertically distributed in the physical database design, with all the benefits of using the global dictionary to record the location of the data tables. The expensive multi-site searching access mechanisms will not be necessary.

If the "bending" could not be achieved then the summary maps could be used to indicate which multi-site searching strategy should be used. For example, figure 4.30 shows that the bulk of the accesses are data retrieval on tables A, B and C. Choose replication.

In addition to producing the summary access path described above it is also necessary to produce summary maps showing the amount of remote data access between locations. One might have a distributed system but the important question is—how much of a distributed system, between which locations is the distribution and is one location more distributed than another? It is a totally different ballgame if 90% of data access is remote than if only 10% is remote. The degree of remote access as a percentage of total data access can vary from location to location.

The location summary access path maps showed for the government system that for day-to-day issuing of public documents each regional office was self-contained and merely required to pass information about documents issued to the location holding the central reference repository. At the end of each week each regional office passed to the head office summarised accounting information. The volume of remote access was therefore low as a proportion of the total and varied by location to location. Access from the regional offices to the head office was solely data maintenance and from the regional offices to the central repository location split more or less equally between data retrieval and data maintenance. Remote accesses were only some 5% of total accesses. This system was certainly distributed, but not much. The system was also distributed vertically as regards all distributed data. Maximum use of the global dictionary could be made and the system was highly efficient.

At the port company the system was distributed vertically as regards the main and small ports in relation to the head office and horizontally between the main and small ports apart from lorry. However, this horizontal data distribution was of no concern. The fact that there were only two port locations meant that if data was not at one port location it must be at the other. There was no cause therefore to worry about multi-site searching mechanisms.

If the entity is logically distributed vertically then all access to an entity will be either all local or all remote, and hence easy to calculate. If the entity is logically distributed horizontally then some of the accesses to the entity might be local and some remote—it has to be calculated for each entity for each remote location split by data retrieval and data maintenance accesses. A large, tedious but necessary task.

Figures 4.31 and 4.32 show the summary access path maps at the main and small ports and figures 4.33—4.35 the volume of distributed remote access between them. The data being transmitted between the ports is information about pallet movements being or planned to be shipped between them. Figure 4.33 shows transactions being raised at the main port to send data to the requisite small port concerning pallet movements to be expected shortly. Figure 4.34 shows transactions raised at a small port to enquire on the main port about planned pallet movements. Figure 4.35 is the reverse of figure 4.33 and is the small port informing the main port about pallet movements to expect. Other inter-site summary maps were produced, for example between the main and small ports and the head office.

A number of points arose from interpreting just these three maps:

- The system was only distributed to some 0.5% of all its data accesses.
- Only some of the entities required remote access.

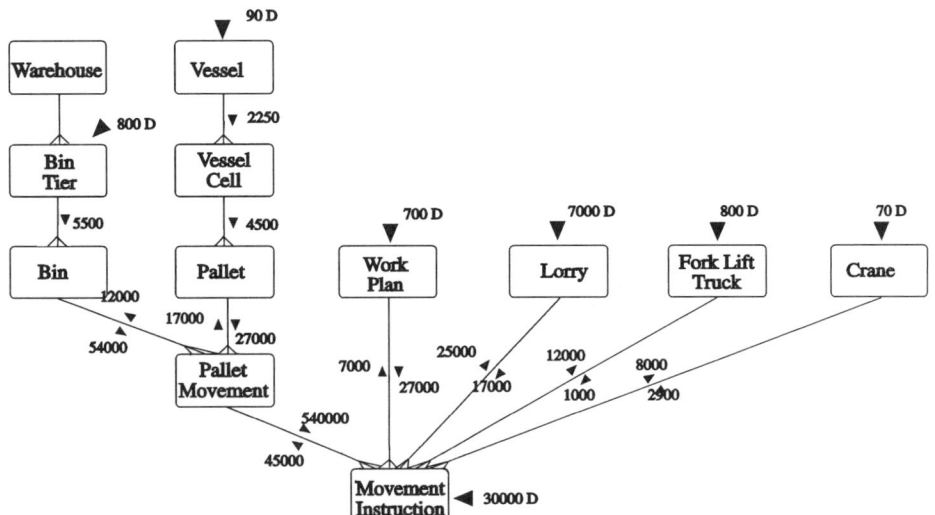

Figure 4.31 Main port summary access path map

Distributed Systems

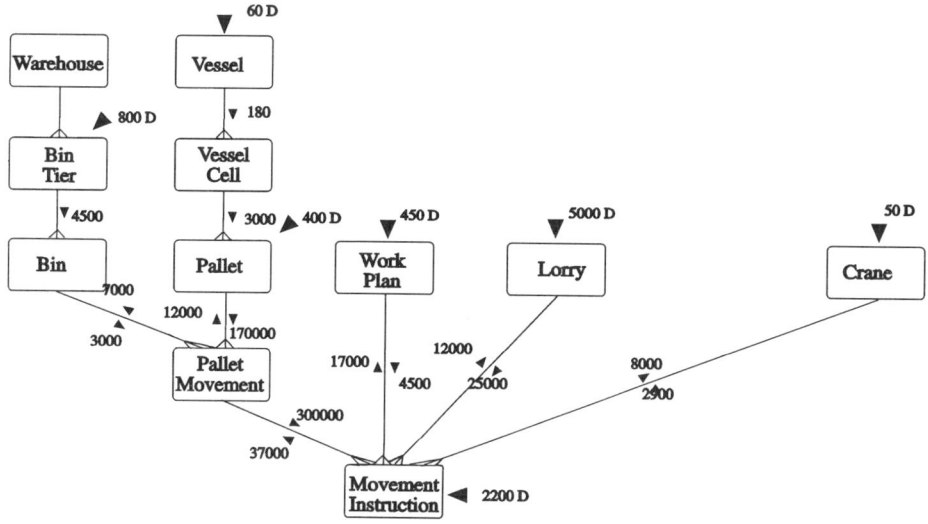

Figure 4.32 Small port summary access path map

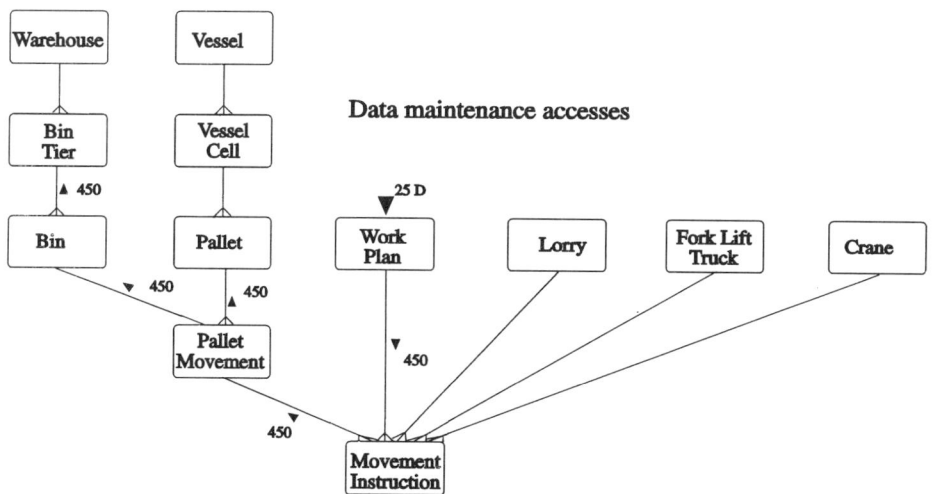

Figure 4.33 Data transfer: main port to small port (1)

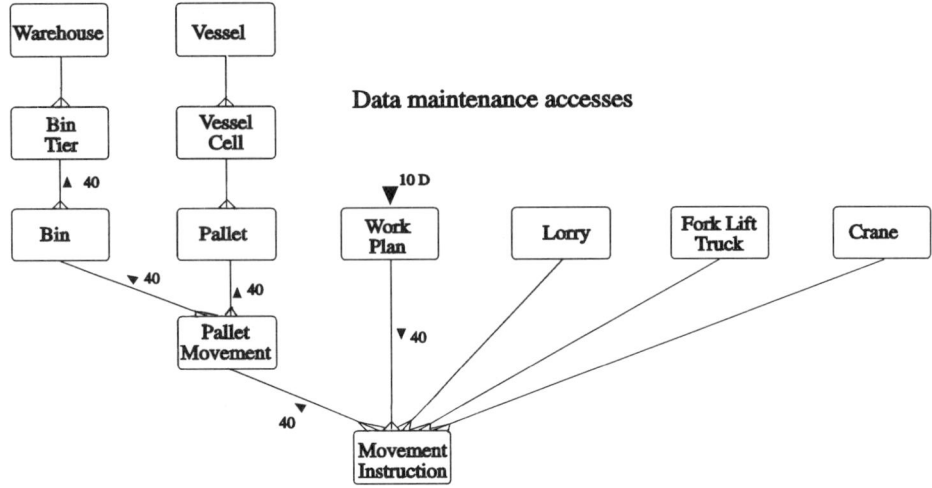

Figure 4.34 Data transfer: main port to small port (2)

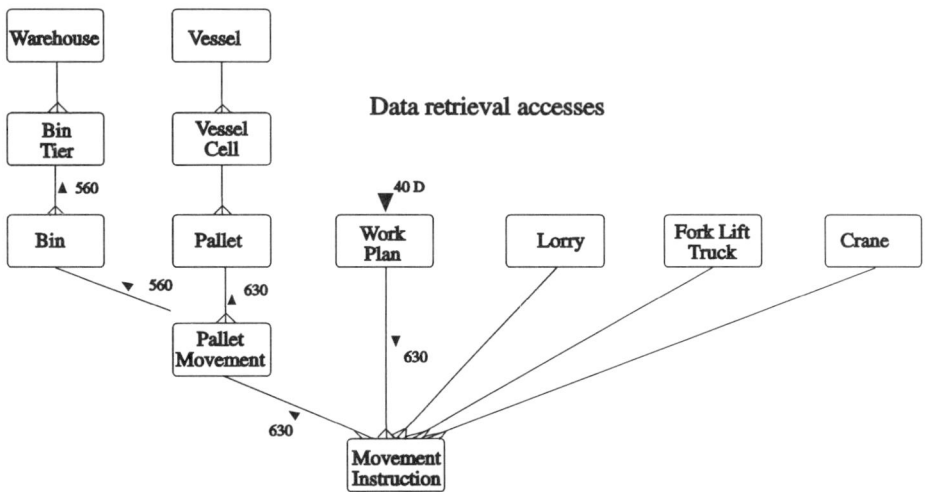

Figure 4.35 Data transfer: small port to main port

- The distributed transactions were of the type that could wait for a response time of up to one hour.

Given these findings, the enormous cost of a distributed system could not be justified. Data was therefore transmitted between sites on the basis of periodic snapshots in the implemented system.

4.1.5.7 Calculating logical data communication volumes.

This is a simple but very productive task. The remote I/O is now known from the various summary access path maps showing inter-site accesses. If all the data in the entities is to be transmitted then multiply the number of logical accesses to the entities by the length of the entity. If only some of the data in the entity is to be transmitted take a percentage figure as appropriate. Endless permutations of "shuffling" the communication lines between the processors around the locations can now be made.

Figure 4.36 shows how the logical telecommunication overheads were rapidly computed for the government system depending upon the design decisions taken. Three of the regional offices had exactly the same volume of business. The number of data accesses between locations was known from the summary access path maps, as was the fact that all data in an entity was transmitted when remotely accessed. The total access to an entity was multiplied by the data length of the entity and divided by the line speed to give average line loading. Peak periods could be identified and the loadings easily calculated by taking the relevant portion of data transmitted during the peak period. A 2400 bit stream line was assumed.

The "complete" strategy assumed that each site has its own processor. The "partial" strategy assumed that Derby could be serviced from Manchester

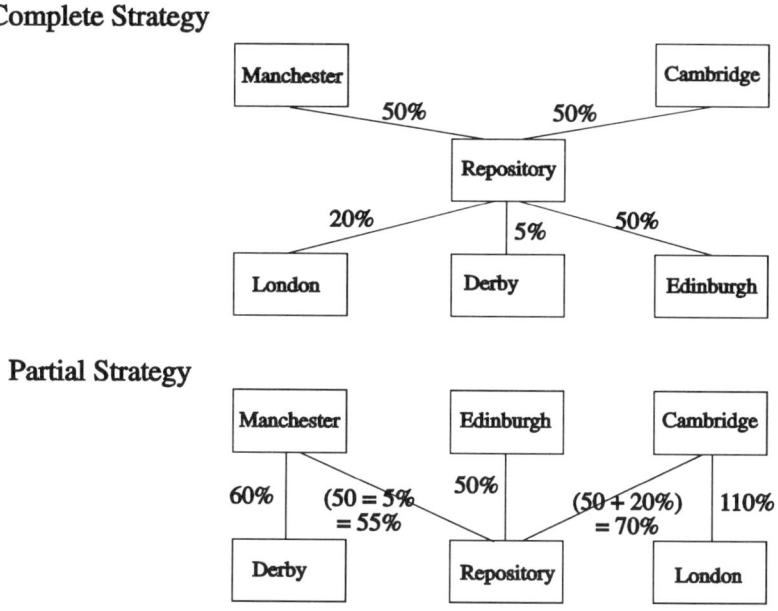

Figure 4.36 Data communications

and London from Cambridge. All data access—local and remote in the complete strategy—at these two regional offices became remote access and their accesses to the central reference repository had to be added to their respective "host's" accesses. It was a simple case of adding remote accesses together depending on the distribution of processing power in the offices. Endless permutations were tried and easily calculated. In the partial strategy it is clear that a higher speed telecommunications line is required between Cambridge and London.

4.1.5.8 Dialogue design

No change is required to the technique other than to be aware of possible variations in local requirements. The element boxes in the LGDEs may vary by location, as may the LGDEs themselves. In most cases one can record which of the element boxes relate to which location as appropriate. If the differences were large and the resultant chart large it may be necessary to draw a different structure chart per location.

The supporting documentation must also reflect the differences in the dialogue inputs and outputs.

4.1.5.9 Distributed processing/client-server

The author is finding that there is an increasing need to use the facilities of distributed processing and client-server file handlers to reduce the volume of data being transmitted between sites. The ability of distributed processing to transmit the processed data between sites is proving to be the solution to the slow speed of the wide area telecommunication networks and the means by which heterogeneous and incompatible file handlers are able to communicate. The ability of client-server technology to place processing where it sensibly should be executed is also enabling small front-end processors to connect with the back-end larger processors. Networks of computers each doing their most appropriate processing can now be supported.

4.1.5.10 Physical design

Given the concept of logical design = physical design, the physical design has already been produced, with the one proviso of breaking this "law" (the only time it is advised) when the data is distributed horizontally across many sites. Furthermore, given that distributed systems are nothing more than co-operating centralised systems, the techniques of database and program design are unchanged.

Distributed Systems

Notwithstanding the comforting statements about the preservation of skills and learning curves the physical designer of distributed systems must take into account extra facilities, such as:

- the extra remote I/O and local processing overhead of multiple message pairs between locations for distributed locking/timestamping and database updates;
- the implications of remote I/O and data transmission overheads generated by the simplest (send all the remote data unmerged to the triggering location) to the most sophisticated (pre-calculate data transmission and send minimum data to maximum data) distributed query optimisers;
- whether to use distributed database access or optimise further with such alternatives as distributed processing and remote database access;
- the data transmission overheads between sites on hardware recovery;
- whether application program "deadly embrace" is supported, and if so by timestamping or multi-phase lock messaging with timeout. For either approach consider how to use transaction classes and conflict graph analysis to minimise the possibility of conflict and optimise it where it occurs;
- whether the global dictionary is repeated *in full* at each site or whether the local dictionaries only have references to the remote sites as regards the data tables held there. If it is the latter consider the *ad hoc* SQL parsing overheads to access remote data;
- how to minimise horizontal data distribution in the physical design;
- how to limit the potential runaway overheads of *ad hoc* distributed queries.

4.1.6 Redefining the logical design

As described in section 2.2.1.1, the concept of logical design = physical design is valid for all data processing environments from the strategic down to lines of application program code, *except* for distributed database. The problem occurs in distributed systems where the data is horizontally distributed. Horizontal data distribution has been shown to have a major adverse impact on finding the location of data. Although the logical design might show that part or all of a distributed system is horizontally distributed the logical design specification cannot be used as the basis of code generation as in the other data processing environments, as a performance disaster would ensue. The logical design has to be "re-interpreted" as described in this chapter, particularly as in sections 4.1.4.8 and 4.1.5.6, and *modified logically* before physical design. This requirement is unique.

4.2 REALTIME PROCESSING

Ed Yourdon defined realtime systems as being "immediate output of current input". This delightfully succinct definition is delightfully accurate, as it clearly identifies the difference between realtime and online processing. Online processing matches the above definition as regards output—a rapid immediate output response is made to a trigger on the computer system. It does not, however, match the above definition as regards current input. The online output display is based on accessing the database for data, data that will have been inserted at a prior point in time, possibly days, weeks and months previously. Making use of the Yourdon definition, online processing could be said to be "immediate output of previous input".

Notwithstanding these differences, there is much that is similar between realtime and online processing. Realtime systems deal with data, which must be structured and accessed. Realtime systems therefore require the techniques of entity relationship modelling, relational data analysis and entity life history analysis. The accessed data is displayed on screens—the dialogue design techniques are therefore used. Data requires manipulation and therefore process models must be specified. *All of SSADM is relevant to realtime processing.* As with distributed database, one need unlearn nothing.

Realtime systems have a number of additional unique features—continuous data flows, event recognition, causally related events and event synchronisation. Techniques to support these features are necessary. Two widely used structured design methods with which the author is familiar have been developed specifically for realtime processing—the Yourdon and JSD methods. There are, of course, other such methods, such as MASCOT. Given, for obvious reasons, that SSADM does not provide techniques or guidelines for these unique realtime features, it has been a question of which of these two methods to integrate into SSADM when undertaking realtime projects.

The JSD method proved incompatible, although clearly this is much less of a problem with version 4. The method does not produce dataflow diagrams, a generic technique which is widely understood. The specification of logic is based on entity life history like structures, but there is an inconsistency. Data maintenance events that update the database are grouped by entity, such that the resultant application programs are entity based. Data retrieval events that are queries on the database are not so grouped, with the query application programs being event based. However, the main problem is that not all the characteristics unique to realtime systems are explicitly recognised. No advice, for example, is provided for handling continuous dataflows and event recognition.

The Yourdon method, by contrast, provides a near perfect match for SSADM. Extensive use of enhanced dataflow diagrams is made, with the

Realtime Processing

other techniques for realtime processing adding value to the information implicit in a dataflow diagram. The method could therefore be easily integrated into SSADM. The Yourdon use of the Chen diagramming notation for data modelling was discarded, to be replaced by the SSADM ERD notation. No conflicts between the two methods therefore occurred. Techniques to support continuous dataflows, event causation and synchronisation were added to SSADM, the second as enhancements to the dataflow diagram technique and the latter on a Yourdon technique fully compatible with dataflow diagrams.

The final selling point for integrating Yourdon into SSADM is its conceptual soundness. It explicitly recognises realtime features and separates them from conventional data processing. The control mechanisms necessary to support realtime are separated from the data processing mechanisms of batch and online processing. There is therefore no confusion between the different but dynamically interwoven environments of control processing and data processing.

The only area where the Yourdon method fails is that no technique to handle the problem of event recognition is provided.

4.2.1 Realtime concepts

There are a number of significant differences between realtime and batch and online processing. These differences are based on differing concepts. Those relating to batch and online processing are described in section 2.2. Those specific to realtime are as follows.

4.2.1.1 Events can be causally related

The identification of events is more problematical in realtime systems than in batch and online processing because they can be causally related. Event A can trigger event B, both functioning together as a continuous process. Events are therefore less distinct from each other.

Assume the following scenario, where realtime software is embedded within the equipment it controls. A warship is sailing the high seas and comes under a missile attack. The attack is recognised (event 1) by the warship's radar control program (program A) which triggers program B (magazine control program) to issue ammunition automatically from the warship's magazine and downdate the database record of ammunition in stock (event 2). Program B now triggers program C (gun control program) which loads the gun with the ammunition and updates the status of the gun to loaded (event 3). Program C triggers program D (warship fire control program) to aim and fire the gun and update the status of the gun to fired (event 4).

These events are causally related in that event 1 triggers event 2 *et seq*. They also occur logically at the same point in time and it is only the slowness in the operation of the equipment that physically separates them. This causal relationship poses the question—what is a realtime event? Given that they all occur logically at the same point in time, events 1–4 could be grouped together as one event.

4.2.1.2 *Events can occur asynchronously*

The above is somewhat simplified in that it ignores the fact that many missile attacks could be occurring simultaneously in a non-predictable manner. Indeed, attacks of other kinds, such as aircraft and submarine attacks, could also be occurring at the same time. The causally related events described above could therefore be triggered by multiple events of different types occurring simultaneously but in a totally asynchronous manner. Control logic to handle synchronous sequencing of causally related asynchronous events is therefore required. This control logic is unique to realtime systems.

4.2.1.3 *Events may not be recognised*

The warship comes under attack. The question is—is it a missile, an aircraft, another warship, a submarine or a group attack of any combination of the above? In batch and online processing a transaction code is always appended to data being input, so the operating system/teleprocessing monitor can recognise the event for what it is. For realtime events such luxuries do not exist. An event recognition mechanism is required.

4.2.1.4 *Information flows may not be discrete*

A realtime system often requires to monitor flows of information that are continuous. For example, the system is monitoring temperature gauges and switches equipment off or on as the temperature rises or falls across a threshold. The flow of temperature data is continuous. With batch and online systems all dataflows are discrete and time distinct from one another. They are clearly recognisable and do not require a threshold boundary mechanism to act as a trigger. Discrete dataflows implicitly represent both the content of the data and the occurrence trigger of the flow at a specified point in time, to which a computer system must make a response. Continuous dataflows do not have a built-in triggering mechanism. These require to be designed into the realtime system, as well as a mechanism to have a continuously running program to support the continuous dataflows.

Realtime Processing 235

All the above concepts are to do with processing. Of the three logical components of a computer system—data structure, data access and data process—it is the data process component that requires enhancement to support realtime systems. The data structure and data access components remain unchanged from batch and online processing.

4.2.2 The impact of realtime concepts

4.2.2.1 *Separation of control from data*

It is noticeable that the Yourdon method recognises that application programs are event based. To preserve this concept the method has skilfully separated the "control" access and process logic for event synchronisation from the "data" access and process logic processing the data about the event. Once the control logic has completed synchronisation the control program calls the data process program(s).

The normal coding of application program processing data at the event level, as in SSADM, is wholly preserved. Their realtime causal relationships and synchronisation are handled by separate control programs. SSADM is therefore entirely appropriate, without change, to support the data processing component of a realtime system.

This control mechanism requires to be able to receive multiple control messages over time for multiple asynchronous events, possibly for multiple entities. It therefore requires to be able to:

- hold itself in a wait condition until the critical event happens (which may be the last event in an unordered sequence of non-critical events);
- switch the application programs processing the realtime event on, off or into a wait condition;
- monitor the state of all its dependent control and data programs;
- pass control data to the dependent control and data programs.

At no time should the control program be accessing or processing user data associated with the realtime events. That is the function of the data programs called by the control program.

The state of a control program reflects the condition the program is at at an instance in time when synchronising realtime events that together effect an entity occurrence. Using the port authority as an example the program synchronising the events affecting a lorry could be in a "Lorry Idle" state. When a certain event or set of events occurs the state could change to "Lorry Busy". When another event or set of events occurs the state could revert to "Lorry Idle".

Control programs, unlike data programs, "talk" to each other dynamically. There is another control program for Lorry Movement Instruction. This control program is dependent on the lorry control program. The piece of equipment being monitored in realtime is the lorry. The lorry is represented as an entity in the logical data model for the container port—see figure 4.15. The lorry movement instruction is issued for a lorry under certain conditions. The entity lorry movement instruction is a detail entity of, and dependent on, lorry. The matching control program can therefore be dependent on the control program for the master lorry entity. When the lorry is matched with a fork lift truck then, because both entities are always processed together for this business requirement, the lorry movement instruction control program is invoked by the lorry control program. Unlike online processing, realtime programs can call each other.

A variety of techniques are available to monitor the control program state. Yourdon uses State Transition Diagrams (STD), optionally supported by State Transition Tables/Finite State Machines.

4.2.2.2 Pitching of control and data programs

A question which any realtime method requires to answer is—at what level do you pitch the control program? Clearly it cannot be pitched at the event level as in batch and online processing, because it can be controlling multiple events simultaneously. Furthermore, these events can be totally unpredictable as to their sequence one with another or indeed whether they will occur or not.

Yourdon once again offers a hint when describing realtime software controlling individual items of equipment. A typical example could be a radar antenna which is part of an air traffic control system. The realtime software is part of and embedded in the equipment supporting the radar antenna. The machinery is recorded as an entity in the logical data model. It is therefore a natural consequence that realtime control programs should be at the entity level, synchronising the asynchronous events that the entity, such as the radar antenna, has to cater for.

Pitching the control program at the entity level when undertaking logical design has proved to be very successful on projects undertaken by the author. For the port authority case study there was a control program for the lorry, fork lift truck and crane entities, as the movement of these pieces of equipment was being monitored in realtime. Indeed there was, conceptually, a control program controlling the realtime events for each occurrence of lorry, fork lift truck and crane. The state of each lorry, fork lift truck and crane regarding pallet movements needed to be monitored, with asynchronous events affecting each occurrence in an unpredictable manner. While the logic for each control program was common for each entity type, i.e. the logic for controlling the fork lift truck was the same for all fork lift trucks, the requirement was to control the asynchronous events for each individual fork lift truck.

By having a control program as described above there is the sweet situation where the realtime events effecting an entity occurrence are synchronised as a separate task before the actual data processing of the event. Once the realtime events are synchronised the control program calls the data program accessing the entity for conventional application processing.

A clear distinction can therefore be made between realtime and online programs based on their role and mechanisms used. *Realtime control programs should be entity based and fulfil the one function of synchronising a set of asynchronous events that, when brought together in a rational order, update the entity.* Because the events are unpredictable and can simultaneously occur over a period of time, control programs require to be able to switch themselves and each other on, off or into a wait condition. They therefore require to monitor their own states. Additionally, because the events can be causally related and affect many entities, control programs can talk to control programs.

By contrast, traditional online data programs should be event based. Since events are standalone application data programs do not talk to each other. Events normally occur in a predictable sequence or within known limits. Being standalone they also occur discretely at an instance in time. Application programs do not therefore require on/off or wait condition switching mechanisms or the concept of program state.

The question of realtime events is answered. The attack event involved four pieces of equipment: the radar, the magazine, the gun loading mechanism and the gun firing mechanism. *Realtime events* are not just triggers as in conventional online systems, they *are triggers at the entity level.*

4.2.3 Integrating the Yourdon method: logical design techniques

This section will not re-describe the Yourdon method. Rather it will detail how it was integrated into SSADM.

4.2.3.1 Dataflow diagrams

The Yourdon dataflow diagram uses different symbols (for example, data processes are circles rather than rectangles) and additional symbols (control processes). The additional symbols reflect the realtime component. Both types of symbols were converted to SSADM format. The symbol set is:

- Control stores.
  These stores contain data about an event that has occurred or control commands (program stop, start, wait on an event etc.) between control processes and between control processes and data processes.

- Event flows.
  Event flows report an event or give a command at a discrete point in time and have no data content. An event flow is an interrupt "pulse" type message between control processes, data processes and control stores. There are three types of event flows: a signal, an activation and a deactivation. A signal merely reports that an event has occurred. An activation is a command to start a process. A deactivation is a command to stop a process. Activations and deactivations contain no data.

- Dataflows.
  Dataflows can be discrete or continuous. Discrete dataflows are associated with a variable or set of variables that is defined at discrete points in time. A discrete dataflow corresponds exactly with a dataflow in SSADM dataflow diagrams. Continuous dataflows record the flow of a data value or set of values that flow continuously over time, for example a temperature gauge. The flow occurs between an external entity, data process or data store.

- Control process.
  This is a process that accepts only event flows as input and produces only event flows as output. Input event flows may only be signals; output event flows may be any type: signals, activations or deactivations. It is the control process that contains the control logic to handle the sequencing of causally related asynchronous events. Control processes activate and deactivate data processes. A deactivated data process abandons any work in progress and restarts rather than resumes its function when next activated.

- Data process.
  This is the standard process for batch and online processing, as recognised by SSADM. It can receive signals, discrete and continuous dataflows and activation and deactivation commands. It cannot issue activations and deactivations. It can issue signals to a control process, control store or data store and discrete dataflows to a data store or external entity. Note that this last sentence states that there is no flow relationship to other data processes. This is totally in line with the concepts of event level processing and standalone events, such that data processes and resultant application programs do not talk to each other.

- Data store.
  This is the standard data store, as recognised by SSADM.

Realtime Processing

The author has always drawn DFDs in a three-ring manner, with the data stores in the middle, surrounded by the processes, surrounded by the external entities. For realtime DFD a five-ringed approach is used, with the control processes and control data stores within the standard data stores, as illustrated in figure 4.37. With these five rings, the realtime part of the application system is illustrated separately from the online part of the application system. An example of a realtime dataflow diagram based on the port authority application is illustrated in figure 4.38. The example reflects directly the ring structure and thereby adds clarity to interpretation. Note that the control processes and control stores are distinct from the data processes and data stores around them. The control processes are concerned with recognising an event and with synchronising an event with other events that are simultaneously but asynchronously occurring against an entity. An example is the control process Control Lorry. Various events are affecting the lorry—that the lorry is unavailable, the lorry is ready and a fork lift truck is matched.

When the synchronisation has been accomplished by the control process(es) then the appropriate data process(es) can be triggered for traditional data processing. Control Lorry triggers Pallet MI, which undertakes the appropriate data processing, in this case creating a machine instruction for a fork lift truck to move a pallet. Note also the new "C" type control stores, and the control flows only between control processes. Continuous dataflows are not represented as they were not relevant to the port authority.

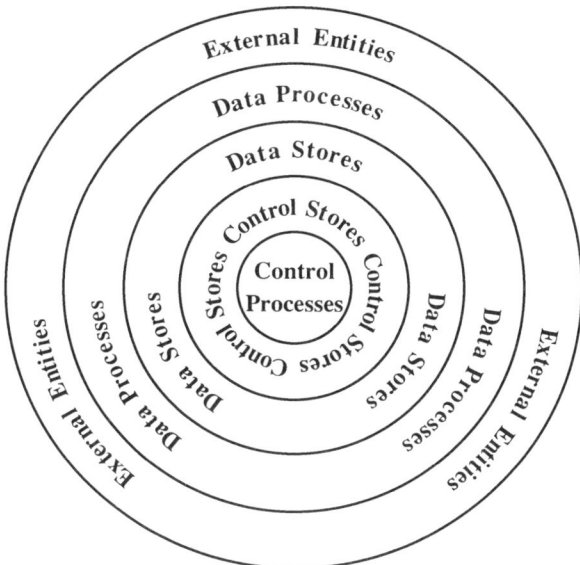

Figure 4.37 Realtime DFD rings

Distributed and Realtime

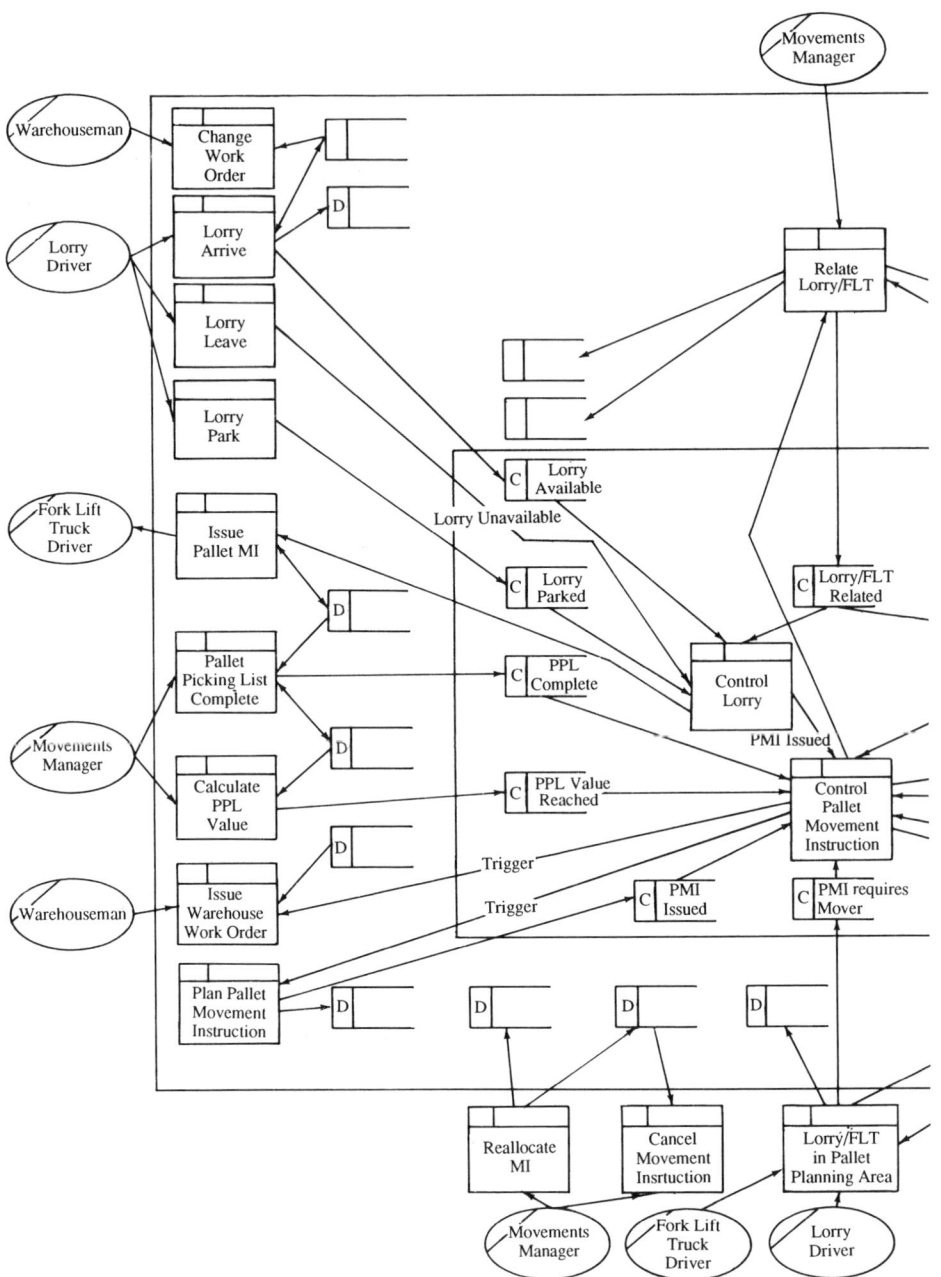

Figure 4.38 Realtime dataflow diagram

Realtime Processing

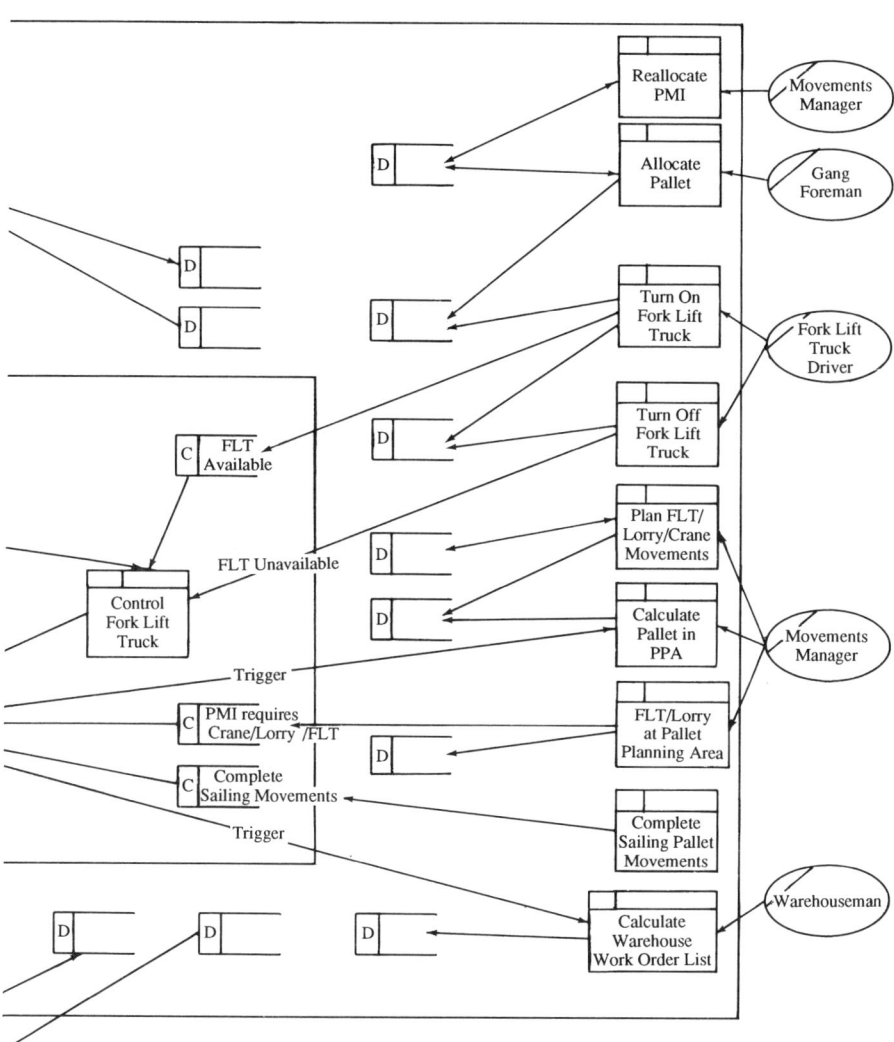

Figure 4.38 *Continued*

The dataflow diagram illustrates certain features of realtime systems not found in traditional batch and online processing:

- A control process supports a particular equipment type, for example a fork lift truck.
- Control processes talk to control processes for the same equipment type and to control processes supporting related equipment types. For example, the entity pallet machine instruction is related to lorry.
- Control processes can trigger data processes.
- Control processes are only triggered by other data or control processes.
- Data processes can trigger control processes.
- Data processes can update "C" control stores but not read from them.
- A control process must trigger a relevant data process, i.e. a data process that modifies the entity type the control process supports.
- External entities cannot trigger control processes.
- Control processes cannot access "D" type data stores.
- The control stores act as message repositories between data and control processes. As part of their event synchronisation task, control processes may require to "wait on" another event before proceeding. The messages about previously occurring events require to be stored temporarily until the "wait on " event occurs. For example, the Control Lorry control program requires to wait on for the event Lorry Arrive when the Lorry state is unavailable—see figure 4.39. Only then can the Wait on Lorry Available message be deleted.

4.2.3.2 *The specification of control processes*

The specification of the access and process logic of the control processes was undertaken by the use of state transition diagrams (STDs). Representative STDs for the control processes Control Lorry and Control Pallet Picking List on the realtime dataflow diagram in figure 4.39 are given in figures 4.40 and 4.41.

Figure 4.39 shows that a lorry alternates between being available and not available. This is reflected in the supporting control program. The initial start point is that the lorry is available. A condition/event Lorry/FLT Related occurs, with three actions/effects—the data process Issue Pallet MI is triggered, the Control Pallet Picking List process is called with a PMI Issued message and the state of the lorry is changed from available to being used.

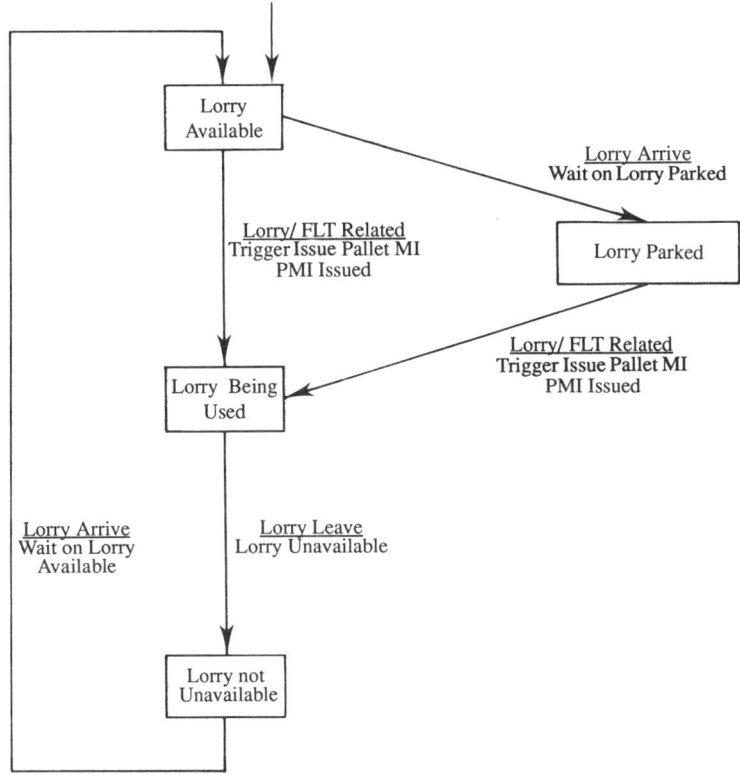

Figure 4.39 Control Lorry state transition diagram

The lorry could also change from being available to being used when the event Lorry Arrive occurs. In this case the action is to wait on another event of Lorry Parked. The lorry changes from being used to not available when the event Lorry Leave occurs. The lorry then reverts back to being available when the event Lorry Arrive occurs and the action is to wait on Lorry Available. In both cases where the action is to wait on the only effect is to change the state of the control program.

4.2.3.3 Continuous dataflow sampling

This is not difficult. All that is required is additional logic in the Process Models. Sampling can easily be handled by using a timed sampling mechanism to test whether a threshold has been crossed or not. If a heating radiator is being monitored to maintain a constant room temperature then

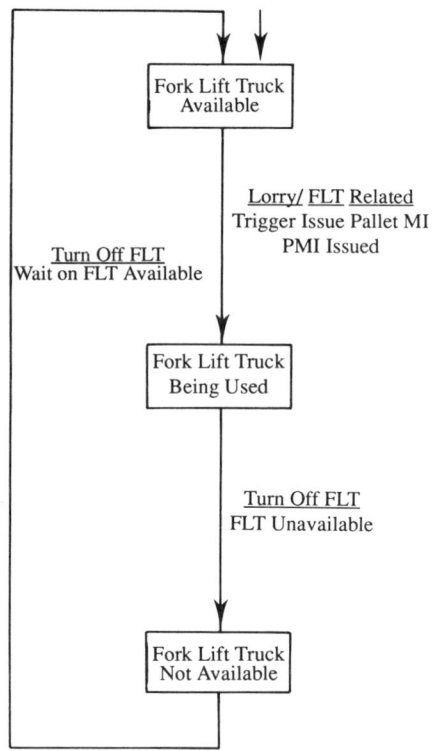

Figure 4.40 Control Fork Lift Truck state transition diagram

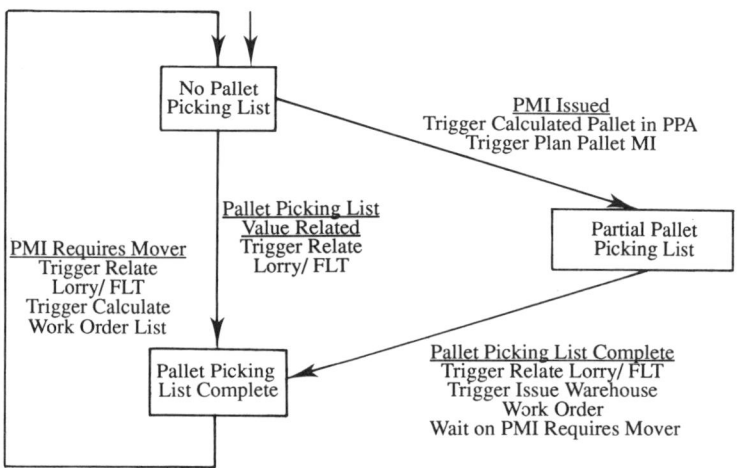

Figure 4.41 Pallet Movement Instruction state transition diagram

the realtime application program re-triggers itself, by issuing its own transaction code to the message queue, every n seconds to sample the room temperature. If the threshold is crossing on an upwards temperature path then the logic is to switch the radiator off. If the threshold crossing is on a downward temperature path then the logic is to turn the radiator on.

4.2.3.4 Event recognition

Event recognition was the one problem area, because the Yourdon method, as described in the Ward and Mellor book *Structured Development for Real Time Systems* and confirmed by consultants at Yourdon UK Ltd., does not contain an event recognition technique.

Event recognition is a task for the control processes. Given that control processes are entity based, it occurred to the author to use the entity life history technique (ELH), also entity based, as it had been used in the SSADM structured design method.

In its previous incarnation ELHs used the posit, quit and admit facility. An event occurs which is not initially recognised for what it is. A posit assumption is made that an event updating the entity is recognised. Using the warship example, the position is that the entity warship is under an attack event that is not initially recognised for what it is. It is therefore posited as a guided missile attack (the state of the warship control program is updated to missile attack). Processing continues as normal on this assumption. A set of causally related events are triggered on the missile attack assumption— an anti-missile missile is loaded (another event updating the state of the warship control program) and trained (another event updating the state of the warship control program). At this point a quit occurs as the originally posited event is now recognised as a civilian aircraft, the events that occurred since and dependent on the posit must be undone (detrain and unload the guided missile launcher) and an admit be made to proceed on the correctly recognised event.

The quit could occur anywhere between the events recognise attack and train launcher. A record of any pre- and post-states of the database changes must be made from the time a posit is made, with the database table rows locked with an exclusive "X" lock, until the posit is confirmed or rejected. If the posit is quitted then the database must be returned to its pre-posit state condition. All changes made to the database from the time that the posit is assumed and the recognition that the posit is valid or wrong must be treated as a single logical unit of work, by being front-ended with a BEGIN statement and back-ended with a COMMIT statement. This ensures that the posited actions are either all successful if the posit is true or all backtracked if the posit is false. When all the pre-posit state changes have been restored the "X" locks can be released and the recognised event be executed as part of normal processing.

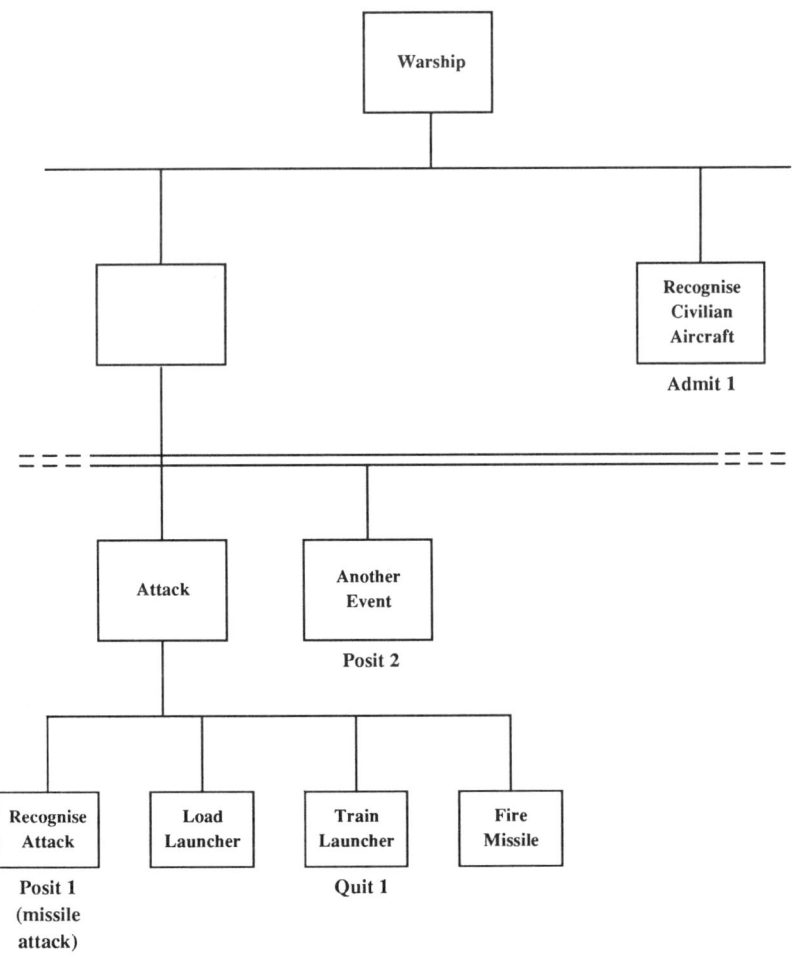

Figure 4.42 Event recognition

A representative ELH in figure 4.42 using posit, quits and admits illustrates this trail of near disaster. Note the use of parallel lives to handle asynchronous events.

Having drawn the warship event recognition structure the effects of the event (say train launcher) were added to the matching state transition diagram for the warship control program as actions for the corresponding condition event (trigger data process Train Launcher and change control program state) and recorded as a new program state (launcher loaded to launcher trained).

4.2.3.5 Combining the control processes

The control processes linked by control flows require to be combined to generate a single control application program. The link is the control flow between two control processes. This means that the related STD per control process also require to be combined. Since control processes are entity based two entities (master-detail) are involved. As described earlier, such related entities are, for example, lorry and lorry movement instruction. The three control processes Control Lorry, Control Fork Lift Truck and Control Pallet Movement Instruction need to be combined as whenever Control Lorry or Control Fork Lift Truck are triggered they in turn trigger Pallet Movement Instruction. The action Trigger Issue Pallet Movement Instruction in the Control Lorry process from the event Lorry/Fork Lift Truck Related links Control Fork Lift Truck and Control Lorry to Control Pallet Movement Instruction. This can be seen in the combined STD in figure 4.43.

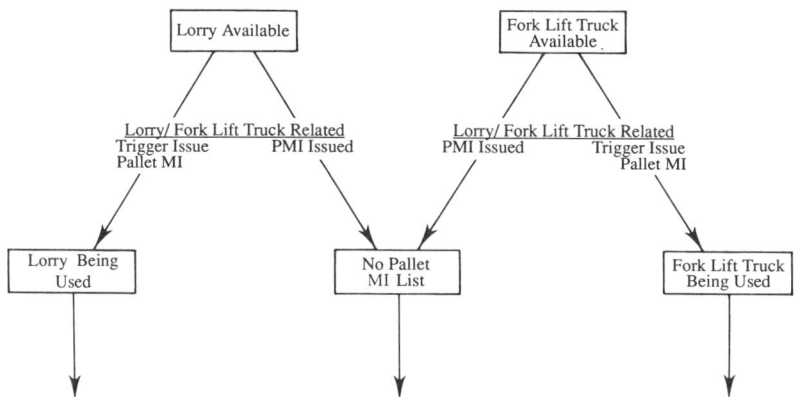

Figure 4.43 Combined state transition diagram

4.2.3.6 The remaining techniques: physical design

The remaining logical design techniques of SSADM, such as Logical Data Modelling, Relational Data Analysis, I/O Structures, Dialogue Design and Process Models, are untouched.

The SSADM techniques of First-Cut Database Design and Physical Design Control are untouched. The program design techniques, of course, require to support the data processes as represented in the figure 4.38 dataflow diagram and the database design to the "D" type data stores. The realtime component was added by:

- using the combined STDs as the basis of specifying the logic of the realtime control programs and converting the combined STD into action diagram code. If it was necessary to split a control program across processors then the technique described in the Yourdon method was used;
- defining the control stores as standard data tables;
- removing concurrency from the STDs. The STDs pay no attention to the concurrency of processes and assume multiple processes can run simultaneously. Unfortunately this is not always true. A processor can only execute a single program "task" at a time. Tasks can be truly concurrent with each other if they run on different processors or on a multi-processor capable of running several tasks simultaneously. The state changes in a STD (which may be a combined equipment type level STD from several entity level STDs) must be converted into a sequence. In figures 4.39–4.41 the three entity level STDs are already sequenced. The combined STD in figure 4.42 is not. The event ST VC Matched triggers two processes simultaneously—ST MI Issued and Issue ST Machine Instruction, the former a control process and the latter a data process. The technique for removing concurrency is described in the Ward and Mellor book.

5

ADDITIONAL DATA PROCESSING ENVIRONMENTS FOR SSADM—EXPERT SYSTEMS

5.1 THE COMMON LINK WITH OBJECT ORIENTATION

Object oriented and knowledge based expert systems are data processing environments that will have an increasing impact on application design and development. Many leading industry thinkers believe that object oriented design will be the next "leap forward" for the 1990s as relational technology was for the 1980s.

The author is convinced that the future of object orientation will be more significant and much more beneficial than relational database. For various reasons, the detail of which are not relevant to this book, the author is not a devotee of relational technology, particularly as regards non-procedural data access to the data. The need to know in SQL the tables in which the data attributes are held and the relationships between the tables when undertaking a multi-table join is not non-procedural access, never mind the claims to the contrary. This is but one of the deficiencies of relational file handlers. Much is claimed for relational technology, such as the non-procedural access, but, in the author's opinion, little of substance is delivered. Such is not the situation with object orientation.

As regards expert systems the author was a considerable sceptic but, as a result of studying the subject and the supporting technology, is now a total convert to the practical benefits of the technology to both users and system developers. This long term confidence is based on the fact that much of expert systems technology is also used by object oriented databases and programming languages. Furthermore, expert systems technology can also fully support the expert systems concepts and is therefore stable.

The reasons why these apparently distinct trends (certainly distinct and certainly trends if one accepts the marketing hype) are considered jointly is because they use much common technology.[1] It has been said that an expert system "is an object oriented system without methods" and that an object oriented system "is an expert system without an inference engine". This is a somewhat simplistic explanation. Basically, both technologies use the additional data structuring components of class and aggregation, but expert systems do not use the object oriented logic components of polymorphism, encapsulation, dynamic binding *et al*.

The technical facilities and their relationships to the trends are illustrated below. It shows that there are many facilities that are common, although an expert system uses only a subset of object oriented technology. Products called GOLDWORKS from Goldhill Computers Inc. (GOLDWORKS was formerly owned by Artificial Intelligence Ltd.) and GENERIS from Instrumatic Ltd. contain facilities for both expert and object oriented systems, and are therefore real life physical proof of the high degree of technical overlap of these currently distinct data processing environments. The technical facilities and their relationships to the trend are:

Object Orientation		Expert System
×	* Semantic Nets/Class Models	×
×	* Class Objects	×
×	* Aggregation Objects	
×	* Abstract and concrete objects	×
	* Business objects	
	* Rules	×
×	* Procedures/Methods (private and public)	
×	* Conflict Resolution	×
×	* Property Inheritance	×
×	* Polymorphism/Late Binding	
×	* Encapsulation	
×	* Message Passing	
×	* Genericity	
×	* Overloading	

---

[1] This means that there is some inevitable overlap when discussing the two technologies. Every attempt has been made to keep it to a minimum. The readers will have to be patient as some of the facilities are explained under object orientation in the next chapter.

The facilities that have been marked as being used by these new data processing environments are those that are specific to supporting the underlying concepts on which the environments are based. Both environments make use of each other's facilities. Expert systems make use of the facility of procedures/methods of object oriented design, because it has been found useful to use under certain conditions. However, procedures/methods are not necessary to support the concept underpinning expert systems. Object orientation can use the facility of frames, but frames likewise are not necessary to support the concepts underlying object orientation. The facilities that are not appropriate to supporting the underlying concepts of the data processing environments are not marked against them.

Expert systems and object oriented design both use many other facilities, but these additional facilities are those initially created for the centralised data processing environment. Such facilities include file handlers, dictionaries and logging/recovery routines. As described earlier, the centralised data processing environment, and its facilities, is generic to all the other data processing environments. It so happens, for example, that expert and object oriented systems use file handlers, dictionaries and logging/recovery routines, and yet these facilities are not regarded as specific to the concepts underlying these data processing environments. These facilities are also not identified in the above table.

The technology of expert systems and object orientation will be described separately, each followed by an assessment of their potential impact on SSADM. Expert systems are considered first as they have been established for a longer time and the technology has "bedded down".

The author does not consider that this assessment of these two technologies is exhaustive. The subjects of expert systems and object oriented design are still in their infancy, notwithstanding all the work that has already been done. The physical technology is new and only now gaining appreciation and acceptance of its potential. As usual, and not unexpectedly, the logical design techniques are some way behind the physical technology. Structured methods for both trends are either in development (GEMINI) or new release (KADS and STAGES for expert systems and HOOD, Shlaer/Mellor, Coad/Yourdon, Booch, and Rumbaugh *et al* for object orientation) with frequent and major upgrade versions being issued.

This chapter is designed, following a description of the technologies, for the author to add his "two pennies' worth" of suggested enhancements to SSADM. The suggestions are in line with the policy stated at the beginning of chapter 3. No attempt is made to detail how the enhancements could be incorporated into SSADM where worked case study examples to prove the enhancements are not available.

5.2 Expert System Concept

Expert systems have the unique ability to store and access knowledge as well as data. This means that they can represent expertise regarding a particular application domain, for example how to service a car or how to diagnose an illness. Expert systems can therefore be relevant to different types of computer applications from those which have traditionally been developed. Expert systems are particularly suited to those applications that require expertise to solve problems. Traditionally data processing has been for those applications where dumb data is presented to users for them, as *homo sapiens*, to interpret. Expert systems, by contrast, store knowledge and can therefore offer advice as well as present data. This is their unique feature. *It is the concept of knowledge and resultant advice provision that separates expert system processing from traditional batch and online processing.* They can therefore "assist" and "advise" in such tasks as diagnosis, planning, design and interpretation.

Notwithstanding the many, sometimes esoteric, definitions of knowledge given by the intellectual community, in an IT environment the definition is much simpler. *Knowledge is a meaningful association of two or more facts or of a fact(s) and a command(s).*

A fact is composed of three elements: a subject, a relationship and a property. These three elements are known as a triple. A triple example could be "Today (subject) is (relationship) Friday (property)". The subject and property, which could either be data or logic, are dumb information when considered in a standalone context. In the triple example today and Friday are dumb data. It is only when the dumb information is brought together as a triple that it becomes a meaningful fact.

Another fact could be that Friday is a pay day. The property of the first fact is the subject of the second fact. These two facts can therefore be combined to create some knowledge, that one gets paid today because today is Friday. Other dumb data could be the days of Monday and Tuesday. It is not until Monday and Tuesday are triple linked with a total of seven different days and associated to form the concept of a week that the individual facts become more meaningful. When related together these facts become knowledge—there is a thing called a week, it is composed of seven days and one gets paid once a week.

Clearly the facts have relevance to a specific application, in this case the domain of time in a Western context, whereby a week begins on Monday and ends on Sunday. In a different cultural context a week can begin on a different day and even be of a different timespan. The technical term used by some expert system products to describe a fact is an assertion.

The facts can be related meaningfully together to form knowledge either via rules or semantic nets. The rules are usually in the form of "If condition A then conclusion B". The rule could be "If today is Monday (one fact) then tomorrow is Tuesday" (another fact) or "If today is sunny (a fact) then go sunbathing" (a command), the meaningful association between the facts and command being the "if...then clause".

There can be multiple conditions and conclusions to a rule. The conditions can be connected by ANDs and/or ORs, but the conclusions can only be connected by ANDs. There could thus be a rule that states "If condition A and B or A and C then conclusion D and E". An interpretation of the rule states that if either the conditions A and B are true or A and C are true then the conclusions D and E can be drawn.

The rules can be chained together by the conclusion(s) of one rule matching the condition(s) of another rule, such as "If condition A then conclusion B" and "If condition B then conclusion C". A complete network of such rule knowledge can thus be created.

It is also possible to record facts as semantic descriptions—"employee is a person" (one fact) and "person is a mammal" (another fact), the meaningful association between the two facts being a common subject/property symbol, in this case person. The subject of one fact is the property of another fact. A complete network of such knowledge descriptions can be modelled in the object relationship descriptions in semantic nets.

Rule facts are the knowledge representation mechanism that enable expert systems to provide advice. *Rule facts are the active component and can be regarded as knowledge logic.*[1] *Semantic facts are the passive component, play a supporting role to the rules, and can be regarded as knowledge data.* Note that the classical

---

[1] With the wide range of facilities now available the term logic has acquired several meanings. In the context of this book there are two major classifications—procedural logic and declarative logic. Procedural logic is that which has been used since the beginning of computing and is conventionally associated with the code in application programs. The logic is positional and composed of individual statements of action. It is positional in that the statements of action only have relevance in the context of their position to other statements of action that precede and succeed them. Furthermore, the logic has to be complete and in the correct order for it to work. The statement of action has no value in its own right.
There is also a sub-division into access logic and process logic. The access logic is the commands that are used to access the information-base. Here again there is a sub-division into record-at-a-time access (Read, Write, Update and Delete) of the pre-relational file handlers and the set access (such as Select, Union and Difference) of the relational file handlers. The process logic is for the pre- and post-massaging of the database accessed data. The commands are sequence, selection, iteration and branching.
Declarative logic is composed of non-positional statements of truth in the form of rules. As the rules are statements of truth they have value and can therefore be used in a standalone context. The rules are non-positional because they are related implicitly together by symbolic matching, with the suffix of one rule matching the prefix of another rule. The software can thereby search for the next rule in the chain of logic no matter where the rule is located. Thus the code can be entered in any order.
Declarative code does not have to be complete. There can be gaps in the logic and the inference software that processes the rules will attempt to "fill the gap". The GENERIS example in section 5.10.1 of declarative logic includes such a gap and yet produces a full and complete information output result. Clearly the greater the gap, the greater the imprecision of the result. There also comes a point where the gap in the logic becomes too large.
The rule also includes access and process logic constructs. It thus means that only one construct need be used for the writing of logic instead of the 13 or more constructs that are required for procedural logic.

physical division of a computer system into data and logic, as illustrated in figure 2.3, is still preserved.

Both type of facts can be used in a standalone manner to provide knowledge. More usually the two types of facts are used to support each other. The knowledge is ascertained by the inference software. The inference software enables the knowledge logic to access the knowledge data to infer yet further knowledge.

An example of the other representation of knowledge (relationship of a fact(s) to a command(s)) can best be illustrated by an example of a rule. The command component has to be in the conclusion of the rule. The rule is "If the date is 31 and the month is January, then issue monthly salary payment". The command is the instruction to issue a payment. Payment could be a call to a procedure, thus providing a link between declarative and procedural logic.

So we have a situation where knowledge can be:

- if a fact(s) is(are) true then some more fact(s);

- if a fact(s) is(are) true then do something.

Knowledge can be universally applicable, specific to an application domain or specific at any level of detail—the buzzword for level of detail is abstraction and is discussed in section 6.6.1. Universal knowledge is a meaningful association of facts that are known to be always true. For example, (income + assets)—(expenditure + liabilities) = wealth. Such knowledge is universal. A piece of knowledge specific to an application domain could be that overtime pay = standard pay 1.25.

By being able to deduce knowledge from facts expert systems are able to support new types of questions. By accessing dumb data in a database traditional computer systems have been able to answer such questions as "How much have I spent?" It is up to *homo sapiens* to ascertain the significance of the information presented. The storage of knowledge could be, for example, in the form of a rule that states "If known monthly outgoings are less than £1000 and the month day is less than 10 and monthly income is greater than £3000 then likely to have money in the bank", purely because one is at the beginning of the monthly pay cycle. One could therefore pose the question "Should I spend?" Certainly is the advice, if the three conditions to the rule are satisfied. By being able to access data *and* knowledge one can pose a much more sophisticated question such as "Should I spend money and how much?" The response to this query will be advisory. If the three conditions are satisfied the advice could further be "Yes, you can, and you have this much in the bank and these are your bills outstanding and these are your standing orders still to be paid". The remaining monies are £*nnn*.

5.3 Knowledge Acquisition and Construction

There are two complementary ways in which knowledge can be acquired. First, topics must be studied formally, as in a school or when one attends lectures and reads textbooks. As a result of such study knowledge is grouped as perspective and general laws. Successful students emerge from courses in accounting or mathematics with a firm grasp of the terms and laws that constitute the formal theories and accepted principles of their discipline. Principles and laws are useful in explaining and justifying why a solution succeeds or fails, but they are often of little help in finding a solution in the first place. General laws usually fail to indicate exactly how one should proceed when faced with a specific problem.

Knowledge can also be acquired by means of experience or by learning from a mentor. In this case the results are different. Domain/application-specific facts are learnt first. Experience, or a mentor, usually teaches the students to rely on rules-of-thumb to perform tasks or solve problems. Students also acquire competence by learning domain/application-specific theories. Thus, accountants who learn from a mentor behave as if they know accountancy theory even though they may not. Knowledge acquired from experience results in heuristics. Heuristics are rules-of-thumb that prune the thought process to a manageable size. They tend to focus the attention *on a few key patterns*. Compiled heuristic knowledge is well organised and indexed. It gives an edge when requiring to solve numerous daily problems.

The sequence of learning usually starts at school and from reading books. This baseline is then applied in a real world and is enriched with the rules-of-thumb that are obtained from mentors and experience. The prescriptive knowledge laws are then gradually converted into heuristic knowledge rules.

There are two methods of expert system knowledge construction. The deductive, or more correctly knowledge elicitation, method requires a knowledge engineer skilled in the use of an expert system product and an expert user skilled in the application domain to build the knowledgebase about the application domain incrementally, knowledge piece by knowledge piece. Knowledge elicitation systems require human creation of the knowledgebase. The inductive method, by contrast, can use software to generate knowledge rules based on database examples. The basic principle is that the software generalises from the particular. For example, an algorithm could be written that if 95% of 1000 records read show that managers managing departments with departmental turnover greater than £250 000 have a status of senior, then one can conclude and generate a knowledge rule that states "If manager manages department and department has turnover greater than £250 000 then status is senior".

Clearly, while inductive rule generation is a great labour saving device, there is a risk in such a mechanistic approach, as it may not take account of exceptions that often occur in an application. An exception could be that a senior manager has been appointed to rejuvenate a department in distress, with a turnover of only £100,000. It is usually therefore found that inductive software is a front-end to generate a set of rules, which can then be vetted by the knowledge engineer and expert user.

5.4 Expert System Components

The basic software architecture of an expert system is illustrated in figure 5.1. There are three main components—the knowledgebase, the inference engine and the man/machine interface. The knowledgebase stores the various ways of defining knowledge, such as rule facts and semantic facts *and* the dumb data traditionally associated with user data stored in the database. The data component of the knowledgebase could well be stored in a separate file handler, typically relational, with the definitions of knowledge held separately in specialised expert system software.

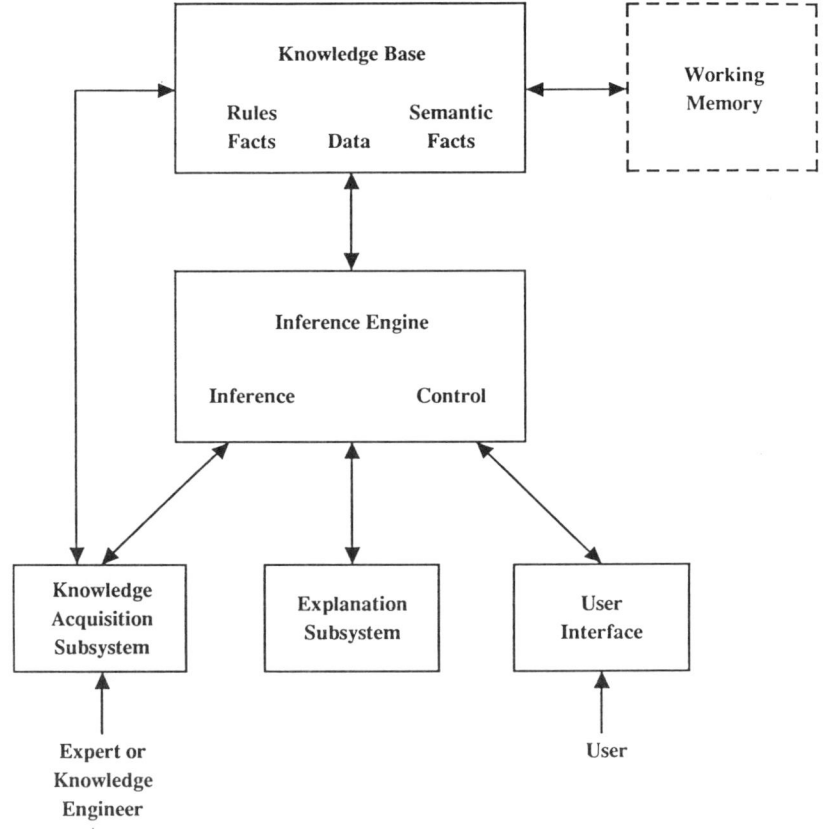

Figure 5.1 Expert system architecture

Expert System Components

The inference engine is the software that executes against the knowledgebase. The engine has two prime functions: inference and control. Inference is primarily composed of three main tasks, rule processing, usually by a technique called *modus ponens* (reasoning with uncertainty) and conflict resolution. As the reasoning is being executed during a consultation process with a user further facts/assertions are constructed and held as temporary working storage in the processor main memory or disk.

Modus ponens states that when A is known to be true and if a rule states "If A then B" it is valid to conclude that B is also true. Stated differently, when we discover that the conditions of a rule are true we are entitled to believe the conclusions. An example of *modus ponens* could be if you know "If today is Monday then tomorrow is Tuesday" and as today is Monday you can conclude tomorrow is Tuesday.

A second implication of *modus ponens* (and many knowledge based systems depend entirely on this rule) is that certain implications that are valid cannot be drawn. For example, another rule of knowledge states that if B is known to be false, and if there is a rule "If A then B", then it is valid to conclude A is false. This rule is called *modus tollens*. An example of *modus tollens* could be "If Joe Bloggs has a tattoo with a romantic message, then Joe Bloggs has a girlfriend". However, a given fact in the database data of the knowledgebase is that Joe Bloggs has no girlfriend. There is therefore a new fact, by implication, that Joe Bloggs does not actually have a tattoo. This conclusion, as seems obvious to *homo sapiens*, cannot be reached by most expert systems. The common mechanism, therefore for deriving new rule facts and semantic facts is *modus ponens*. It is a simple intuitively appealing way to conduct reasoning.

In conventional programming we expect that all required information is provided before computation can take place. It makes no sense to try to balance an account with missing numbers. In knowledge programming this is not necessarily the case. It is believed that some of the financial transactions are not accurate or even lost. An inference engine must therefore be able to handle incomplete or unknown or erroneous information.

Incomplete information can sometimes be handled by uncertainty factors. Unknown information is handled by allowing rules to fail when information necessary to evaluate the conditions of these rules is simply unavailable. The results, of course, depend on the exact nature of the conditions. If the conditions are connected to each other by AND then all must be evaluated as true before the rule conclusion can succeed. In this case if the user or knowledgebase answers "unknowing" to any part of the conditions the rule fails. If, however, the conditions are connected by OR then one piece of unknown information need not preclude the rule from succeeding. In the second case the rule may succeed even though the user or knowledgebase answered "unknowing" to a question related to only one condition of the rule. Furthermore, a condition could be satisfied even though it is less than certain, for example, the condition that Joe Bloggs has a girlfriend is only 80% certain. This condition uncertainty would propagate to the conclusion.

The final inference process is conflict resolution. As a user builds a knowledgebase, using whatever knowledge representation mechanism is appropriate, so the user may, perhaps wittingly, make statements that are or appear to be contradictory. For example, there could be a rule defined by a civil servant that states, based on a law passed by parliament, that persons above 80 years of age are not allowed to buy a car. But at the same time another rule defined by a sales director states that anybody over 18 years of age can buy a car. Such conflicts need to be resolved.

The control function of the inference engine is concerned with rule processing (either forward or backward chaining) and ascertaining the rule access mechanism (usually depth first or breadth first) with monotonic or non-monotonic reasoning. The rule access and chaining mechanisms are described later in this session, with worked examples.

Inference engines can be distinguished as to whether they support monotonic or non-monotonic reasoning. In a monotonic reasoning system all values concluded for a property (property is an expert system buzzword meaning an attribute/data item) remain true for the duration of a consultation session with the user/knowledgebase. Facts that become true remain true and the amount of true information in the system grows steadily or monotonically. These temporary facts (or assertions as they are sometimes called) are stored in the working memory component of the expert system software.

In a non-monotonic reasoning system facts that are true at one time may be retracted later. Planning is a good example of this problem type that demands non-monotonic reasoning. In the early stages of a planning problem it makes sense to make an assumption to go a certain way. Later as information continues to come in it may turn out that an early decision was wrong. The subsequent decisions and consequences need to be retracted (shades of the event recognition problem in a realtime system). Changing the value of a single attribute to retract a conclusion is not difficult. Tracking down all the implications that are based on an initial assumption is difficult and can be a resource-hungry process.

Most expert systems marketed today support monotonic reasoning, but allow only carefully controlled types of non-monotonic reasoning.

The third main component in expert system software is the man/machine interface, for the construction of the knowledgebase through knowledge acquisition and for user query/consultation facilities for which advice will ultimately be offered. In knowledge acquisition the expert user describes the expertise appropriate to an application and the knowledge engineer records this as rule and semantic facts as appropriate.

The querying of an expert system is different from traditional processing.

With traditional processing a query is raised and a reply made and that is the end of it. In expert systems there is a conversational consultation with typically a query and answer session. This is because of the underlying technology. At the end of the consultation chain of rule condition based queries and answers (which may be invisibly undertaken automatically by the inference engine obtaining the answers for the rule conditions from the database component of the knowledgebase) advice is offered.

5.5 THE KNOWLEDGEBASE

The knowledgebase is composed of two kinds of information—the dumb data as entered by users and stored in a database and factual knowledge entered by the expert user and knowledge engineer. The data *and* knowledge are usually stored and accessed by a conventional, typically relational, file handler.

There are three main ways in which knowledge is currently defined—rules, semantic nets and frames. All three are structured as triples. These three facilities are described below.

5.5.1 Rules

The most widely used mechanism of the first generation type expert systems, such as CRYSTAL, is to represent knowledge in the form of rules. Rules are the basic execution instructions for the inference engine. The rule structure is usually in the form of "If condition(s) then conclusion(s)". A typical rule could be "If today is Monday then tomorrow is Tuesday". The structure is sometimes reversed. "Tomorrow is Tuesday if today is Monday."

Each rule is a declarative statement of some universal or application domain truth. *Each rule therefore has value in its own right. A rule can therefore be standalone.* A rule can also be linked with other rules relevant to an application domain to form a ruleset. A ruleset is a group of rules for a particular application domain task (a problem-to-solve process in centralised data processing) for which domain relevant advice is sought. A domain could be how to service a car. Domains can be decomposed into many tasks and sub-tasks and hence rulesets within rulesets. Car servicing tasks could be how to change a tyre, check the oil and maintain the electrics, each requiring expertise and advice.

In order to ascertain if the condition(s) or conclusion(s) is true, the inference engine will access the knowledgebase first and if the answer cannot be ascertained there then usually the condition(s) or sometimes the conclusion(s) is posed as a question to the user. If the user cannot provide the answer a default value is used, that is, a rule without a condition.

5.5.2 Rule types

There can be four types of rules—inference rules, action rules, daemon rules and truth maintenance rules. An inference rule is a query against the knowledgebase and makes no changes to the knowledgebase. They are therefore passive, merely imparting knowledge "if condition fact(s) the conclusions fact(s)". A worked example of an inference rule is given in section 5.10.

An action rule is a rule where the conclusion is an executive statement to do something. The form of the rule is therefore "If condition(s) facts then conclusion command". For example, an action rule could be "If the date is 31 and the month is January, then issue monthly salary payment". This conclusion command could well be a procedure call to a conventional application program, which in this case not only issues the monthly salary payment but also updates the database to record the event. Inference rules and action rules can be grouped into rulesets.

Daemon rules are like "gremlins" continuously running (in a virtual context) in the background waiting for something to happen. A daemon rule is similar to a database trigger found in more modern relational file handlers. They are an expensive form of knowledge execution, for they are tested as to whether they should be "fired" on every fact evaluation of the knowledgebase. They are free-standing rules and do not relate to any ruleset. They are globally applicable to all the application domains. Their prime role is to interrupt a domain ruleset access sequence in order to undertake some action.

Daemon rules are usually identified with a slightly different syntax. The IF precursor to an inference or action rule is typically replaced by a WHEN precursor. For example, "When weather is sunny then display sunny weather options screen". Any inference or action rule that has a conclusion that the weather is sunny would automatically invoke the daemon rule, which would be fired and display the sunny weather options screen. The remainder of the ruleset is then processed, if appropriate.

Truth maintenance can also be supported by using the rule facility and, again, the rules are free standing and continuously running, but this time passively. No changes to the knowledgebase and no interrupt to the ruleset processing are made. A typical truth rule could be "When there are no clouds and it is daytime then the sun is shining". It is obvious that such a rule is universally recognised to be true. Like daemon rules, truth rules are usually applicable to an entire enterprise application.

Daemon and truth rules are controlled entirely by the inference engine and, unlike the inference and action rules, require no human intervention.

The Knowledgebase

5.5.3 Ruleset access strategies

There is one ruleset for each goal/task for which advice is sought. Each ruleset will have one to n rules. The rules in the ruleset can be mapped as a hierarchical structure as illustrated in figure 5.2. Starting at the top it shows that the condition(s) of rule A point to the conclusion(s) of rules B and C, whose conclusion(s) separately point back to rule A. The condition(s) of rule C points to the conclusion(s) of rules B and F. The conclusion(s) of rule C points back to the condition(s) of rules F, A and B.

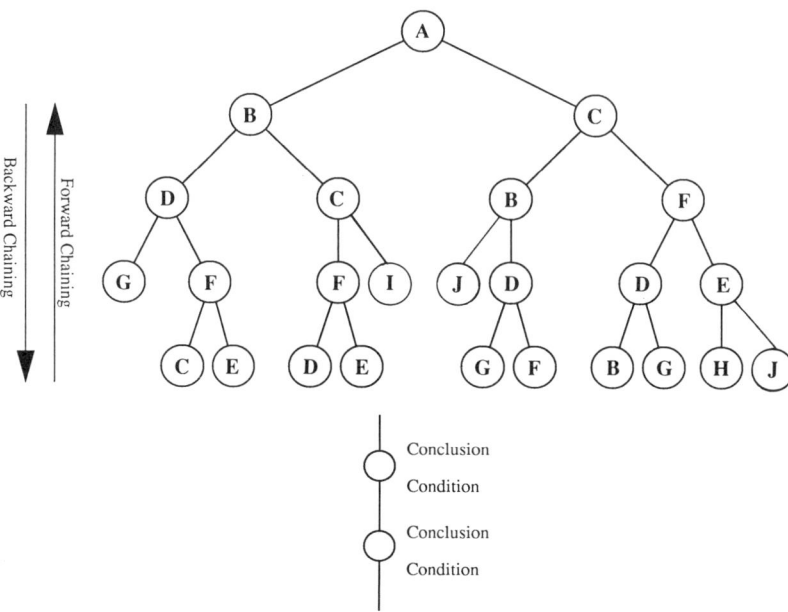

Figure 5.2 Ruleset structure

Backward chaining is down the structure from the single goal rule (the rule with the conclusion matching the goal) and forward chaining is up the structure beginning from any rule with conditions matching the initial set of assertions/facts. It is therefore evident that, although backward chaining has the advantage of starting from a single entry point, in this case rule A, there is the disadvantage that multiple "dead end" rule paths may be accessed before the proper advice is presented. It may be that the rule paths that contain the advice appropriate to the consultation with the expert system are A, C, F, D, G but the rules to the left of the path in the diagram have to accessed first. Forward chaining has the disadvantage of multiple rule entry points, up to one per initial assertion, but the strong advantage of fixed and efficient access paths towards the goal once the entry point(s) has been established. There is no chance of "dead ends".

Note that the hierarchical structure does not preclude the rules within a ruleset being a network. For example, rule F is related to "higher" rules C and D as well as to "lower" rules C, E and D.

What the hierarchical structure of the rulesets also shows is the access overheads in the ruleset, for both forward and backward chaining. Beginning with backward chaining the conclusion(s) of rule A points to the condition(s) of rules B and C. Rule C points to rules B and F and so on. It also shows that forward chaining can begin on any of the rules C, E, D, G, F, B, H and J. If the entry point was solely to rule H then the rule chain would only be E, F, C and A. If, however, the entry point was E the two rule access path options would be E, F, C, B, A and E, F, D, B, A.

There are a number of search strategies, such as breadth first, depth first, hill climbing and best first. The two most widely used are breadth first and depth first. If the search strategy is breadth first access starts at the first rule appropriate to the goal and then access to the remaining rules is by searching all the rules at the next level down in the ruleset hierarchy before proceeding to the rules in the next lower level in the hierarchy. Depth first, by contrast, accesses the first rule as above but then proceeds down the leftmost rule path and progressively migrates to the rightmost rule path. In breadth first access the sequence would be A-B-C-D-C-B-F-G etc. With depth first access the sequence would be A-B-D-G-F-C-E-C-F etc. Both access strategies are illustrated in figure 5.3.

The most commonly used access strategy is depth first with backward chaining.

5.5.4 Ruleset access mechanisms

A ruleset defines knowledge as rule facts for a particular task in an application domain. For example, there could be n rulesets each of n rules to describe the knowledge and expertise required for the tasks of servicing a car. There could be any number of rulesets within the application, for example how to change a tyre, how to diagnose an electrical fault, how to tune the carburettor. A complete hierarchy of rulesets can be obtained. Wherever advice is required in an application domain there must be at least one ruleset.

Within the ruleset access strategy as defined above rules can be chained in a forward and backward direction. The basic mechanism is that rules in a ruleset are chained together via symbolic values with the conclusion(s) *pattern matching* the condition(s) of another rule. Using this symbolic pointer principal rules can be chained in a ruleset in a forward and/or backward direction with equal facility. If the goal conclusion is known in advance (I

The Knowledgebase

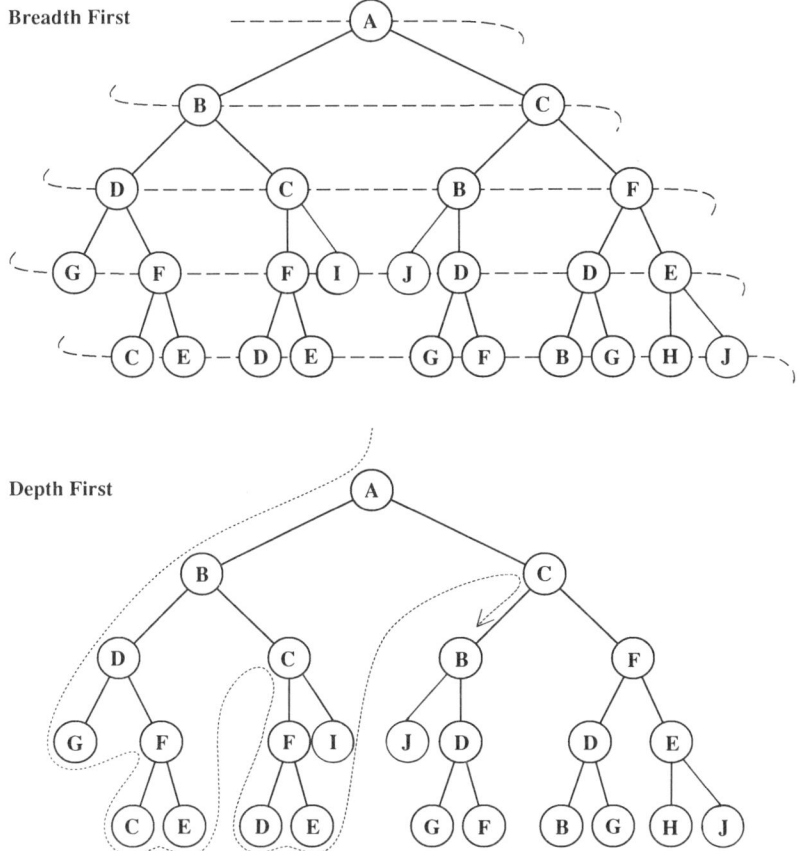

Figure 5.3 Ruleset access strategies

want to buy a computer) for which condition based advice is required (do so if conditions A + B + C are satisfied) then conclusion based backward chaining rule access is used. When the end of the chain of rules in the ruleset is reached advice is provided. Backward chaining systems are sometimes called goal directed systems. If the goal conclusion is not known in advance and requires to be constructed based on a set of conditions (I have a temperature, sore eyes, runny nose and headache, what illness do I have?) then condition based forward chaining rule access is used. Forward chaining systems are sometimes called data driven systems.

In a forward chaining system the condition(s) of the rules in a ruleset are examined to see whether they are true or not, given the information available. If they are true then the conclusions are added as temporary assertion facts to the initial set of facts known to be true (temperature, sore eyes, runny nose and headache), and the system examines the rules again. As the facts are evaluated they are cancelled from the list of outstanding facts. When the end of the chain of rules in the ruleset is reached any outstanding assertion facts (i.e. outstanding conclusions) are displayed as advice. Based on the initial temperature, sore eyes, runny nose and headache facts you have flu.

The basic mechanism used by expert systems in forward chaining is by posing questions based on an initial set of condition(s). If the answer to the condition(s) is yes (assuming the conditions are linked by AND or to a condition if the conditions are linked by OR), the condition(s) is true, and the rule matching all or part of the condition(s) rule is "fired" and the conclusion(s) is drawn. The conclusion(s) is then used to search for any other rules in the ruleset where the condition(s) matches the conclusion(s) of the rule just processed. To ascertain the answer of the condition based questions, the expert system software will first access the knowledgebase to the application domain. If the knowledgebase can answer the question by providing relevant information (for example, yes customer 12345 does have outstanding orders) then the rule is fired and the conclusion is successfully drawn. If the knowledgebase cannot answer the question then the condition(s) is posed as a question to the user. If the user cannot provide an answer a default rule value is used, that is, a rule composed only of a conclusion. Depending on the reply given, the condition(s) is found to be true or false and the rule is fired or not fired. When the end of the chain of the rules in the ruleset is reached the goal/objective of the ruleset is achieved and appropriate conclusion based advice is offered.

The same basic mechanism is used with backward chaining, but the questions posed are conclusion based. The initial question posed is the goal for which advice is required. The inference engine matches the goal to the conclusion of the rules in the ruleset, and then tries to match the conditions of these rules to the knowledgebase or to the conclusions of other rules in the ruleset. At the end of the rule chain condition based advice is provided. Variations of this basic mechanism exist from expert system product to expert system product.

It is possible to switch from forward chaining to backward chaining and vice versa. An example of this will be seen shortly.

An example of rule forward chain firing in a ruleset is illustrated in figure 5.4. The first rule in a ruleset is used to access the knowledgebase to ascertain whether its condition(s) (with forward chaining from an initial set of facts) are true. If the knowledgebase cannot provide the answer the condition(s) are displayed as a question to the user. The first question would be "Is it a sunny day?" If the answer is yes the condition of rule 1 is true, the rule is fired and its conclusion drawn to go sunbathing. This conclusion is then matched against the other rules in the ruleset to see if another rule has its condition matching the conclusion. In the first ruleset the answer is no, so the original conclusion is true and conclusion based advice is displayed to go

The Knowledgebase

By
1 Answers to questions
2 Rule chaining mechanism

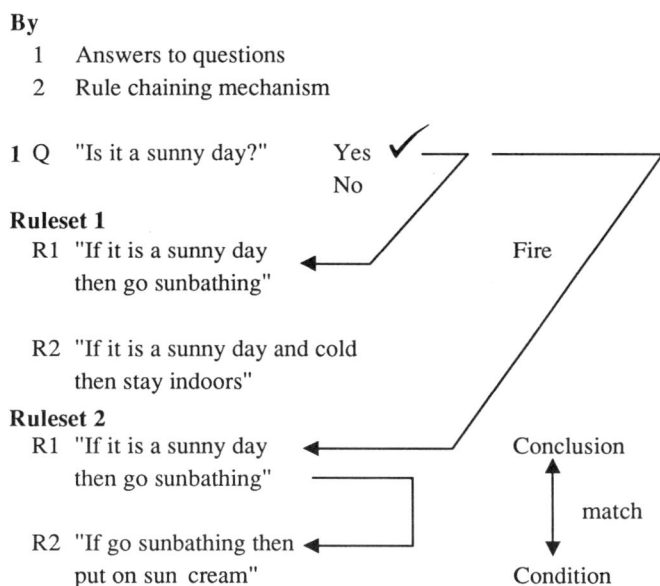

Figure 5.4 Rule firing

sunbathing. In the second ruleset, it can be seen that a second rule matches the conclusion of the first rule. If the condition to the second rule is not found in the knowledgebase a second condition based question would therefore be displayed on the screen, which would be "Do you intend to go sunbathing?" If the answer is yes then the next rule would be accessed. Because this second rule is the end of the ruleset chain, the conclusion advice would be displayed to put on sun cream.

Rules in a ruleset can be standalone and not symbolically chained to the other rules, that is, the logic may not be complete. Where this occurs the inference engine creates a temporary "agenda" list of rules relevant to the task being processed and then tries to match them symbolically by progressively grouping the conditions and conclusions as appropriate and using the semantic net and knowledgebase. Worked examples of this agenda matching are described when considering the GENERIS and GOLDWORKS products.

Rules are stored on disk as variable length records. This means that they can be treated as normal data. *Rule logic is stored and accessed like data.* To store a rule in free form in full on disk would involve a large amount of disk storage. Rules are therefore usually compressed or tokenised, that is, they are given a set of meaningfully coded numbers. Separate tables containing the code descriptions are required to convert the rules to meaningful form when being displayed in question or advice form to the user.

It is important to appreciate that the chaining of rules does not have to be single path, where rule A points to rule B points to rule C. A given rule can point to multiple other rules, each of which can be in different rulesets. A ruleset can therefore provide multiple pieces of advice, one per rule path. Given that many if not all the rules in a ruleset are accessed when the goal/objective is queried it is important that appropriate rules in the ruleset are clustered on disk in order to optimise performance. However, within the ruleset cluster, the order/sequence in which the rules are stored is completely immaterial as far as the execution of logic is concerned. Given that the rules are linked by symbolic values and are standalone statements of some truth, the rules can be "shuffled" in any order and, unlike the procedural logic of traditional application programs with their flow of control, the declarative logic of rules will still execute correctly. The rule access mechanisms function within the rule access strategies.

5.5.5 Rule storage and access

It will clearly improve efficiency on disk if the rules are stored in the sequence they are accessed on disk within a ruleset and clustered on disk as a ruleset. If the rules are only chained together by the matching of symbolic values each access to a rule will require scanning of the ruleset. This could well lead to considerable inefficiency, particularly if the ruleset occupies a large number of disk pages. It is therefore necessary to be able to chain the rules in a ruleset using a direct pointing mechanism. This can be done by using direct address pointers between rules or by indexing the rule condition(s) and or conclusion(s).

In order to gain initial entry access to a ruleset it is important that the "first/last" rule in the ruleset (depending on the ruleset access strategies and mechanisms) is indexed. Backward chaining is initiated from a goal, so that the only single entry point value (Should I buy a computer with a printer?) need be searched for, usually from a single goal based index pointing to the conclusion(s). By contrast, forward chaining can begin from an initial set of facts which may, probably will, vary from consultation to consultation and therefore may or may not be indexed. For example, patients pose a set of ailments to the doctor. The ailments vary from patient to patient and the expert system may not have been designed for all possible ailments. The entry point set of condition facts to the ruleset (in this case the facts are the ailments described earlier by the patient, such as headache and sore eyes) may not be indexed at all, in which case the entire ruleset requires scanning, or only some of the ailments may be indexed, in which case all the rules to the indexed ailments will require to be accessed as a start point to access those ailments not indexed. If the non-indexed ailments cannot be found from the indexed ailments then scanning of the ruleset will be required. Forward chaining systems are therefore more difficult to predict when producing a design. Performance could be adversely affected and would certainly be variable.

Rules can thus be chained either using the symbolic mechanism of conclusion(s) matching condition(s) or linking the conclusion(s) and condition(s) through indexes or direct address pointers. One can see that the access mechanisms to the ruleset are exactly the same as to data with relational file handler mechanisms, with the rulesets being tables of data and the rules in the rulesets being table rows.

The initial start point of the chaining mechanism will be the goal (backward chaining) or initial set of facts for which a goal is to be constructed (forward chaining). If the initial entry point(s) is not indexed then the ruleset requires scanning to find the initial rule. A worked example of backward rule chaining is illustrated in figure 5.5. The example illustrates backward chaining to obtain advice regarding the purchase of a computer with a printer. This goal would be phrased as a question to the user "Do you wish to purchase a computer with a printer?" Assume the answer is yes. If the goal is indexed the index would act as the entry point to the ruleset. As the example uses backward chaining the index would point to the rule with its conclusion matching the goal. If the ruleset is not indexed then, of course, the ruleset requires to be scanned.

Goal Statement

    Purchase a computer with a printer

Rule 1

    If a manager's personal assistant
    Then conducting management meetings

Rule 2

    If conducting management meetings,
    Then writing management memos necessary

Rule 3

    If writing management memos necessary
    Then producing reports is a requirement

Rule 4

    If producing reports is a requirement
    Then purchase a computer with a printer

Figure 5.5 Rule chaining (backwards)

It can be seen that the conclusion of rule 4 matches the goal. The index would therefore point to rule 4. Assuming the knowledgebase does not contain the answer to the condition of rule 3 the condition is then posed as a question to the user "Is producing reports a requirement?" If the answer is yes then the condition is used to search the remaining rules in the ruleset to see if another rule has its conclusion matching the condition of rule 4. Depending on the access mechanism, the ruleset requires either to be scanned again or accessed by indexes or direct address pointers pointing from the condition of rule 4. It can be seen that the conclusion of rule 3 matches the condition of rule 4. Assuming the knowledgebase does not contain the answer the condition of rule 3 is posed to the user "Is writing management memos necessary?" The cycle of rule matching is completed when the end of the chain of rules in the ruleset is obtained. If the answer to all the condition based questions (after the initial goal based question) is yes then condition based advice is offered "Purchase a computer with a printer if producing reports is necessary, if writing management memos is necessary, if conducting management meetings and if a manager's personal assistant". This advice is obtained by running in a forward direction from the end of the backward chain those rules that were used in the backward chain.

The structure of a ruleset using the indexing and direct address pointing mechanisms is illustrated in figure 5.6. The index in this example is used purely for the initial entry point to the ruleset. The index value could be for the goal for backward chaining or set of factual conditions for forward chaining. In this example once the index entry to the ruleset has been obtained the rules are chained using direct address pointers. Indexes could be used just as easily. From the goals/conditions there will be a pointer value pointing to the rule in the ruleset matching the goal value(s) or condition value(s) as appropriate. The index goal pointer will point to a rule, the conclusion(s) of which match the goal value(s). The index condition(s) pointer (there may be multiple indexes each supporting a single condition—thus, in the example there could be three indexes pointing to the conditions D, E, and F, point to a rule, the condition(s) of which matches part or all of the initial set of condition facts. The pointing procedure then follows the normal mechanism of pattern matching the conclusion(s) of one rule to the condition(s) of another rule or vice versa until the end of the chain of rules in the ruleset is reached.

5.5.6 Rule I/O

Give that rules are stored as variable length records, rule access obviously involves I/O. As with normal data records, logical I/O occurs when the rules are accessed in the processor main memory buffer pool and physical I/O when it is necessary, because the rules are not in the buffer pool, to access the

The Knowledgebase

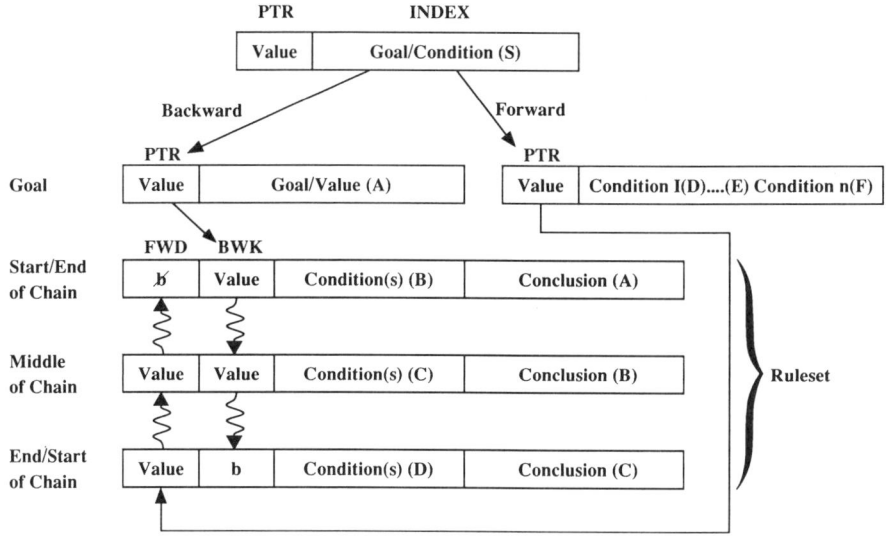

Backward Chaining from Goal Value to Rule Conclusion
Forward Chaining from Condition Value(s) to Rule Condition(s)

Figure 5.6 Rule chaining and indexing

disk. The amount of I/O is easy to calculate, as it is exactly the same as for calculating normal access to user data stored in table rows in a database.

Assume that a ruleset has ten rules and each rule takes half a kilobyte of data. The ruleset is also not indexed, so that scanned access is required. Assume further that the page size is one kilobyte, so that on average two rules can be stored per page. The ruleset therefore occupies five pages. The initial scanned access to pull the ruleset from disk into the buffer pool requires five pages, hence five physical I/Os. Subsequent scans are only against half the ruleset on average but, with the ruleset in the buffer pool, the I/Os are logical. Each rule in the ruleset requires a scan. Since a scan involves accessing half the ruleset and hence 2.5 logical I/Os, the total scanning overhead for the ten rules is 25 logical I/Os.

To access a ruleset using an indexing strategy, with the index pointing from the goal to achieve the start point for backward chaining or set of initial condition facts for forward chaining (and assuming that there are indexes to each fact in the initial set of facts), would generate on average one physical I/O to read the index to main memory (assume the index upper layers are already in the buffer pool), some five physical I/Os to read the ruleset in physical order into main memory and subsequently two logical I/Os (one to access the index, one to access the rule pointed to) for each access to the ten rules.

In this worked example the scanning approach requires five physical I/Os and 25 logical I/Os and the indexing strategy requires six physical I/Os and 20 logical I/Os. This ratio between the two access mechanisms will obviously vary depending on the access strategy and access mechanisms in each ruleset and the access requirements of the initial query. A basic rule-of-thumb is that the greater the number of rules per page and the narrower the ratio of one rule to another (usually one to one), the greater the advantage of ruleset scanning—the reverse is true for pointing/indexing.

5.6 Semantic Nets

The structuring of data in such a way that it can support knowledge is discussed under Object Orientation in section 6.6.1.4.

5.6.1 Property inheritance

This facility is an essential component of object orientation and is discussed in section 6.6.1.3.

5.6.2 Conflict resolution

It is because of the inheritance facility that situations will arise where there is a conflict in the value of the information in the knowledgebase. There can be conflict in the data. There could be a fact that says "a" and another fact in the same class hierarchy that says "b". Different experts have different opinions about something. One expert says the world is round, another says with equal conviction it is square. Inevitably there will exceptions/conflicts to general laws.

The mechanism for solving any conflict between data is the particular overriding the general. The more particular information is always to be found in the entitles lower in the semantic net. The lowest and most particular point in the semantic net is the instances of the class entities.

The conflict can be between the instance and the class of an object and between classes of objects. Figure 6.7 shows both types of conflict for data, and the use of the particular versus the general to provide resolution. As regards class-to-class conflict the sub-class object is more particular than its super-type object in a class hierarchy. In figure 6.7 the class entity SEO has a salary of between £12 000 and £18 000. This is more particular than the salary level of £5000 to £35 000 of the Administrator's salary. The SEO class salary level therefore overrides that of Administrators. By contrast, Middle

Management will inherit the pay property of Administrators and thus have a salary of between £5000 and £35 000, because they do not have salary levels for their class. As regards the instance of the entity being more particular than the class of the entity assume that Smith has a salary of £22 000. Smith the instance is more particular than SEO the class and Smith gets his/her salary and not that of an SEO.

5.7 RETRACTION

One of the problems that can occur in forward chaining, where the advice which is being sought requires to be partially constructed from an initial set of facts, is that an initial assumption(s) made as part of the consultation process is based on the initial facts. The initial facts could be misleading, such that the assumption could prove to be wrong. Any actions taken on the initial assumption(s) require to "undone". A patient complains to a doctor that he/she has a headache, runny nose and sore eyes. In the processing of these initial facts one by one it is initially assumed that the patient has flu. Other facts ascertained from the knowledgebase as part of normal inference processing identify that the patient lives in the countryside, has bee-keeping as a hobby and works part-time driving a tractor at harvest time and generally gets these symptoms from spring to autumn, facts that disprove the initial assumption of flu, to be replaced by the conclusion of hay fever. The flu assumption has to be cancelled and any action taken on the assumption (a bulk anti-flu vaccine order has been raised, the warehouse containing the vaccine is now, having supplied the order, below the minimum stock level, and £100,000 of new vaccine has been authorised to be purchased...!) must be retracted/back-tracked. The retraction and back-tracking is the same problem as event recognition in realtime systems. The backtracking mechanism is described in section 4.2.3.4.

5.8 UNCERTAINTY

The inference mechanism must be able to support the processing of logic and data which may be uncertain.

5.8.1 Uncertain logic

Traditional data processing software does not include facilities for handling uncertainty. Data is definitive or null. Expert systems include many ways of supporting uncertainty in knowledge logic, such as classical probability, Bayesian probability, certainty factors and fuzzy logic. All of these use the assignment of a numeric value to a given proposition to indicate its likelihood. The numeric value is a certainty factor (CF). A proposition could be a condition or conclusion of a rule.

Certainty factors lie between 1 and −1 and represent the measure of belief in a given proposition (a positive CF between 0 and 1) and the measure of disbelief in a proposition (a negative CF between 0 and −1). For example, a measure of belief or disbelief that it will rain.

In the case of classical probability the propositions are independent and the numeric value will always lie between 0 and 1, 0 indicating impossibility and 1 indicating certainty. The two independent propositions could be the probabilities of it raining or buying a house.

Bayesian probability is an extension to allow the treatment of conditions to the probabilities, the example of the probability rain/sky is red being the probability that it will rain given the condition that the sky is red. The two propositions, sky is red and rain, are not independent of each other.

Certainty factors are designed to cope with accumulation of evidence about a single proposition. For example, two rules happen to draw the same conclusion proposition but with different certainties. Multiple certainty factors for separate occurrences of the same proposition are combined in a way which allows evidence to accumulate gradually on both sides, i.e. rule 1 to rule 2 or rule 2 to rule 1.

Fuzzy logic is used to indicate not so much a probability of a proposition being true or not true but a measure of agreement with an indefinite proposition. An indefinite proposition could be "The car is expensive". This is a relative statement of belief. To a rich person a car of a given price may appear cheap; to a poor person a car of the same price may appear expensive.

5.8.1.1 Classical probability

There are many ways by which classical probability can be managed. The propositions are independent as they have no causal relationship. For example, if the independent propositions A and B relate to two throws of the dice, A is the event "A six is thrown on the first throw" and B is the event "An even number is thrown on the second throw", then

$p(A) = 1/6$
$p(B) = 3/6$
$p(A \& B) = 1/12$

Three common ways for the treatment of independent probabilities are:

- Condition(s) to a rule may be less than certain. Uncertain conditions lead to uncertain conclusions. For example, a rule could be "If A (CF 0.5) then B". This will lead to a conclusion that if A is true then B is true with a measure of belief of 0.5.

Uncertainty

- Conditions to a rule are evaluated differently, depending on the number and type of clauses and logical connectives they use:
  — If uncertain conditions are connected by ORs, take the maximum CF value. For example, a rule could be "If A(CF 0.2) or B(CF 0.8) then C". This will lead to a conclusion that if A or B are true then C is concluded with a measure of belief of 0.8.
  — If uncertain conditions are connected by ANDs, take the minimum CF values. For example, a rule could be "If A(CF 0.2) and B(CF 0.8) then C". This will lead to a conclusion that if A and B are true then C is concluded with a measure of belief of 0.2.
- Where there are uncertain conditions a "threshold of acceptance" should be specified. If the CF factor of a condition is below the threshold then the condition is not accepted as being true. In MYCIN, an early expert system product, rules succeed if the condition(s) certainty value has a threshold value greater than 0.2. The rule succeeds if:
  — The condition is a single clause and has a confidence factor (CF) greater than 0.2; or
  — The condition contains clauses connected by ORs and the maximum CF value for each of the clauses is greater than 0.2; or
  — The condition contains clauses connected by ANDs and the minimum CF values for each of the clauses is greater than 0.2.
- Conclusions of a rule themselves may carry separate certainty factors. Any uncertainty of a rule conclusion is propagated to other rules in the ruleset by multiplying their combined uncertainties together.

    An example of rules with uncertain conclusions propagating uncertainty to each other is illustrated in figure 5.7. There are four rules in the ruleset, which if all are fired will produce a combined conclusion certainty of 0.9 0.8 0.9 0.9 and thereby provide advice to purchase a computer with a printer with an overall certainty factor of 0.57. The system is not massively confident in its own advice!
- If both the condition(s) and conclusion of a rule are uncertain then they are multiplied together to produce a rule certainty. For example, "If X(CF 0.5) then Y(CF 0.5)". The conclusion result would be a CF of 0.25.

5.8.1.2 Certainty factors

Facts may be concluded by more than one rule. A combining function blends the certainty factors.

When the inference engine is processing the knowledgebase, a set of common facts/assertions may be concluded by different rules which by themselves each contain degrees of certainty. For example, the value for Joe Bloggs' job type may be concluded in one rule as "consultant" with a

---

Goal Statement

Purchase a computer with a printer

Rule 1

If a manager's personal assistant
Then conducting management meetings
Confidence 0.90

Rule 2

If conducting management meetings,
Then writing management memos necessary
Confidence 0.80

Rule 3

If writing management memos necessary
Then producing reports is a requirement
Confidence 0.80

Rule 4

If producing reports is a requirement
Then purchase a computer with a printer
Confidence 0.90

Figure 5.7 Rules with uncertainty

certainty, a measure of belief, of 0.6 and later concluded again with another rule as "consultant" with a certainty of 0.4. The mechanism to handle this is to combine these less than definite certainties. The revised certainty is then added to another common fact/assertion with a certainty factor, each addition increasing the measure of belief. The algorithm to calculate the addition of multiple common certainties is:

$$mb(A + B) = (mb(A) + mb(B)) - ((mb(A) \times mb(B))$$
$$= (0.4 + 0.6) - 0.24$$
$$= 0.76$$

A and B are the 0.6 and 0.4 values of the common proposition of Joe Bloggs being concluded as a consultant.

The order in which the information is combined does not matter. Combining 0.6 with 0.4 is the same as combining 0.4 with 0.6. As this process of adding revised certainties to other assertions/facts continues, the confidence in the conclusions rises. If a definite conclusion occurs then the combined certainty becomes definite. Indefinite information on the other hand will never accumulate to yield a definite conclusion.

5.8.1.3 Fuzzy logic

Fuzzy logic is an example of multi-valued logic: instead of logical propositions being either true or false and independent they have a numeric value, again lying between 0 and 1, and a causal relationship. The difference between fuzzy logic and classical probability lies in the interpretation placed on the numbers and on the way they are combined.

It is usual to combine fuzzy propositions in terms of their values, not their probabilities:

$v(A + B) = \min(v(A), v(B))$ as shown in the shaded area of figure 5.8.

This differs from the treatment of ordinary independent probabilities. Here probabilities are used:

$p(A + B) = p(A), p(B)$

For an example of fuzzy logic, assume two distinct but related propositions—a car is very expensive (proposition A) and a car costs about £30 000 (proposition B). Figure 5.8 shows a probability curve of A that says that if a car costs about £30 000 then 50% of people regard it as expensive but if it costs more than £40 000 then everybody believes it to be expensive. With a price of about £13 000 then nobody believes the car to be expensive. The proposition of B states that if the car costs £20 000 then nobody thinks it costs about £30 000 (nobody is fooled) but that a car can go up to about

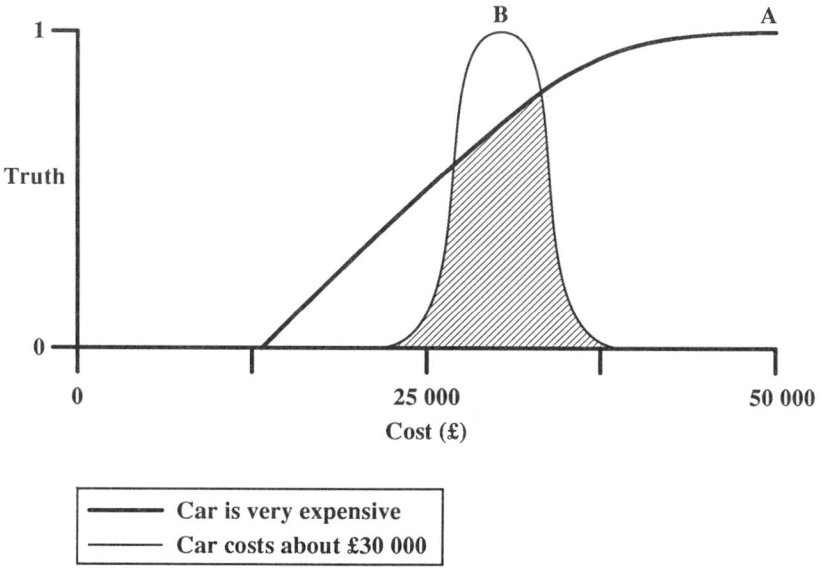

Figure 5.8 Fuzzy logic

£40 000 in the price before nobody can be persuaded that the car is about £30 000. In between these price ranges a varying percentage of people agree the car can be about £30 000. The shaded area in figure 5.8 is the truth of the combined propositions. Thus if a value is produced within the shaded area any dependent rule conditions would be regarded as satisfied.

Note that where the two propositions are being ANDed the minimum result of the propositions A and B is taken, hence the shaded area below the line of the two A and B curves. If the two propositions are ORed then a maximum result is produced.

5.8.1.4 Bayesian probability

Bayesian probability adds conditions to the independent propositions, as represented in two dice. Consider figure 5.9. Dice are being thrown but, unlike the earlier example of two independent throws of the dice for classical probability, in the context of a condition. The condition is the value of 10. The combination of two figures of 6 + 4, 4 + 6 and 5 + 5 combine to produce 10. These three are equally probable. The probability of throwing 6 on the first throw given the value of 10 is 1/3.

$p(\text{6 on first throw} \mid \text{total is 10}) = 1/3$

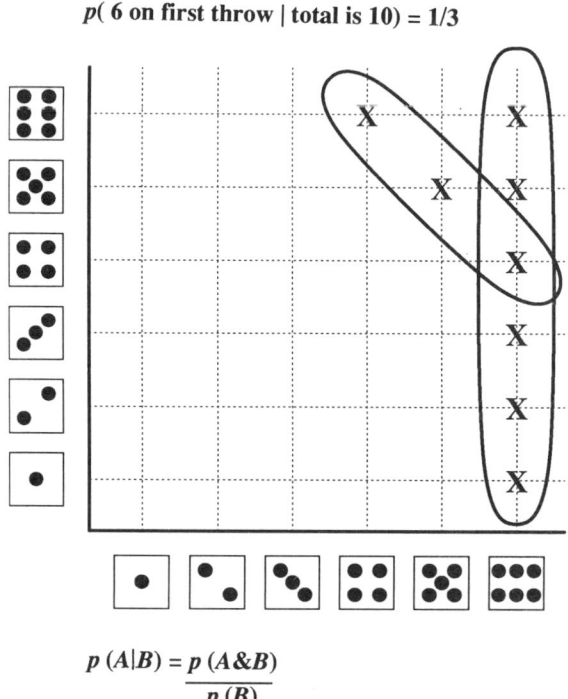

$$p(A|B) = \frac{p(A\&B)}{p(B)}$$

Figure 5.9 Bayesian probability

5.8.2 Uncertain data

It must also be appreciated that data stored in the database component of a knowledgebase can also contain a certainty factor. For example, it is perfectly legitimate for the expert user to insert data into the database with a degree of certainty less than definite. For example, the user could say that, to the best of his/her knowledge, a person has a salary of £10 000 with a certainty factor of 0.8.

5.9 Dictionary and Database Synchronisation

The schema description of the expert system semantic net in the dictionary is more closely integrated with the database data than with normal relational and earlier pointer chain technology type database management systems. Given property inheritance from the database instance tables/concrete objects to the dictionary class table definitions, when a change is made to a database table all "inherited" tables in the semantic net class hierarchy to the table being changed need to be correspondingly "X" locked for obvious synchronisation reasons. The schema description in the dictionary is therefore treated no differently from user data stored in the database. It is necessary to synchronise locking of the dictionary with the database.

5.10 Worked Examples

A number of worked examples show how expert systems can be used as intelligent query languages and advice providers. The examples are based on GENERIS from Instrumatic Ltd[1] and GOLDWORKS from Goldhill Computers Inc. The examples illustrate a variety of features, such as how the components of a knowledgebase are linked together, the criteria on which the access strategy to the rules and the rulesets is based, forward chaining versus backward chaining and the relationship of one ruleset to another ruleset.

5.10.1 New query languages (with GENERIS)

New topic based information access query languages are now available that provide genuinely non-procedural access to information of whatever type. The procedural table languages, such as SQL, are technically obsolete. Not only that, but the languages include "intelligence", such that they can:

---

[1] GENERIS was initially developed by Deductive Systems Ltd, which has been taken over by Instrumatic Ltd.

- provide advice in response to the query: an example of this was given in section 5.2;
- infer/create yet further information in response to the query.

A worked example of how the GENERIS inference mechanism works, how the individual components of the knowledgebase interrelate and how it is able to infer information now follows.

The GENERIS inference mechanism is based on four components—the database component containing the dumb data, rules, property inheritance through the semantic net facility and local computation. The first three components work in conjunction with each other. The knowledgebase is illustrated in figure 5.10. Note the use of the age property in the database to contain logic, in this case procedural rather than declarative. This is an example of an expert system product taking advantage of technology developed to support the normalisation of logic in object oriented systems. The two rules are inference rules and, in this case, are standalone. They are not chained together, in contrast to the rules in the example in figure 5.5, nor grouped as a ruleset. It will therefore be necessary to build tables of temporary results and hold these in the working memory component of expert system software. The query is that the user wishes to "Display the colour and age of elephants". This is very definitely a good statement.

In this example GENERIS is being used as an intelligent query language gleaning as much information from the knowledgebase as possible rather than an expert system offering advice. Notwithstanding this prime function GENERIS is quite capable of being used as an expert system.

The inference mechanism begins by parsing the keywords in the query, in this case colour, age and elephants. The knowledgebase is then accessed to find the relevant relationships between the keywords. Colour is in the database as a property of the animal instance table and in the semantic net as a class of colour. The conclusion of both rules 1 and 2 (GENERIS puts the condition after the conclusion) are also both related to colour. There is thus a three-way relationship between the three components of the knowledgebase, in this case via colour. Animals also have occupations as defined in the database and the condition of rule 2 has a clause concerned with occupation of circus performer.

The appropriate rules 1 and 2 can now be fired through their conclusion association with colour (backward chaining is being initiated). The sequence is that rule 1 accesses the database to find all animals that live in the jungle and ascertains that it is Jumbo. The result is stored in a temporary working memory table of inferred information—see figure 5.11. Given that the condition of rule 1 is true (that there is an elephant living in the jungle) the conclusion of the rule is true, such that Jumbo must therefore be muddy. Rule 2 accesses the database on the value of circus performer and ascertains that there are two elephants that satisfy that condition and also puts the two

Query "Display the colour and age of elephants."

Rule 1 - Animal has colour muddy if animal lives in jungle

Rule 2 - Animal has colour pink if animal occupation circus

Animal	Birthday	Age (computation)	Occupation	Colour	Location
Nellie	5/8/79	...logic...	Circus Performer		
Fred			Circus Performer	Black	
Elephant				Grey	
Jumbo				Dull	Jungle

--------------------------------Semantic Nets----------------------------

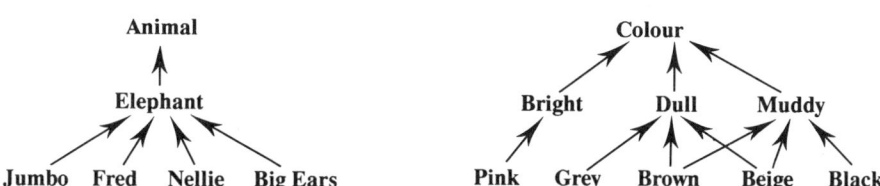

Figure 5.10 The GENERIS knowledgebase

Temporary Table		Result		
Elephant	Colour	Elephant	Colour	Age
Fred	Pink Black	Fred	Conflict (pink/black)	Unknown
Nellie	Pink	Nellie	Pink	9
Jumbo	Dull Muddy	Jumbo	Brown Beige	Unknown
Big Ears	Grey	Big Ears	Grey	Unknown

Note: the speckled shaded area is the only information that would be returned with a relational file handler

Figure 5.11 The temporary and final results

elephants, Nellie and Fred, into the temporary working memory table and infers that Nellie and Fred, based on the conclusion of rule 2, are pink. The database information regarding colour is that Jumbo is dull and Fred is black. Combining the inferred and database information produces a conflict on Fred—the inferred information says he is pink and the database information says he is black. The conflict cannot be resolved (both sets of information are at the instance level and there is no class hierarchy with the sub-classes overriding the super-classes) and the users are presented with the conflicting colours for them to resolve. Nellie is pink and Jumbo is muddy (from rule 1) and dull (from the database).

To this add the semantic net information. Muddy and dull intersect to state that Jumbo is brown or beige. Using inheritance in the semantic net Big Ears is an elephant and it is known from the database that elephants are grey. Big Ears is therefore grey. Information regarding colour is now complete.

Computation must now be executed to ascertain if any more information can be gathered/computed. The age of the elephants is more difficult to ascertain. The only elephant for which age can be computed is Nellie. For all the other elephants the age is unknown. The output response to the query is illustrated in figure 5.11.

This simple example of using a knowledgebase of rules and semantic net facts as well as data illustrates a number of facilities that relational technology cannot support. It shows that when data is entered into the database by a "standard" user the information may actually be wrong when set against the knowledge of an expert as defined in the knowledgebase. For example, the user says that the elephant Fred is black and a circus performer, the expert that circus performing elephants are pink. An expert system could automatically correct any database errors on the basis that the expert knows better than the ordinary user. GENERIS happens to leave this example of conflict open.

A relational file handler would have shown little of this information. It would have shown only the colour of Fred as black but without showing the conflict, the colour of Jumbo as dull where actually he is brown and beige and the age of Nellie as a birthday rather than an age. In fact the only accurate information that would have been displayed is the name of the elephants with poor Big Ears totally ignored. All the other information—Jumbo being brown and beige, Fred's colour pink and 9 years of age and Big Ears being grey—would not have been presented with relational technology.

5.10.2 GOLDWORKS

GOLDWORKS incorporates facilities for predefining the access algorithms to portions of a knowledgebase and therefore goes some considerable way to providing multiple access mechanisms to the knowledgebase. The structure of the GOLDWORKS inference engine is illustrated in figure 5.12. It is one

Worked Examples

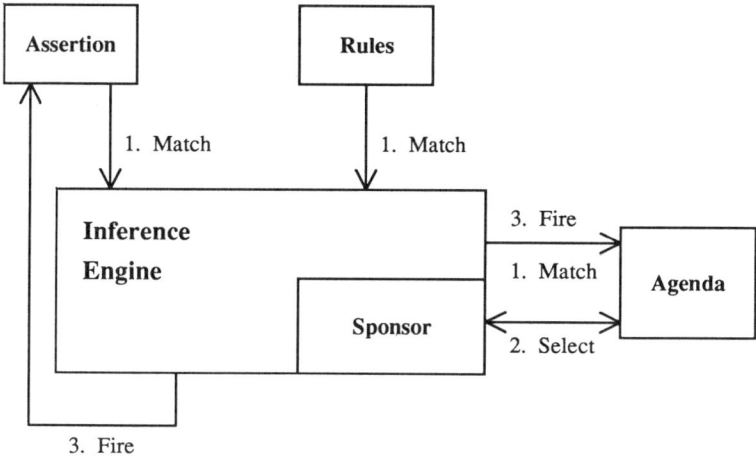

Sponsor — Controls rule selection mechanism in an agenda and groups rules in an application domain

Assertions — Dynamic statements of truth (may have uncertainty)

Rules — Cause and effect logic - antecedents(s) ⟷ consequent(s)

Agenda — Dynamic stack of rules waiting to be fired for query

Figure 5.12 GOLDWORKS inference engine

of the most sophisticated available. The sponsor controls the rule access mechanism in the agenda and controls when rules assigned to the sponsor may or may not fire. Sponsors are used primarily to control the mechanism for the firing of forward chaining rules during the consultation process.

Assertions are dynamic statements of facts that are ascertained to be true during the life of the consultation. The assertions could be initially defined as a set of condition facts when requiring to start a forward chaining consultation. For example, in an illness diagnosis system the assertions could be that the patient has an earache and temperature. Additional assertions may be obtained from the rules and semantic net in the knowledgebase by the inference engine during the consultation process. The consultation will be either to the knowledgebase or, if the assertions cannot be obtained there, with the user. The assertions are held in the working memory component of the expert system software. An agenda is a prioritised queue of rules that are matched on outstanding assertions and waiting to be processed. The rules, of course, are the cause and effect logic statements. They must be allocated to a sponsor and may be allocated to a ruleset. Each sponsor and ruleset is relevant to an application domain task for which advice is sought. The sponsor and ruleset facilities are independent of each other, but can be related explicitly if required.

The sponsor is a mechanism that enables rules to be grouped together into sponsors, with the sponsors related to each other as a hierarchical structure, and executed within the structure in the sequence of top to bottom, left to right, that is, depth first.

When you define a rule you assign it to a sponsor. Thus you can group rules under a different sponsor according to their purpose and function, i.e. their task in the application domain. This is illustrated in figure 5.21, where an overall sponsor structure is split into six task groupings. The application domain is the servicing of a car, within which there are certain tasks requiring advice relating to electrical, mechanical and other servicing. Within the electrical task there are further sub-tasks relating to servicing the car lighting and heating. Each task and sub-task is supported by a sponsor group.

Sponsors enable the access strategy to the rules within their scope to be accessed in the optimum manner, either breadth first or depth first. If one assumes that the domain expert knows that when a car is serviced the electrical tasks are undertaken before the mechanical tasks, which are undertaken before the other tasks, then it will be best to search the sponsors in that sequence, i.e. depth first. The appropriate access strategy for each sponsor can be predefined in advance, based on an understanding of how the application works. Further refinement would be desirable but is not currently supported. For example, within a sponsor grouping, assume that the most frequent sequencing of rule access to obtain advice in the heating ruleset is between rules 5 to 6 to 15 to 16 to 17, therefore choose breadth first searching, whereas in the lighting ruleset the rule access sequence is usually 7 to 18 to 8 to 19, therefore choose depth first searching. Within the electrical task the rules are accessed in a highly variable manner, so therefore choose the default. It would optimise performance if the rule access strategy could be defined per sponsor.

The agenda is the dynamic list of rules waiting to be fired during the consultation. It is a prioritised queue, onto which agenda items are placed according to their priority, which is determined by the priority of the agenda item rule and the breadth or depth first ordering of the agenda—see figure 5.14. An agenda item is a rule waiting to be fired. An item is actually a pointer to the rule. The ordering of the agenda items that share the same priority is determined by the type of agenda specified when the sponsor itself is defined. The system places agenda items among other agenda items that share the same priority using either depth first or breadth first ordering. The default ordering is breadth first. In breadth first ordering when agenda items are added to the agenda they are placed behind other agenda items

Example (GOLDWORKS) 1.

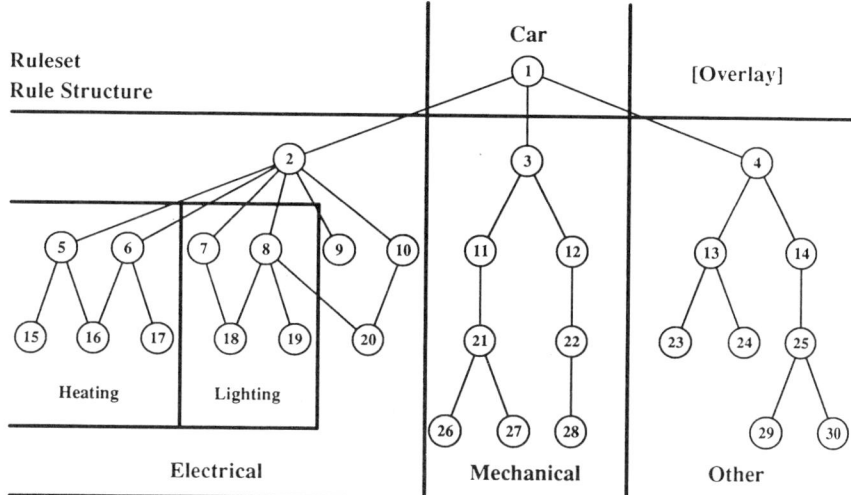

Figure 5.13 Sponsor rule allocation

sharing the same priority. In depth first ordering the agenda items are placed in front of other agenda items sharing the same priority.

In contrast to the sponsors, rulesets are standalone groupings of rules, with no hierarchical structure and no control mechanisms for their sequences' execution. Each ruleset is mutually exclusive. The execution of a ruleset has to be invoked explicitly, typically from an application program. Ruleset firing is therefore intrinsically *ad hoc*. A ruleset and the individual rules within a ruleset can be invoked from a ruleset—their relationship, if required, is many-to-many. Thus the sequenced processing of the sponsors and the *ad hoc* processing of the rulesets can be combined to mutual benefit. It could be, for example, that there is a ruleset for providing advice on when and how to rebore a car engine. Under certain conditions the mechanical sponsor in figure 5.13 can invoke the rebore ruleset. The sequenced invocation by the sponsor does not compromise the standalone nature of reboring a car engine, which in the normal course of events is not part of car servicing.

Sponsors offer sequential partitioning of the rulebase whereas rulesets offer mutually exclusive partitioning of the rulebase.

The forward chaining mechanism used by GOLDWORKS is that the initial assertions are used to match against the rules which could be fired from the list of agenda items. The rules to be fired are selected from the list of agenda items in the sequence entered onto the agenda. The sequence is set by the rule priority and agenda search specification. Once a rule has been fired from the agenda items, the result of the conclusion is treated as an

Rule Structure within a Sponsor

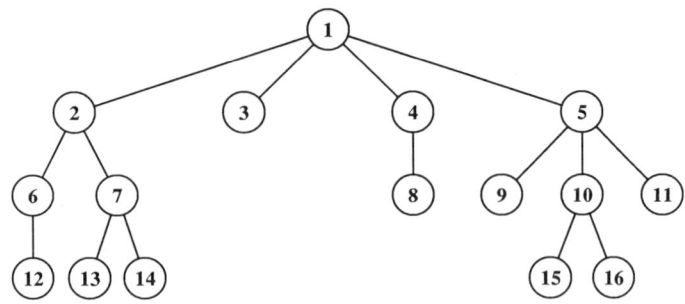

Agenda Sequence (Breadth First : FIFO)

After 1	After 2	After 3	After 4	
2 →	3 →	4 →	5 →	
3	4	5	6	Etc.
4	5	6	7	
5	6	7	8	
6	7			

Agenda Sequence (Depth First : LIFO)

After 1	After 7	After 3	Etc.
2	7	4 →	
6	13	8	
12	14	5	
7	3	9	
⋮	4	10	
	⋮	15	
		⋮	

Figure 5.14 Rule firing example (GOLDWORKS)

assertion and added to the list of assertions being dynamically constructed, those assertions which were used to fire the rules are marked as processed and the agenda item pointing to the rule which has just been fired is cancelled. As the rules in the agenda items are fired so the assertions are constructed from the conclusions of the rules and added to the knowledgebase. These assertions can, in turn, cause more agenda items to be created, continuing the forward chain. The assertions are then used to match against the rules and, if a match is found, the rule is fired. The process is repeated for all the agenda items. When all the agenda items are processed the conclusions of the final unprocessed assertions are presented as advice.

Worked Examples 285

Each agenda is controlled by a "sponsor" and each rule is linked to a sponsor and, optionally, to a ruleset. The agenda items that point to the rules can be searched in a breadth or depth first manner, as specified in the sponsor. If a conflict exists with multiple rules waiting to be fired the rule chosen is that with the highest priority. Sponsors exist as a hierarchy controlled by a "top sponsor". There is one top sponsor for the overall application. The top sponsor fires its agenda items first and then, if there are further sponsors in the hierarchy, to the next sponsor in the sequence of depth first, left to right order. When all the sponsors have processed their agenda items, control returns to the top sponsor.

A worked example of forward chaining using GOLDWORKS illustrates the inference mechanism processing rules, assertions and agendas. Advice on the choice of wine for a meal is being sought. The rules are defined in figure 5.15 and the sequence of rule firing to ultimate advice provision is illustrated in figure 5.16. Note that the rules cannot be symbolically matched via their conclusions and conditions as they stand. The inference engine software therefore requires to create a temporary table of rules waiting to be processed (the agenda items) and progressively build up the symbolic relationships between the rules during the consultation. The simplistic rule structure in figure 5.2 of a rule nicely pointing to a rule(s) may be broken.

Rules

1. If Meal is Poultry then Wine should be White
2. If Wine should be White and use Light Body Wine then select a Chablis
3. If Meal has a Delicate Taste then use a Light Body Wine

Figure 5.15 GOLDWORKS example rules

An initial set of assertions/facts are made, possibly the user stating that he/she is about to eat a meal of poultry and the meal has a delicate taste. At point-in-time 1 these initial assertions match the conditions of rules 1 and 3 and references to rules 1 and 3 are therefore put into working memory as agenda items. At point-in-time 2 the top rule in the agenda items, i.e. rule 1, is fired and its conclusion that the wine should be white is added as an assertion. Rule 1 is cancelled from the agenda item list. The first two assertions have already been processed in order to put rules 1 and 2 onto the agenda item list. The conclusion of rule 1 does not contain sufficient information to point to the condition(s) of another rule, so that no further additions are made to the agenda item list. At-point-in time 3 the top item in the agenda item list is ascertained to be rule 3, rule 3 is fired, its conclusion to use a light-body wine is added to the assertion list and rule 3 is cancelled from the agenda item list. The two unprocessed assertions—wine should be white and use a light-body

P1 Initial Assertion/Facts
- Meal is Poultry
- Meal has a Delicate Taste

Conditions of Rules 1&3 match assertions - therefore put into working memory as agenda items.

Agenda: A Item #1: Rule 1
         B Item #2: Rule 3

P2 Rule 1 is fired and conclusion added to assertion and cancelled from Agenda Item List

Assertions:
- Meal is Poultry   ⎫
- Meal has Delicate Taste ⎬ Processed
- Wine should be White

Agenda: B Item #2: Rule 3

P3 Rule 3 is fired and conclusion added to assertion and cancelled from Agenda Item List

Assertions:
- Meal is Poultry   ⎫
- Meal has Delicate Taste ⎬ Processed
- Wine should be White   ⎫
- Use a Light Body Wine   ⎬ Match Rule 2

Last two assertions match Rule 2 conditions so Rule 2 conclusion added to agenda

Agenda: C Item #3: Rule 2

P4 Rule 3 is fired and conclusion added to assertions and cancelled from Agenda Item List

Assertions:
- Meal is Poultry   ⎫
- Meal has Delicate Taste ⎬
- Wine should be White   ⎬ Processed
- Use a Light Body Wine   ⎭
- Select a Chablis

Agenda: Empty

Agenda is empty so unprocessed assertion(s) offered as advice

Figure 5.16 GOLDWORKS example forward chaining

wine—are matched against the ruleset. The two conditions of rule 2 are found to be a match and rule 2 is added to the agenda item list. At point-in-time 4 the top item in the agenda item list, rule 2, is fired and its conclusion—select a Chablis—is added to the assertion list, and rule 2 is cancelled from the agenda item list. The agenda item list is now empty and the conclusions of the remaining unprocessed assertion(s) is offered as advice—that is, select a Chablis.

5.11 Enhancements to SSADM

The more the author has delved into expert systems technology, the more it has become apparent that knowledge is stored and accessed like standard data using a standard file handler, typically of a relational type. *Once the facilities of*:

- semantic nets (knowledge data), rules (knowledge logic) and how they are related together by the inference software are understood;

and it is appreciated that:

- knowledge data is merely a limited extension of a database dictionary/repository;
- knowledge logic is stored and accessed like data;
- knowledge data and logic and database data are related as required to infer yet further knowledge;

expert systems technology is well understood. Expert systems technology is low risk. What is true of the technology is also true of the design techniques. This low risk is one of the main reasons why expert system products are well developed, in contrast to distributed database.

As with all new trends of information technology, of which knowledge/expert systems is but one, the underlying technology remains very much the same with new layers of software added on top.[1] This is in line with policy followed by the vendors of these new technologies to preserve as far as possible existing technology. This policy was followed, as we have seen, for distributed and realtime systems and is likewise followed for expert systems. The benefits of this approach are that much of the existing technology and design techniques are preserved. *Thus all of the SSADM techniques are relevant to applications requiring knowledge engineering. Once again one need unlearn nothing. The SSADM techniques merely require enhancement by addition, in this case for the recognition, representation and processing of knowledge.*

There are few structured methods currently available for the design and development of knowledge based expert systems. One of the principal methods available is called STAGES—Structured Techniques for the Analysis and Generation of Expert Systems. STAGES has been developed at Ernst & Young. STAGES also has the advantage that it is designed to be compatible with traditional structured methods, such as SSADM. Another method is KADS—Knowledge Acquisition and Documentation System from Touche Ross. No method has yet achieved significant market penetration.

---

[1] A perfect example of this is illustrated and discussed in section 5.10.

Given the newness of expert system methods, the STAGES techniques will be described in some detail. Suggestions as to how SSADM can support knowledge engineering and what enhancements are required to its techniques are therefore based on the material from STAGES, as well as the author's own understanding of the underlying expert systems technology. The applicability of the STAGES techniques to SSADM is obvious to the reader.

The following areas need to be addressed in order to use SSADM in knowledge based systems:

- Provision of guidelines as to how to select applications requiring knowledge and to identify which areas within the application use knowledge.
- Provision of a set of structural standards regarding the stages, steps and tasks that must be undertaken in an expert system project.
- Provision of a set of procedural standards regarding knowledge engineering techniques, specifically:
  — how to model knowledge in the form of rules and semantic nets, describing when to use which mechanism and how to integrate them and their deliverables;
  — how to decide which knowledge access mechanism to use;
  — how to decide which knowledge access strategy to use;
  — how to design a knowledgebase physically and to calculate and tune its performance;
  — how to record knowledge uncertainty and which mechanism to use;
  — how to interpret the major role of prototyping in the creation of an expert system design;
  — how to validate and verify an expert system prior to release in an operational environment;
  — to provide a set of documentary standards;
  — to describe the project structure appropriate to an expert system application.

The last four points are not addressed in this book.

5.11.1 How to identify expert system applications

There is no hard and fast rule, no fixed mechanism, to say this application is suitable for an expert system solution and that application is not. The unique task of expert systems is to offer advice about an application domain based on the knowledge of that domain. The types of application for which advice is most suitable are those involved with diagnosis, interpretation, planning

Enhancements to SSADM

and design. The first task therefore is to classify the purpose of an application and see if it falls within one of these four categories. If it does then one needs to assess whether the application has certain characteristics, such as:

- the problem is recurrent. This means it is suitable for computerisation;

- the problem requires expertise that is both stable and in short supply but available. There should be a shortage of expertise, otherwise there would be no justification for the expert system; however, expertise must exist and be available, otherwise the system could not be constructed. The diagnosis of rare deseases would be a typical application. Certain types of problem are, by contrast, always or often in a state of flux, such as those resulting from scientific research. It is only when the research findings are accepted as valid that the scientific facts can be converted into knowledge;

- the application must be of a manageable size with clear and precise boundaries. Expert systems tend to generate a lot of rules, which in turn proliferate the number of logic paths to be tested. It can soon become impossible to test rigorously all possible combinations.

Once the identification and justification of an expert system has been accomplished one can decompose the application to ascertain which business areas, functions, events and problems-to-solve contain knowledge/expertise from which advice can be offered. The approach adopted by STAGES is to use modified versions of entity modelling (MLDMs) and dataflow diagramming (MDFDs) to identify where knowledge is required and, once identified, to specify the knowledge logic in knowledgebase maps (KBMs) and knowledge data in MLDMs.

The MDFDs are decomposed in the standard manner, preferably as in section 3.2.1. Each process at each level of decomposition is viewed as to whether:

- it is to be implemented in its entirety as an "E" type expert system process, a "D" type data processing process or a "M" type manual process;

- the datastores are "D" type datastores usable by both traditional data processing systems and expert systems or "E" type datastores usable only by expert systems.

The process boxes and datastores which indicate a need for knowledge imply a set of knowledge distinct from other processes and datastores. Should there be a need for shared knowledge between expert system processes and datastores the common part of the knowledge should be decomposed into a further process and datastore.

As described earlier, the knowledgebase contains dumb user data as well as the other facilities for representing knowledge, such as semantic nets and rules. This dumb data is a crucial component of the knowledgebase, providing many of the answers to rule conditions from which deductive inferences can be made. It is for this reason that expert systems can access "D" type datastores.

MLDMs support the concept of class and property inheritance. They are also a variant of an entity model designed to show that the structure of data can be dynamic, depending on the context of the data. Consider the modified logical data structure in figure 5.17. The modified data structure model indicates that a company can be a bank, a retailer or a manufacturer. That these are exclusive possibilities is indicated by the arc. Each possibility gives rise to a different data structure. This structure is dependent on the content of the application as defined in the knowledgebase. A company is a bank if.... Many would recognise the retailer, manufacturer and bank as no more than entity sub-types, which can be better handled by the sub-class facility as shown in section 6.6.1.1. MLDs do, however, go further in recording the context/business rules that distinguish the entity sub-types as expertise.

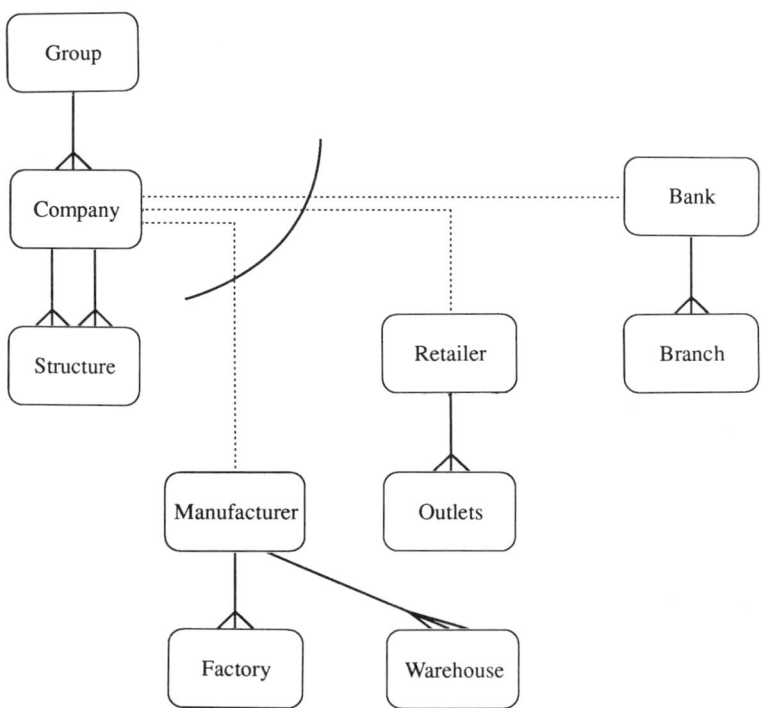

Figure 5.17 Modified data structure diagram

Enhancements to SSADM 291

Knowledge engineering techniques can now be applied to the areas of the application requiring knowledge.

5.11.2 Provide a set of structural standards

This is an aspect on which the author will not comment. The policy followed in this book is that comment will only be made on the basis of practical experience and illustrated with case study support. Details of how STAGES is structured can be obtained from Ernst & Young.

5.11.3 Provide a set of knowledge engineering standards

Knowledge engineering techniques represent knowledge in a form which can be computerised. The techniques need to produce two logical design deliverables that are currently used to represent knowledge, namely semantic nets and rules. In STAGES the knowledge data relates to the "E" and "D" type datastores, and is diagrammatically represented in MLDMs and knowledge logic to the "E" type processes in the MDFDs, and is diagrammatically represented in knowledgebase maps (KBMs).

Both knowledge data and logic are structured as meaningful groupings of triples. Knowledge data is stored as inter-entity and inter-attribute semantic descriptions in a semantic net MLDM and defined as tables of data in the dictionary schema description. Knowledge logic is defined as rules, which are stored as variable length records, similar to data. Knowledge logic rules can thus also be structured like a data model, rules with relationships to other rules, records with relationships to other records. Such a structure is represented in figure 5.21. These structures can be re-expressed as KBMs.

5.11.3.1 Knowledge data

Knowledge data is nothing more than dictionary table data definitions with extra facilities. The extra facilities are inter-entity and inter-attribute semantics, class and instance entities, data property instantiation where the data value is relevant to all instances and property inheritance in the class hierarchy. These facilities convert a normal entity model of dumb data into a semantic net of knowledge facts as described in section 6.6.1.4. The dumb database table rows are also part of the knowledgebase and linked to the dictionary through property inheritance of the defined class hierarchy.

These extra facilities can easily be added onto the normal SSADM data modelling techniques. The policy followed by the author is to:

Additional Data Processing Environments for SSADM—Expert Systems

- Conduct data modelling using the LDM technique without change.

- Identify the common data properties in the entities implicit in the LDM and abstract into classes hierarchies through specialised sub-classes or generalised super-class entities of the existing relationally based entities. The rules for this are discussed in section 6.8.1.2.

- Add inter-entity business semantics (free form), class semantics ("is a") and inter-attribute aggregation semantics ("has a"). The inter-entity business and class semantics can be easily written onto the relationships between the entities. SSADM already records business relationship descriptions between entities, so users of the method will be familiar with the task. The conditions under which inheritance aggregation and business semantics are used is described in section 6.6.1.

- Define the entities. The entities on the logical data model are implicitly defined at a class level. Some of the entities may not point to instances of the class entity because they are abstract. The definition of the class template (this would include such information as the name of the entity, its data properties, their names and format) is stored in the dictionary and the instances of the class entity are stored in the database, although also defined in the dictionary. The entity description form therefore needs two further columns, one to indicate whether the entity is abstract or concrete with instances and the other a property/attribute instantiation column.

If the entity is abstract then the entity abstract indicator needs to be ticked and all the attributes describing the class entity require to contain a value. In the next chapter, such an abstract class entity in the class hierarchy example in figure 6.4 is Man, which contains a generic property of number of legs with a value of two. If the entity is also a concrete entity it will hold both generic class *and* instance properties in its definition. In this case the class/instance indicator needs to be ticked against each individual property as appropriate. If the data property can contain different values for the instances of the entity then it will not contain a value.

The attributes and any class values they contain must be defined in the entity description document.

The upgrading of the logical data model to a semantic net is thus an easy if laborious task.

STAGES goes further to identify areas where expertise is required in the data model. Data models are drawn of the relationships between attributes where expertise about the business context of the attributes is required. An example is given of reference numbers identifying nostro financial deals (banks dealing with banks) and the expertise required for reconciling unmatched deals. "Detective" expertise is applied to whittle away gradually the unmatched deals to identify missing characters, extra characters, wrong characters, transposed characters.... An "attribute" data model showing the relationships between deal reference numbers is drawn. The idea of attribute models reflect the intra-object relationships of a semantic net (see section 6.6.1.4).

- Record uncertainty. If the data property value at the abstract or concrete class level is less than certain record a certainty factor against it.

5.11.3.2 Knowledge logic

This is defined in the form of rules. No SSADM technique as it stands is appropriate for knowledge logic specification as SSADM is familiar only with procedural logic. The first task is to identify which parts of an application require to support knowledge logic. STAGES uses MDFDs as the prime means of this identification. As the processes are decomposed they are progressively identified as requiring to support expertise. Such processes are marked as "E" type processes. Each such process has an implied set of rules, which are recorded in a KBM.

The author suggests making each E type process a ruleset. The use of process models for modelling such declarative logic would not be appropriate. The condition of a rule could refer to one entity and the conclusion to another. How would such logic fit in a process model? The rulesets become tables and the rules table rows. One is thus organising knowledge logic as one organises knowledge data. In data storage and access terms there is no difference.

STAGES makes greater use of the dataflow diagramming technique in MDFDs than SSADM. SSADM does not use dataflow diagrams in such a positive way, but merely as a means of modelling the business structure and grouping the processes into sub-systems and thence into systems. The STAGES differences are:

- The "E" type processes are of two kinds—those that require expertise to "control" the execution of other, possibly lower level, E type processes that contain the "inference" logic more usually associated with the application expertise. Both type of processes require KBMs. The control processes are most easily identified as those with many dataflows to other processes—a somewhat vague definition. Expertise is required to control the execution of the other processes, which themselves may also use expertise. SSADM only recognises the need to specify logic for the lowest level processes. STAGES supports multi-level logic.

- Process dependency is identified and by this means the order in which rules/rulesets execute. The author has considerable sympathy with the STAGES approach. As described in section 3.2.1 and illustrated in figure 3.23, problem-to-solve processes are causally related and, although occurring logically at the same point in time as the event to which they relate, do nevertheless have a sequence—process 4.1.1.2 is sequenced after/is dependent on process 4.1.1.1. In centralised processing this can be used to identify the sequence of program module calling.

SSADM uses Entity Life Histories to show sequence dependency between event level processes but not process sequence dependency within an event. STAGES is therefore at variance with SSADM by incorporating the concept of time in dataflow diagramming.

This use of process dependency is very useful when designing the layout of rulesets and ascertaining the ruleset access strategy. Consider figure 5.13. It could be that the MDFDs identified "E" type processes for heating, lighting, electrical, mechanical and other tasks when undertaking car servicing and, via process dependency, that heating is undertaken before lighting, which is undertaken before electrical... and that the rules in heating are undertaken in the sequence of 5, 6, 15, 16 and 17. The rule access strategy should therefore be breadth first.

There are some problems with the STAGES' use of MDFDs. As described in section 2.4.2.3 DFDs are not suited to data retrieval business requirements, which are intrinsically standalone and independent. Yet data retrieval business requirements are just as susceptible to requiring expertise as data maintenance requirements. The KBM in figure 5.19 is, for example, for a data retrieval business requirement. A clear statement in STAGES that data retrieval business requirements are identified separately from MDFDs.

The other problem is the use of the E type datastores. STAGES does not identify what data is stored in E type datastores, how the data differs from that in the D type datastores and how the E and D type data is identified and separated. One assumes that knowledge data is included in the E type datastore. Knowledge data is represented as a pair of related semantic descriptions in the semantic net. However, from a knowledge data modelling point of view a semantic net is nothing more than a standard logical data model with class entities abstracted and semantic descriptions between entities and constituent data properties attributes, as described in section 6.6.1.4, to form triples and thereby convert the dumb data in a database into knowledge facts in a knowledgebase. *Knowledge data is the semantic description of the relationship between entities and the data attributes of the entities.*

The inter-entity semantics can be recorded on the ERD logical data model and the inter-attribute semantics in the standard entity description form.

STAGES also uses the MLDs to identify where expertise is required. This is discussed in describing knowledge data. Once the requirement for expertise is identified a KBM is drawn for each E type process.

The KBM is the STAGES technique for modelling knowledge logic. It is an

Enhancements to SSADM

impressive technique. Given that knowledge logic is defined in the form of rules and that rules are stored as variable length records it is not surprising that a KBM is a cross between a data structure diagram and a program flow chart. A most elegant solution. The KBMs allow a full representation of the rules in a ruleset and their interactions, and represent all the variables, constants, operators, relations, facts, questions, defaults and actions in the system as illustrated in figure 5.18.

Symbol	Meaning	Comment
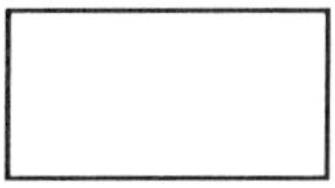 ▭	Variable	Typed identifier
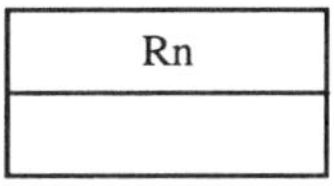 Rn	Rule and principal conclusions	Conditions below Other conditions above
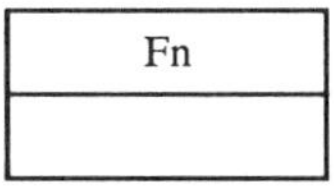 Fn	Fact	Rule without conditions
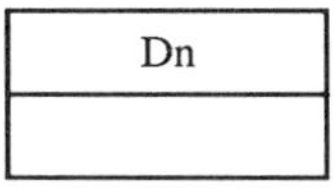 Dn	Default	Rule without conditions
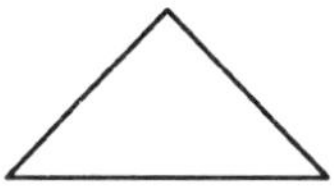 △	Question	
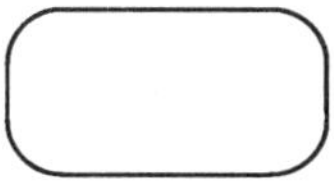 ⬭	Arithmetic relation or operator	Constants inside box, variable below
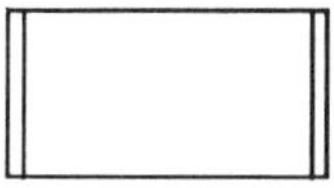 ▭	Action	Procedure or action
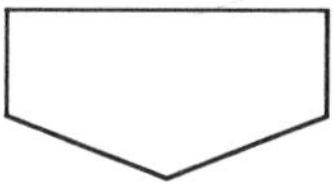 ⬠	Continuation	
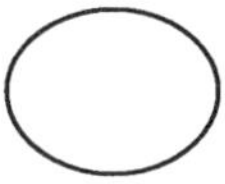 ○	Logical Operator	

Figure 5.18 Symbols used in KBMs

There are certain conventions implicit in the KBM, relating to order and priority:

- Order Convention: so as to define uniquely the order in which the rules should be fired or questions be asked, the map should be read left to right and downwards;

- Priority Convention: the conditions to a rule need to be satisfied before the rule is fired and the conclusion drawn. The sources of information for the conditions are in the following priorities:
  — user data from the database/facts from the semantic net;
  — questions to the user;
  — further rules in the ruleset;
  — defaults.
  Thus, if the user data/semantic net facts return a value of unknown then the appropriate condition is posed in question form to the user and so on down the priority sequence.

- Precision Convention: the top of the KBM should deal with high level, low precision concepts; the lower levels should contain progressively more precise and detailed knowledge. In this way, the declarative nature of the expertise can be maintained and expert users can remain at the top of the map, while non-experts can use an appropriate level of detail and explanation.

The logical operators used in the conditions and conclusions of a rule may include NOT, AND, OR, X OR (A X OR B = A or B but not both), X(XA = A or NOT A).

The KBMs are of two kinds, the types reflecting the purpose for which expertise is required. STAGES identifies "control" expertise to bring order to the execution of application processes and "inference" expertise for the execution of the application processes.

An example of a KBM—in this case an inference process—is illustrated in figure 5.19. The knowledge map illustrates a ruleset with multiple rule paths. Its structure is exactly that illustrated in figure 5.2 and is therefore the direct basis of drawing the rule structure within a ruleset. It illustrates that within a ruleset knowledge can be decomposed into a hierarchical structure, on the lines indicated in figure 5.13 for a particular application.

The purpose of the consultation is to work out the length of the month for the Gregorian calendar and to provide advice as to the number of days within a month. The logic of the knowledge map is that, because by convention one accesses in the sequence of top to bottom, left to right, question 1 is posed to the user first. Question 1 on the leftmost rule path asks the month length. If the month length is provided by the user then the consultation is finished. The sequence of the other priorities, such as posing questions to the user, is

Enhancements to SSADM

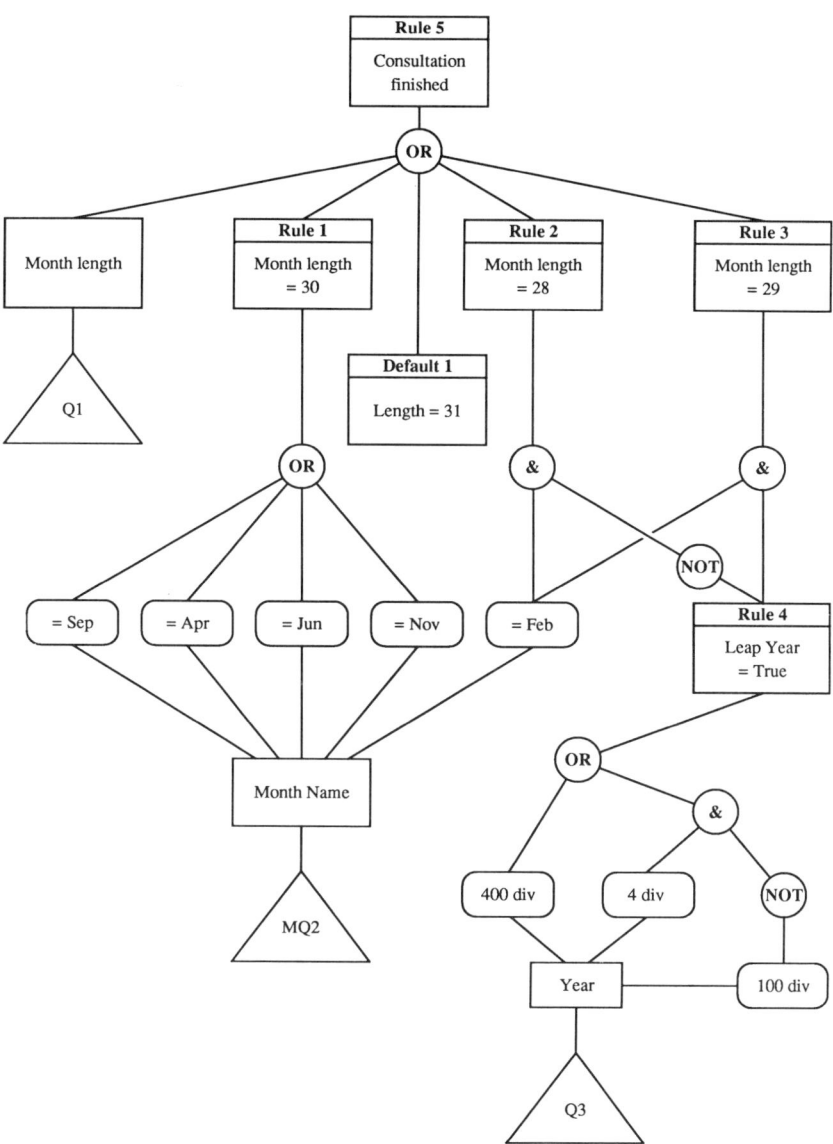

Figure 5.19 Knowledgebase map

obviously irrelevant and there are no rules and default values on this rule path.

If the consultation is not finished the second rule path is accessed, in this case to rule 1. The bottom of the rule path indicates menu question 2, which poses a question regarding month name—is it September, April, June or November? If the knowledgebase provides an answer then the condition of rule 1 is satisfied, the rule is fired and advice provided that the month length is 30. There are no other rules and default values in the rule path for testing against the condition of rule 1. By-passing the default rule, the third rule path is accessed, that is, the rule path beginning with rule 2. The answer to menu question 2 is inspected again to see if the month name is February. If it is then the condition to rule 4 requires to be tested to ascertain if the year is not a leap year. Question 3 is therefore displayed. If the year is divisible by 400 or if the year is divisible by four and not by 100 then the year is a leap year. If the two conditions to rule 2 are true, that is, it is February and not a leap year, the conditions to rule 2 are satisfied and the conclusion is drawn and advice provided that the month length is 28. Rule 3 is fired if the conditions to rule 2 are found not to be true. The fourth rule path, in this case via rule 3, is then accessed. Following the knowledge map, the answer to menu question 2 is inspected yet again for the month name. If it is February then rule 4 is tested again to find out if it is a leap year. The logic for rule 4 has already been described. If the month name is February and the conditions of rule 4 are true then the conditions to rule 3 are satisfied, the rule is fired and the conclusion is drawn, namely that the month length is 29. If rules 1 to 4 are not true then the default rule is fired and advice provided that the month length is 31.

Using pseudo code the above rules would be defined as:

- Rule 1—If month name is September or April or June or November then month length is 30.

- Rule 2—If month name is February and year is not leap year then month length is 28.

- Rule 3—If month name is February and year is leap year then month length is 29.

- Rule 4—If year is divisible by 400 or (divisible by 4 and not divisible by 100) then year is leap year.

- Default 1—Month length is 31.

The KBMs contain all information that is necessary for the coding of rules in a ruleset—the sequence in which the rules are to be accessed, the conditions and conclusions of the rules, any uncertainty of whatever type in the conditions and conclusions, the questions to be posed to the users if the answers cannot be obtained from the knowledgebase, and whether there are

Enhancements to SSADM 299

other rules in the ruleset and if not whether there are any default values as the sources of knowledge.

Given its ability to show the structures of rules in a ruleset and to represent knowledge logic, the STAGES KBM facility is an excellent and powerful technique.

Where the conditions and conclusions of a rule contain uncertainty this requires to be recorded as a certainty factor. It is unfortunate that one of the four mechanisms for handling uncertainty described in section 5.8 is called by the misnomer of certainty factors. This type of uncertainty should be handled automatically by expert system software, as it is based on a fixed mathematical formula. For classical probability a threshold of acceptance should also be specified for the condition uncertainties. If Bayesian probability is required then a condition(s) requires to be specified against the rule condition(s) in order to create conditional uncertainty (no puns intended here!). If fuzzy logic is required then this can be coded as a set of probability statements. The KBMs need to be upgraded to show rule iterations about a common test—in this case two propositions and their combined result. Create one KBM for each E type process and record the rules in the KBM as a ruleset.

What is noticeable with the STAGES approach to rule definition is that the KBMs are effectively pitched at the event level. There is a KBM for each query/update requiring expertise. No attention has been paid to the possibility that, if knowledge logic is to combine with object oriented design, the rules need to be normalised to relate to the objects they describe, rather than the event they support.

On a project the author worked on, the advice given to the SSADM practitioners was to separate the procedural operations of the SSADM command language from any rules that require to be defined for any entity being accessed in the Process Models. It was known that some of the processes to be undertaken were based on rules. These were specified as appropriate, but allocated to the entity leaf boxes on the Process Models or to the event root box if appropriate to the event.

The handling of the rules was to be supported by a relational file handler with the ability to treat the rules as triggers/daemons, so that there were the beginnings of the design of an expert system and the treatment of the rules as if they are normalised.

5.11.3.3 Knowledge access

Both knowledge data and knowledge logic are stored as table rows. The knowledge data is stored as class tables in the dictionary (one table row per table) and instance tables in the database (n table rows per class table). The knowledge logic is stored as variable length records, with the rules being table rows and the rulesets by which action and inference rules grouped as tables with a row for each rule. *Knowledge data and logic therefore have the same underlying storage and access mechanisms.*

The considerations in identifying rulesets, ascertaining the ruleset access strategy of breadth first or depth first and whether forward and/or backward chaining is required were identified in sections 5.5.3 to 5.5.6, with an example based on GOLDWORKS in section 5.10.2.

Some basic principles and task sequences in undertaking knowledge access are:

- Ascertain the application domain for which advice is required. In the GOLDWORKS example the overall application is the servicing of a car, within which there are certain domain tasks requiring advice, such as electrical, mechanical and other servicing. Within the electrical task there are further sub-tasks relating to servicing the car lighting and heating.
- Each task equates to an "E" type process in the MDFDs and each task becomes a ruleset.
- For each ruleset draw the structure of the rules using a KBM.
- From the type of trigger to the KBM decide whether forward and/or backward chaining in the ruleset is required. If the answer (i.e. goal) for which advice is sought is known then adopt backward chaining; if the answer for which advice is sought is not known (i.e. a set of initial conditions is posed) then adopt forward chaining.
- The technique for undertaking knowledge access design in STAGES is a version of that described in section 3.1.1 and is extremely thorough and elegant. The variation is the explicit identification of the probability of rule firing and rule traverse. (In reality this aspect is nothing more than the probabilities of access to different access paths in a standard data model.) The other difference is that the cardinality ratio between the KBM nodes is invariably one to one. Cardinality ratios are therefore not a consideration as they are in conventional transaction access path analysis.

The technique is based on calculating the proportion/weight of accesses to each node of the KBM for each triggering of the KBM transaction. The calculation of weights proceeds as follows:

- *Step 1:* Number each branch of the KBM tree which leads to a node, i.e. to:
  - A rule;
  - An implied test (e.g. if a variable is "unknown");
  - A test.
- *Step 2:* For each numbered branch calculate recursively *PT* (the probability of traversing the branch) and *PF* (the probability that the branch will fire).

PF is calculated as an independent probability and is not conditional on any events except those subsidiary to it in the KBM. PT is calculated differently depending on the node from which the branch descends. If it is a single node, e.g. a rule or a unary operator, PT is defined to be 1. If it is an operator with many arguments, PT is calculated as follows:
— If the operator is AND:
 PT = Π(PF) (for all preceding branches from the node)
— If the operator is OR:
 PT = Π(1-PF) (for all preceding branches from the node)

- *Step 3:* The weight, W, of a branch is calculated as:
 $W = \sum WC$ (for all subsidiary branches) + 1
- *Step 4:* The weight contribution, WC, of a branch is calculated as:
 WC = (PT W)
 This is the mathematical formula of what was described in section 3.2.5.4.

Consider figure 5.20. It represents backward chaining knowledge access. The paths with greatest weighting contribution are those that are accessed most frequently. Using this technique one can quickly ascertain which parts of a KBM are accessed most frequently. In the example it is clear that rules 1, 2 and 7 are the most frequently accessed. It is obvious that the ruleset access strategy should be depth first.

The use of weightings is the opposite of the approach adopted in SSADM. Weighting requires their calculation *towards* the entry point. In the example backward chaining is used. The entry point is therefore rule 1. SSADM uses a counting of the actual number of accesses *from* the entry point. Both approaches are valid. It is a matter of style.[1]

STAGES attempts to assess only the access to the knowledgebase in database optimisation. This is too late. It should be done as part of the logical design in order to identify potential performance problems before they occur. This is not difficult, and, given the concept that logical design = physical design, entirely valid. Just bring the technique forward to an earlier logical design phase of the method.

- Any linkages of the ruleset to the semantic net structure is usually automatically ascertained by the inference mechanism. This was illustrated using the GENERIS worked example in section 5.10.1. Certain products, such as Software AG's NATURAL EXPERT, require explicit linkage.

---

[1] The author is delighted to see that STAGES undertakes access path analysis and ascertains which are the busiest access paths. One wonders if SSADM, when it provides expert systems, either within the method or more likely through an Interface Guide, will abandon the thoroughness of STAGES. It has for database access path analysis against the logical data model. One hopes that this will not be done against the semantic net.

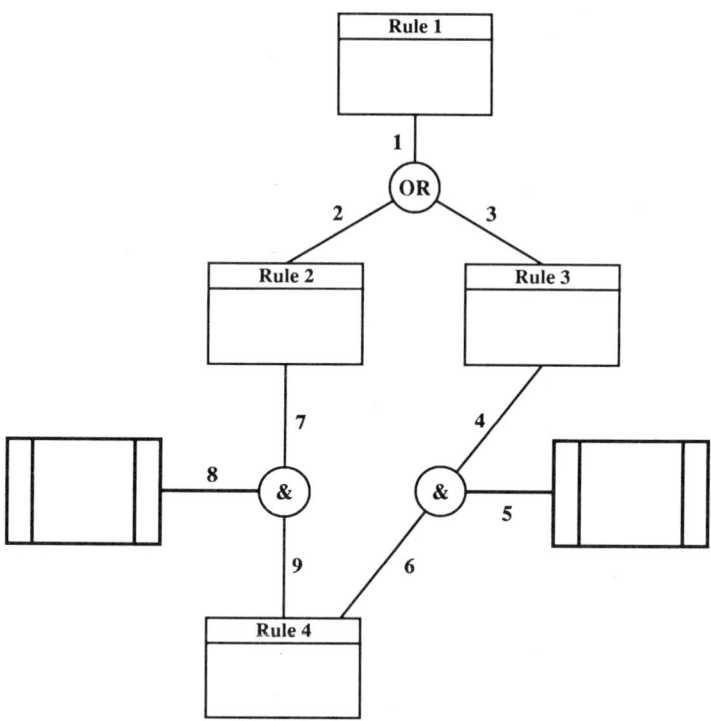

Branch	PF	PT	Weight	Weight Contribution	Subsidiary Branches	
1	0.90	1.00	7.76	7.76	2	3
2	0.81	1.00	6.34	6.34	7	
3	0.09	0.19	2.19	0.42	4	
4	0.09	1.00	1.19	1.19	5	6
5	1.00	0.09	1.09	0.10	6	
6	0.09	1.00	0.09	0.09		
7	0.81	1.00	5.34	5.34	8	9
8	1.00	1.81	1.90	3.44	9	
9	0.90	1.00	0.90	0.90		

Figure 5.20 Knowledge access

Enhancements to SSADM 303

- The calculation of physical I/O overheads in accessing knowledge rule logic in the rulesets was described in section 5.5.6. The rules drive access to the knowledge data in the semantic net structures. Each access to the semantic, be it to the class table in the dictionary or to an instance table in the database, is likely to generate disk I/O, as well as any overheads in accessing ruleset indexes. Before accessing the knowledge data the rules may require to pose questions to the user. This, of course, generates the standard message I/O to the terminals.

One can calculate the performance of a knowledgebase using the standard practices of database optimisation. The SSADM techniques of database optimisation are perfectly valid for this.

6

ADDITIONAL DATA PROCESSING ENVIRONMENTS FOR SSADM—OBJECT ORIENTATION[1]

6.1 THE PROBLEM

It is inevitable that current technology reflects design decisions made years, and possibly decades, ago by the developers of information technology. The long gestation period of technology development contrasts to the short period for human thought. It is therefore not surprising that most current technology for the storage and access of data and logic does not reflect the latest thinking. There are now recognised to be a number of deficiencies with the current technology. As always, with the benefit of hindsight, the design decisions taken then can be improved now.

One general problem we face is that there is a physical separation of data in the database and logic in the application programs. This separation is illogical—after all, both data and logic are information. This illogicality leads to inefficiency. Because the two are stored separately—the data in the

---

[1] One of the hazards of writing about object oriented technology and techniques is the confusion about terminology, particularly two terms. This confusion reflects the evolution of the technology. Object orientation first appeared via object oriented (OO) programming languages. OO programmers used the terms class and objects. In a relational context class means table with a template schema definition of data properties and object means table row. The developers of object oriented file handlers use the terms of class objects and object instances, the class objects being the tables and the object instances being the table rows. Reflecting the background of the author the terms used in this book are class objects and object instances.

database and the logic in the application program—it is necessary to bring them together before processing can take place.

The policy followed by all developers of information technology has been to move the data to the logic. The data is accessed from disk, brought into the processor main memory buffer pool and from there moved to the application program working storage, all involving expensive logical and physical I/O. It is only when the two are brought together, so that data can be processed and presented to the user according to the logic instructions, that the benefit of a computer system is achieved. The merging of data and logic would eliminate this inefficiency (although at a cost elsewhere, as will be seen—nothing is free of charge), as well as provide many other advantages.

6.2 THE SCOPE OF OBJECT ORIENTATION

The scope of object oriented design techniques and implementation technologies is exactly the same as for today is database and programming technology, namely the design and development of computerised application systems. Where object orientation is compared against database technology it is assumed to be relational. Most database users are now relational and relational technology is an almost universally understood standard. And relational technology is the third generation of database technology,[1] as well as being the generation that object orientation is replacing.

Relational technology is composed of the following main facilities:

- a file handler for the storage and access of data;
- a data definition language (DDL)[2] for the description of the database tables and their data properties;
- a command based data manipulation language (DML) query language (universally SQL) for the access of the tables of data;
- a programming language 4GL[3] for the specification of processing logic for the manipulation of data once accessed;
- a cursor based screen painter for the easy formatting of input and output screens and the definition of appropriate built-in functions for screen presentation and data editing;[4]

---

[1] The first generation is the hierarchical database file handlers, such as IMS from IBM, and the second generation is the Codsyl standard, such as IDMS from vendors such as ICL and the erstwhile Cullinet.

[2] This language is used for defining the class structure of data tables and the data properties of the tables of the relational and pre-relational databases in use today. The definition of the data is therefore taken from the programmer and transferred to the database administrator.

[3] The DML contains the sequence, selection and iteration facilities of the traditional programming language for the writing of process logic for the manipulation of data once accessed.

[4] Quite a few relational products combine the 4GL programming language and the screen painter. A leading example of this is ORACLE with SQL*FORMS and SQL*PL.

- a report writer for the rapid definition of the formatting and logic requirements of batch reports.

These facilities provide all the functionality required for designing and developing an application system.

So it is with object orientation. Object orientation contains an equivalent of a file handler, an equivalent of a programming language and an equivalent of a report writer. As described in chapter 3 almost all of the object orientation products are currently restricted to a file handler and a programming language. When the relational vendors respond with their own enhancements of the relational products with object oriented facilities then the full set of relational type facilities for the object oriented development of software applications will be available.

6.3 The Concepts of Object Orientation

There are three concepts that are unique to object orientation. They are:

- The Normalisation of Information.
  Dr Codd of relational fame[1] developed rules for ensuring that the data properties were correctly placed to the tables of data they "belong to", Customer Name "belonging to" the Customer table. But these normalisation rules relate only to data and to the fact that the data properties normalise to the key identifying the object instances. There is no consideration of the role of the data and matching it to the role of the object class. It is now possible to relate these normalisation rules to logic. It is now appreciated that some 80% of logic relates not to the event/business requirements but to the objects/tables. Virtually all information, be it 100% of the data or 80% of the logic, can be normalised to objects. The 20% of logic that remains at the event level is typically the initialisation of the variables, input and output formatting and any pre- and post-processing prior to and post the accessing of the data.

  This concept of normalising information of either type is appropriate given that an object in object orientation is able to support, unlike relational tables of data, both data *and* logic within object instances.

  A formal definition of an object is that "it is an entity that exists uniquely and distinctly in time and space, *containing both data and logic about itself within itself*". The first part of the definition is not particularly significant.

---

[1] Dr E. Codd defined a set of 12 rules that became the foundation of relational technology. The basis of the rules was that data was defined as flat records/tables containing data attributes/properties that described the objects the tables represented. There could be a Customer table with data properties that describe the customer. Backing these rules was a further set of rules for "normalising" data so that the data properties would correctly describe the table. The data property Customer Name would be correctly normalised to the Customer table on the basis that Customer Name "belongs' to" Customer.

Today's database technology supports it. Logical entities and physical tables of data describe "things" of interest to an application about which data, and only data, can be recorded. But objects are more than entities. Objects describe "things" of interest to an application about which data *and* logic can be recorded. *It is the logic component which primarily distinguishes objects from records/tables of database file handlers, which only contain data.* There are other facilities, as will be ascertained, but it is the logic that is the prime distinguishing feature.

The significant point here is that information, be it data or logic, is stored *within* the object to which it logically relates. It is well known that the customer name data attribute normalises to the customer entity. Why should not the logic "All customers with red hair receive a lump sum of £100" also be normalised to the customer object class? The answer is that it can be—it is not logic that relates to any particular event. The logic, like the data, is normalised to the object Customer. *The first concept of object oriented design is therefore the normalisation of information.*

- Application Independent Designs.

  The normalisation of information leads to the second concept, *the creation of application independent information designs*, that is, the design of information generic to a corporation. To date, the only part of information that is corporate is data, hence the creation of corporate databases. The reason for this is that the data has been normalised to the appropriate entity/object and not to a business requirement/event of a functional area of an application. Customer Name is a data property describing Customer, not a business requirement. The data is therefore not specific to an application and its constituent business requirements. The same data objects can be "accessed" by multiple functionally based applications.

  The same is true when the logic information is normalised to an object. If the logic is "All Customers with red hair receive a lump sum of £100" then it, like the data, has nothing to do with a particular business requirement/event but with the Customer object, in the same way as Customer Name. The logic is a logic property describing Customer, not a business requirement. It is therefore likewise normalised to the Customer class object. When logic is normalised like data then it has the same corporate characteristics as data in that it too is application independent—but in this case the logic can be "re-used" (a word much used by object oriented devotees) by multiple applications. *Only the 20% of logic that remains at the event/business requirement level is application dependent.*

- Change by Addition.

  There is a third concept. This concept is perhaps the most significant in that it most distinguishes object orientation from pre-object oriented technology. The concept is that *enhancements made to application systems are through addition rather than through modification.*

The Concepts of Object Orientation

There are two ways in which enhancements to an existing application system can be made—by modification or by addition. Of the two, addition is much the better approach as the existing information is untouched and therefore remains stable.

Pre-object oriented technology enhanced an existing application through the process of modification. The user requests an enhancement to an existing business requirement in the application system. The code of the supporting application program would be modified, recompiled, retested and run. Modification would be the approach adopted for supporting the changed situation. The problem with this approach is that the stability of the existing application program is lost, hence the need for retesting the modification.

Object orientation adopts the strategy of supporting enhancement by the process of addition. There is no modification. When a change is required to the design of the data or the logic it is not a case of altering the existing design but of using the existing information as much as possible and of adding data or logic properties to the existing information only if it is not able to support the new requirements. This process of addition is mainly supported by the creation of specialised sub-class objects to the existing class objects, with the new sub-class object inheriting the data and logic properties of the existing and now super-class objects.

One can also create generalised super-class objects of the existing information. As new properties are added to the object class model new commonality of now common properties between the object classes can be ascertained. These common properties are abstracted as super- or sub-class objects to the existing base class objects. This change by addition to the existing information of the base application is illustrated in figure 6.1.

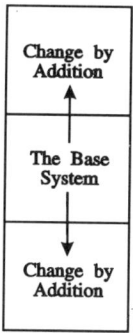

Beneficial Result

Existing information becomes increasingly reused and stable

Figure 6.1 Enhancement by addition

6.4 Does Object Orientation Require a New Way of Thinking?

In discussion with people experienced in implementing object oriented systems the author has been consistently advised that a new way of thinking is required. "You need to throw away much of what you have learnt before about systems design and development and apply new ways of thinking."

It is argued by some (B. Meyer in *Object-Oriented Software Construction*, Prentice Hall) that object oriented design is the antithesis of the traditional analysis and design structured methods, such as SSADM and Information Engineering. These methods typically use the top-down approach of taking a high level process and decomposing it down into its constituent sub-processes. This approach is good at ensuring that the design will meet the initial user requirements, but it does not promote re-useability—the sub-processes may overlap. Re-useable software requires that systems are designed by combining existing elements as much as possible, which is the definition of bottom-up design.

The author is not so sure. It was stated earlier that centralised processing is generic, because the facilities of batch and online data processing supporting the centralised environment are common to the other and more recent data processing environments, such as expert systems and object orientation. Why should object oriented systems be fundamentally different from the generic facilities of centralised processing when the other data processing environments of distributed, realtime, conversational are nothing more than centralised plus a little bit?

As already mentioned, there are two ways in which existing software implementation technology and design techniques can be enhanced—by addition or by modification. The former is much to be preferred if possible, because it preserves the investments in the existing computer systems. *What is becoming clear is that the traditional technologies and techniques for data structure require addition to become object oriented, while those for access and process logic require modification to become object oriented.* Given that data is normalised, all that is required is to build a traditional logical data model and then to add the object oriented facilities, such as abstraction of common data properties and the role of the data to reflect the role of the object class, to convert the logical data model into an object class model.

But even for the logic part requiring modification there is a position of comfort. The technique for normalising logic is, as we shall see, the same as that for data. This, of course, results in modification, as the current logic design paradigm is event based. Application programs are event based. There are business events such as Create Invoice and Pay Invoice. For each there is an application program. But only some 20% of logic is actually event based. There is therefore modification to some 80% of the logic—that is, the logic that can be normalised to object classes. This 80% of logic is redesigned to be object class based rather than event based. *But we are using the technique for the design of data to the technique for the design of logic.* The rules of data normalisation, suitably modified, can be used for logic normalisation. It is therefore not a case of unlearning the skills we have but of applying one of the techniques, the technique of data normalisation, in another context.

It is therefore much more a situation of using existing technologies and techniques and enhancing them by addition. The author cannot think of anything that requires a practitioner of the current structured design methods and relational and other database file handlers and programming languages to unlearn his/her skills. Apply it differently in some parts, yes, but unlearn no.

Object orientation is a major subject, with a substantial array of technical facilities not found in traditional processing. These additional facilities require to be reflected in the design techniques. These facilities include property inheritance in the object class model, encapsulation, property instantiation at the abstract class level, polymorphism of processes that have a common name but different functionality depending on the object instance being accessed, function name overloading, where a process can support different functionality depending on the programming operator being used, genericity, where a process of the same name can support different functionality depending on the parameters passed in the message, logic normalisation and message passing between objects. But none of these facilities negate the pre-object oriented skills.

6.5 An Object—What Is It?

An object is the unit of object oriented application system design. It identifies something of interest to a business application or a system process about which information is required to be modelled and stored. An object could thus be an order in a purchasing application, a policy in an insurance application, a robot in a manufacturing application, a container crane in a container port application and a print routine in a system process. The object contains all the information, both data and logic, that describes the object.

An object oriented application system is a network of intercommunicating objects—each object in itself being totally self contained with all its information, but functioning as part of a total application system. The application system is the total of all the objects and their relationships with each other, plus the application program events, one for each event/business requirement. Many regard, quite correctly, the business requirements as event objects.

An object is, in simple terms, composed of two types of information, as illustrated in figure 6.2. The two types are the data and the logic information components. An object does not have to have both types of information simultaneously and can contain data and logic or just logic. The object must

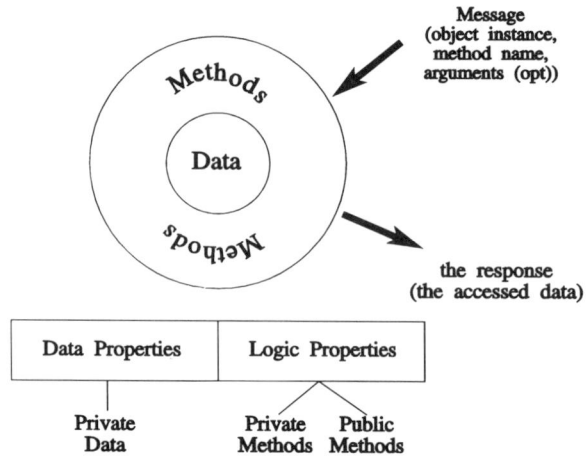

Figure 6.2 The structure of an object

contain logic, because, as we shall see, the data component should not be accessed except via the logic component. The data component is optional.

The data component has been described as the attributes and the logic component described as the services provided by the object. Another term for services is methods. This is the term more usually used in this book. The attributes and their values describe the state of the object at any point in time and the services/methods define what can be done to the attributes of the object.

The data and logic components both describe the object class and are defined in the dictionary schema definition of the object. Unlike the standard schema definitions of relational and earlier database file handler types, the class object definitions of the properties can contain values of information. In the case of the methods the value is the logic. Both components can have many properties (object oriented buzzword for data attributes and methods), which correspond to the fields of a file handler record/table. Thus a method is classed as a property just as a data item. An object can have many data and logic properties.

The methods are of two types—the public methods and the private methods, the former user triggered, the latter system triggered. They are described in detail in section 6.6.6. Suffice it to say at this stage that the public methods are invoked by an explicit message from the user triggering an event/business requirement, while the private methods are invoked automatically by the application when specified conditions occur. All objects must have at least one public method, they do not have to have a private method.[1]

There is, in fact, an invisible third component, the interface layer. The interface layer is composed of messages to the objects and responses to the messages from the objects. Through the facility of encapsulation object oriented technology does not allow objects to know the information that is stored in any other objects. Indeed these other objects do not need to know and do not care what the information is. *The messages provide encapsulation to the methods and the methods provide encapsulation to the data.* The only thing that the users/other objects need to know is the interface message to send to the object to invoke a named method and the response that is sent in return. The interface message mechanism is described in detail in section 6.6.5.

It is now feasible to consider objects as abstract, abstract in that the user of the object does not know the data and logic contents/properties of the object. An object is a "thing of interest" about which information, all information, can be recorded, but what that information is is unknown, is abstract, to the outside world.[2]

6.6 THE OBJECT ORIENTED FACILITIES

6.6.1 Class, composition/aggregation, property inheritance and relationship semantics → information abstraction

These four facilities are unique to object orientation. They each require the other, class and composition (increasingly called aggregation) being unsupportable without property inheritance, property inheritance having no purpose without class and aggregation, and the relationship between the different types of objects, and from this the property inheritance paths, being defined by specific kinds of semantic descriptions. Each without the other would not function—that is why they are described together.

Class, aggregation, property inheritance and relationship semantics together provide information abstraction, this being the single most important feature of object orientation.

---

[1] In C++ there is a three-part distinction—private, which declares that the properties that are visible only to the class itself, protected, which declares the properties that are visible only to the class object itself and its sub-classes, and public, which declares the properties that are visible to all other objects that message the class.

[2] There are two definitions of abstract objects. One is that defined above. The other is of an object class with no instances. An example would be the class Man, there being no instances of Man, just a generic description of the class, such as having a data property of the Number of Legs with an instantiated value of "2". These classes are merely template definitions of the data and logic properties the classes "contain".

6.6.1.1 Class

Class is the facility that models common behaviour between objects. Sets of objects with common behaviour have a common class. Common behaviour is identified as data or logic properties that are common to more than one object. Figure 6.3 shows that the two class objects Employee and Broker had common data and logic properties of Date of Birth and Calculate Salary, so that there is a common behaviour of being Persons.

People have long classified information into classes—and the classification goes from the most general to breakdowns of the general class into sub-classes of increasingly specialisation—a hierarchy of generalisation to specialisation. For example, a very general classification of animals could be into those that are warm blooded and those that are not. The warm blooded animals can be further sub-classified into those that are herbivores and those that are carnivores and the carnivores into those, let's say, that hunt as packs and those that hunt singly to form what are known as class hierarchies. And so on. All these classifications are based on the fact that there are groupings of animals that have common characteristics, common data and common logic operations, that describe their state and behaviour. Warm blooded animals could be lions and cheetahs and eagles. The eagles would fall into the classification of animals that fly, the lions into those that walk and hunt as packs and the cheetahs into those that walk and hunt singly. And there are many other animals that fall into the same classification categories.

Figure 6.4 is a variation and extension of figure 6.3 and shows a model of object classes with the Mammal object class at the top and the Star Performer object class at the bottom. The class structure represents a fully abstracted class object model of the base class object Employee, in that the base class object properties, including logic properties, have been generalised and specialised as appropriate. This facility of a class behaviour relationship is supported by the semantic description of the relationship as being of the "is a" type.

The object classes higher up a class model are more abstract/general in the information they contain than the class objects lower down the class model, which are more particular/specialised. The Mammal is a more general case of Species, which is in turn is a more general case of Person, which is in turn a more general case of ... and so on down the class model. All the classes in figure 6.3 are related by class in that they are all mammalian—they all have a common mammalian class behaviour, the properties of a super-class being relevant to all instances of the sub-class. Warm Blooded is thus relevant to the sub-class object instances of Man down to Star Performer, Date of Birth being relevant to the sub-class object instances of Employee and Star Performer.

The mammal class is relevant to all the sub-classes of the class model illustrated. The sub-class objects of Man, Person, Employee and Star Performer are all mammalian in the mammalian class hierarchy. The Man class is relevant to all the classes of the model except the Mammal. Person,

The Object Oriented Facilities

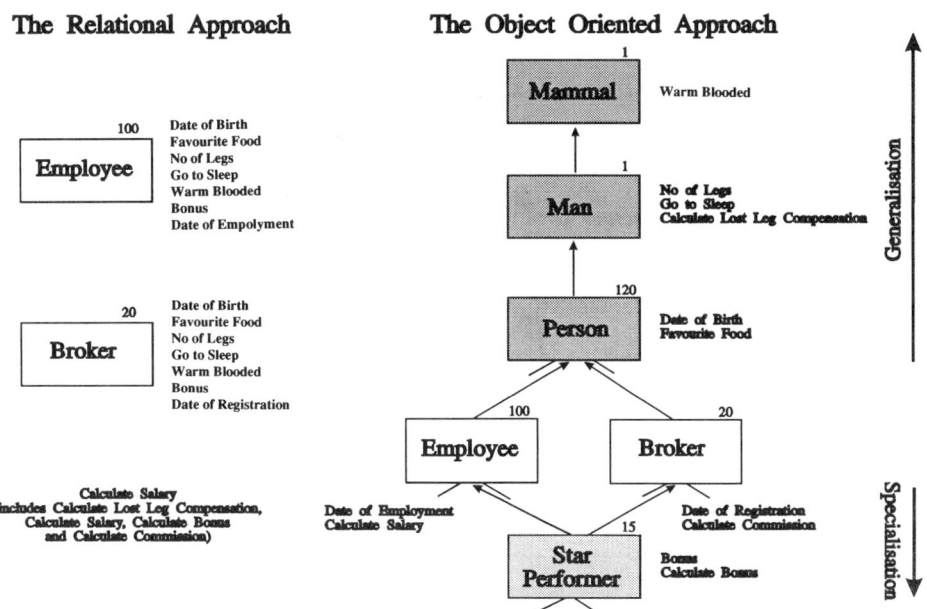

Figure 6.3 Common information → common behaviour → class abstraction

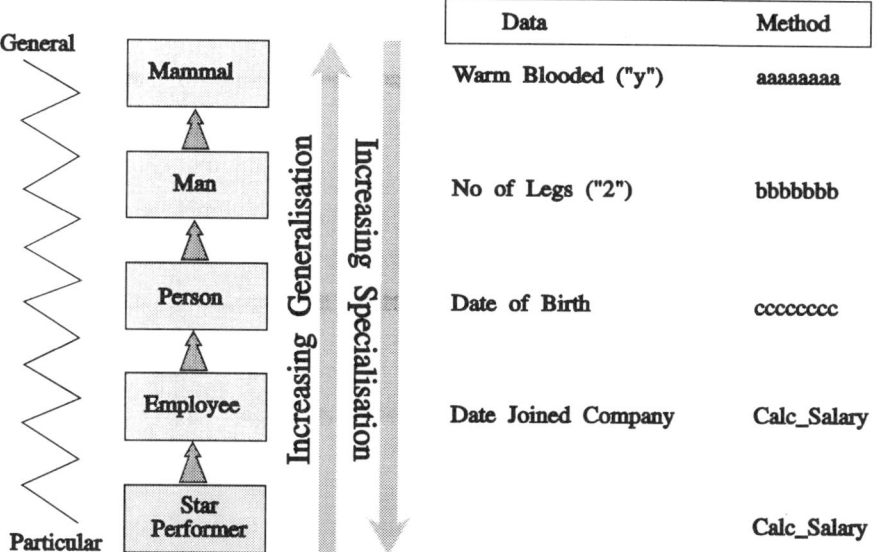

Instantiated class information is generic (ie, applicable) to all instances of the class

Figure 6.4 Abstraction: generalisation ↔ specialisation

Employee and Star Performer are therefore all within the class hierarchy model of Man. And Employee and Star Performer are also within the class hierarchy model of Person. And so on, being progressively more restrictive as one progresses down the class hierarchy model.

There needs to be an understanding of the difference between class and class hierarchy. All objects in an application system are class objects because they have a class template definition of their data and logic properties but not all objects belong to a class hierarchy.

Consider figure 6.5. The objects Customer, Order and Product are class objects, each in their own right, but there is no class hierarchy. This is because there is no common behaviour—it would not make sense to say that an Order "is a" Customer and an Order "is a" Product. And there are no common properties between the objects. The data property of Customer Name is not common to all instances of Order—indeed the data property is not to be found within Order. Order is not a sub-class specialisation of Customers or of Products and therefore does not inherit the Customer's and Product's properties. Nevertheless, Customer, Order and Product are each class objects with their own template definitions in the object base schema. The fact that there is no class hierarchy, no common behaviour, is modelled by the fact that the semantic description of the relationship between the class objects is of the free form type describing the business role of the relationship between the class objects, as shown in the figure.

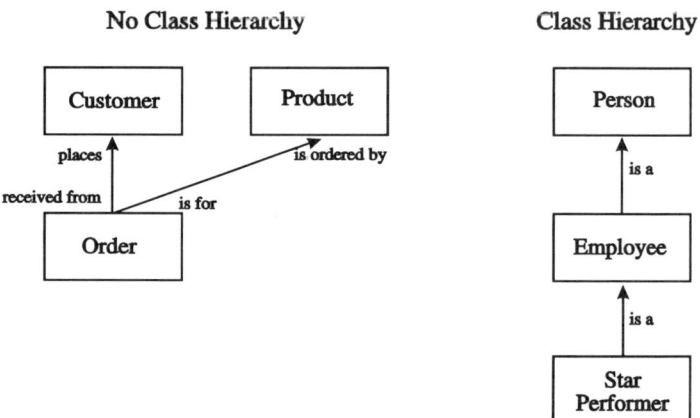

Figure 6.5 Class and non-class hierarchy

The fact that objects are currently only class objects and not part of a class hierarchy does not mean that they can never be within a class hierarchy. At some point in time someone might ascertain that it is possible to generalise and specialise any one of the class objects. There could be a later business need to specialise the Customer class object into those that are creditworthy and those that are not credit-worthy and thus create behaviourally related sub-classes of the base class Customer and a class hierarchy of Customer. By contrast, the class objects of Person, Employee and Star Performer are very much within a class hierarchy, that of Person. There is a common behaviour of employees and star performers of both being persons. And the class hierarchy is modelled with the "is a" semantic of the class object relationships.

As so far described, the increasing specialisation as one descends the model is only at the class level. However, the specialisation can go further—to the instances of the class. In the case of being warm blooded, all the instances of the class objects that can inherit from Mammal can only be warm blooded—they would die if they were not. An instance of Person would be dead if his/her blood was cold. However, this total applicability of class information does not have to be relevant to all the class objects. Consider the Man object. Not all instances of Man will necessarily have two legs. It could be that a particular Man has lost a leg. The instance (Henry has lost a leg) would be a specialisation of the class *at the instance level*, Henry having only one leg.

6.6.1.2 *Aggregation*

This is a data modelling facility that is not used by database technology. It occurs where an object is composed of other object classes. A Car class is composed of the classes Engine and Carburettor. It is the ability to take each data property of a class object and generalise them into super-class objects of the base class. This is shown in figure 6.6 where the base class object Customer has two data properties of Name and Address which have been generalised into two super-class objects.

The relationship that is established is of the data properties to the prime key of the base class object. The "has a" description is that the prime key, Customer Number, "has a" Address and "has a" Name is "composed of" Address and Name. It would be nonsensical to have a "has a" relationship between the non-key properties of Address and Name. There are no other relationships between the data properties, otherwise the class would not be in third normal form (it would be in second normal form).

The composition facility is particularly useful if there is a method that processes a specific data property of a class object. With the ability of object orientation to support the normalisation of logic, it is reasonable to take the normalisation to third normal form and see if there is logic dependency on a non-key data property, if there is a dependent relationship between a data property and a method. The first two rules of normalisation test the data and logic properties against the prime key and therefore against the class as a whole. The rule of third normal form tests a non-key data property against a non-key data property, inter-data dependence, and therefore against part of the entity. And so it is for the normalisation of logic—is a piece of logic dependent on a non-prime key data property.

In the example in figure 6.6 there could be a method to check that the Customer Names are within a certain value range. (The business is in the marketing of named shirts and sends sample shirts out to lists of customers on the basis of named selections. Once the selection has been made there is some additional processing regarding the marketing of the shirts.) While this logic is still generally relevant to the class object Customer, it would be more true to say that the logic, the method, is specifically relevant to the Customer Name property, which requires therefore to be abstracted as a class in its own right. One is correctly normalising the method to the specific property of the base class, abstracting it and creating a new class of a single non-key data property and one or more associated methods. One is in fact testing inter-data/inter-logic dependence, the rule of third normal form for logic. The piece of logic (the method) is dependent on a data property of the class and not the class as a whole. This being the case, abstract it (remove it in normalisation terms) with generalisation. The role of the method has been correctly related to the role of the data property.

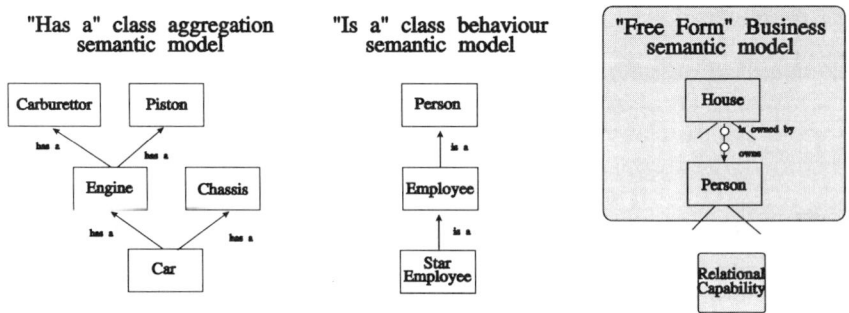

Figure 6.6 The object oriented semantic models

Without this facility of decomposition of a complex object of n data properties it is not possible to take the normalisation of information to the ultimate form. The first two rules of data and logic normalisation have been based on relating the dependence of the values of the data properties and the role of the logic against the prime key of the class object/relation.

The logic that remains as methods in the base class is that which requires to process the key data property of Customer Number as well as other non-key data properties or, and this suggestion is controversial in that the method logic is not being properly normalised, to process several non-key data properties. Again using the Customer example of figure 6.6 the selection process for the shirts is for specified names *and* addresses. In this case the logic is appropriate to several non-key data properties. This method, while functionally the same as the one for the name, is dependent on two data properties, and therefore more of the class as a whole. It is not abstracted.

The Object Oriented Facilities 319

If one created class objects for every combination of methods and the data properties they access within the base class the object class model would become even larger, possibly even unmanageably large. From a normalisation point of view this abstraction into a "has a" super-class object should be so done. From a design purity point of view the author is writing heresy, but there is the inevitable design compromise.

With the facility of class composition one can now take logic to third normal form. The author developed the idea of the normalisation of logic being one of the basic features of object orientation, but could only show examples of this to second normal form.[1] This book is able to expand on this idea and take it to third normal form, via the composition facility. The normalisation of logic is fully discussed in section 6.8.2.2.

The composition of the base classes and their abstracted data properties and related methods is defined with the use of the "has a" type semantic description. To recompose the base class of its constituent properties it is merely a case of following the "has a" relationships and rebuilding the base class through the inheritance of the abstracted properties back to the base class object.

6.6.1.3 Property inheritance

There are two types of property inheritance—class and aggregation.

Class/behaviour property inheritance Class property inheritance is obtained where there is common behaviour between super- and sub-class objects, that is, they share common data and logic properties. An example of a behavioural class hierarchy is described in section 6.1.1.

The inheritance relates to all the data properties. Figure 6.4 shows that all the class objects contain logic properties in the methods. There has to be at least one method as the data properties cannot be accessed except via the methods. Note that there is no data property of the Star Performer class object. Here the bonus is calculated in the polymorphism method called Calculate Salary, the bonus being some $x\%$ of the basic salary, this being a data property inherited from Employee. The value of the basic salary is inherited from the Calculate Salary method in the Employee class object, which issues a response to the message from Calculate Bonus.

A more complete example of data property inheritance is illustrated in figure 6.7. Higher and Middle Management will inherit the pay property of Administrators and thus all three class objects will have a salary of £35 000.

---

[1] This idea was described in the books *SSADM for the Advanced Practitioner* by J. Hares (Wiley, 1990) and *Information Engineering for the Advanced Practitioner* by J. Hares (Wiley, 1991) in the chapters describing how to make the two structured methods object oriented.

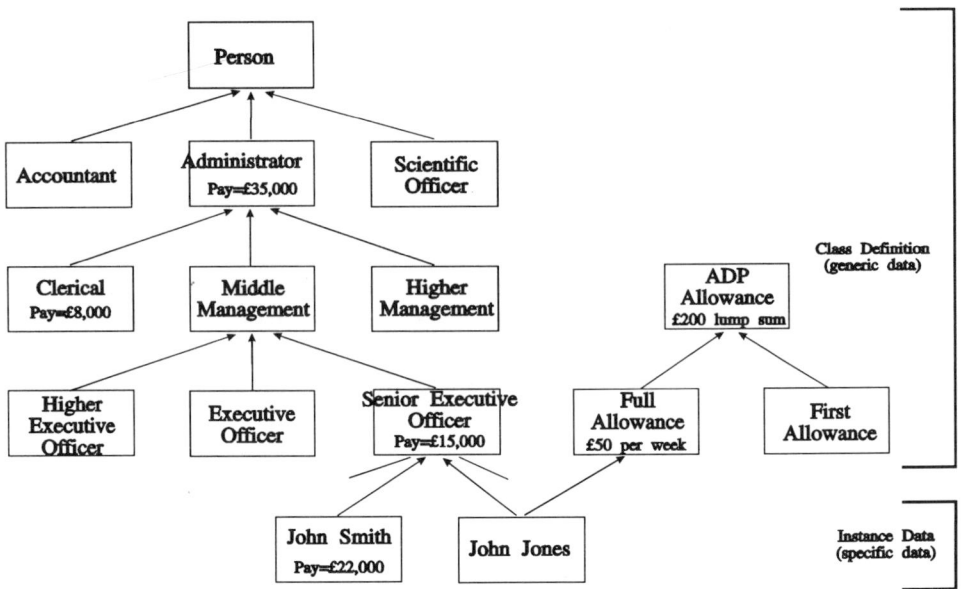

Figure 6.7 Data property inheritance

On the principle that the more particular/specialised information lower down the class model overrides the more general information higher up the class model the Senior Executive Officers will override the Administrator's salary with their own salary of £15 000 and the Clerks a salary of £8000.

The most particular information of all is the instances of the class objects. Note that John Smith and John Jones are instances of the class Senior Executive Officers. All SEOs have a salary of £15 000 because the data property of Pay is instantiated with a value of that sum of money. But of the two named instances John Smith has a pay value of £22 000. This is because he has negotiated a special deal. His special object instance level deal overrides the general class case. John Jones has no value for his salary and therefore inherits the general case salary for SEOs of £15 000. John Jones is a lucky person in that he can also inherit a Full Allowance of £50 per week and the ADP Allowance of a £200 lump sum. Jones has done well as regards allowances, Smith has done well as regards salary. The class structure shows that clerical persons get paid £8000 but that higher management inherit a salary of £35 000.

The diagram shows that accountants and scientific officers will have no salary! The model is a variation of a model produced by a government consultancy body—perhaps there is a message here! If not there is a mistake in the model. If there is no mistake then accountants and scientific officers will have to live off charity.

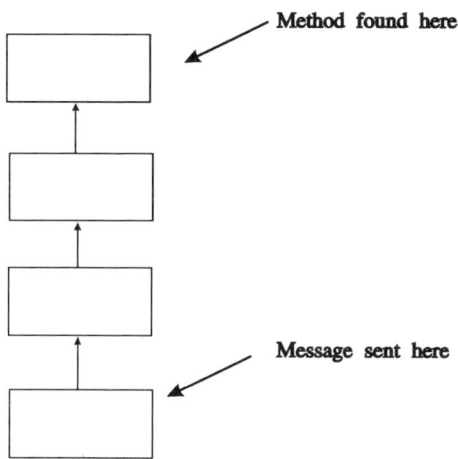

Figure 6.8 Logic inheritance (method may be at a distance from the message)

It needs to be understood that for logic inheritance the "location" of the method being invoked does not require to be known by the sender of the message. The message merely names the method and the method is located within the Class Model by the object oriented software searching up the class hierarchy using the inheritance facility. Figure 6.8 shows this. A message is sent to a class at the bottom of a class object model and the method is a property of a class at the top of the class model.

There are two approaches that can be used to obtain property inheritance. The first approach is to send messages from the sub-class method to the super-class objects in the class hierarchy with the named method in the message. The process continues up the class hierarchy until the named method is found or until the Object object is reached (for Object object see section 6.7.1.1). If no method is found an error message is issued. This is the approach used by object oriented languages that are not compiled before running, that do dynamic/late binding of the methods to the messages at run time. Such languages include SMALLTALK, which would issue the error message "doesnotunderstand" if no method is found. For this approach to work there has to be a message dictionary to list the methods of a class and the messages to which the methods can respond.

The alternative approach to dynamic explicit message searching up the class hierarchy is to compile the class hierarchy into the application program. This is the approach of the strongly typed programming languages such as C++ and Object Pascal. The compiler can resolve the method message call to a simple sub-program call and the compiled class hierarchy is searched. With such typed languages inheritance is not "do it yourself" at run time. It can be predefined.

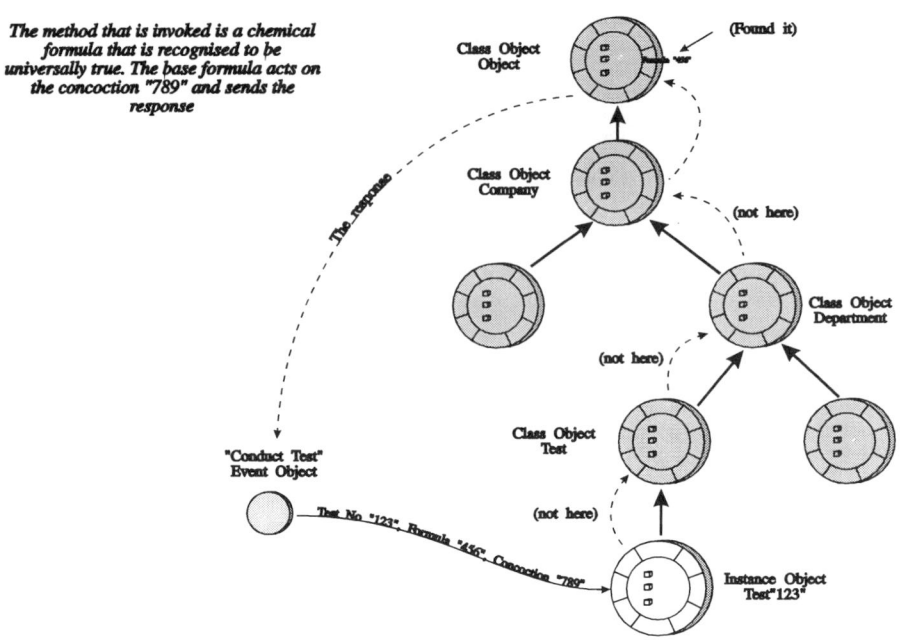

Figure 6.9 Searching for the method (specialised to general)—1

G. Booch (*Object Oriented Design with Applications* Benjamin/Cummings Publishing Inc., 1991) estimates that the dynamic approach for the searching of the methods takes about 1.5 as long as a simple sub-program call.

An example of the searching for the method named in a message using the dynamic messaging approach is illustrated in figures 6.9–6.11. The business application is that of a chemicals company, which mixed standard chemicals in various standard and non-standard ways to produce generic and company-specific products.

In the first example from the event/business requirement "Conduct Test" the message is to the class Test to access the instance "123", invoke the method Formula "456" and pass to it the parameter of Concoction "789". The Test class is at the bottom of the class model. The test instance "123" is accessed and the named method is not found. The class of the instance is accessed and again the named method is not found. The class above Test is accessed and found to be Department. The method is not within the Department class object. The Company class above Department is accessed and again the Formula "456" method is not found. The Object class is finally accessed to find the named method. It so happens that Formula "456" is a standard universally true chemical formula and is therefore at the top of the class model. The method executes and returns a response to the event level class Conduct Test "123".

The Object Oriented Facilities

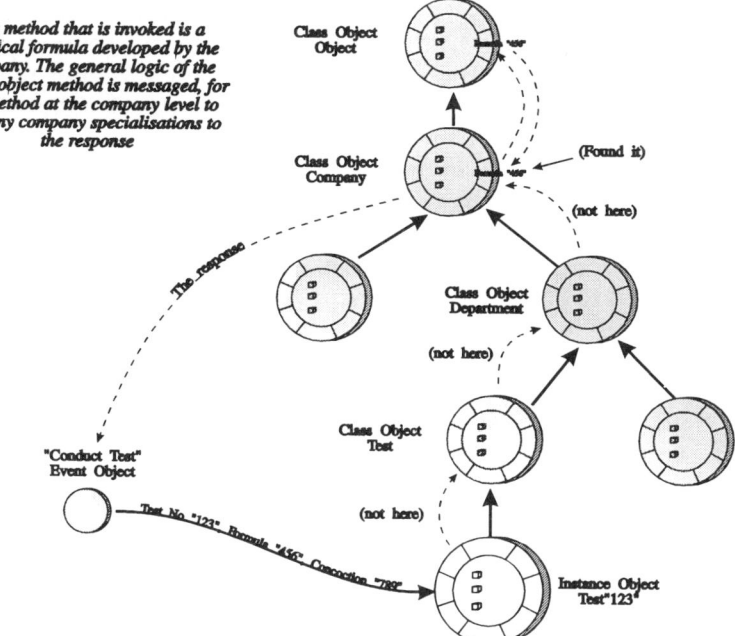

Figure 6.10 Searching for the method (specialised to general)—2

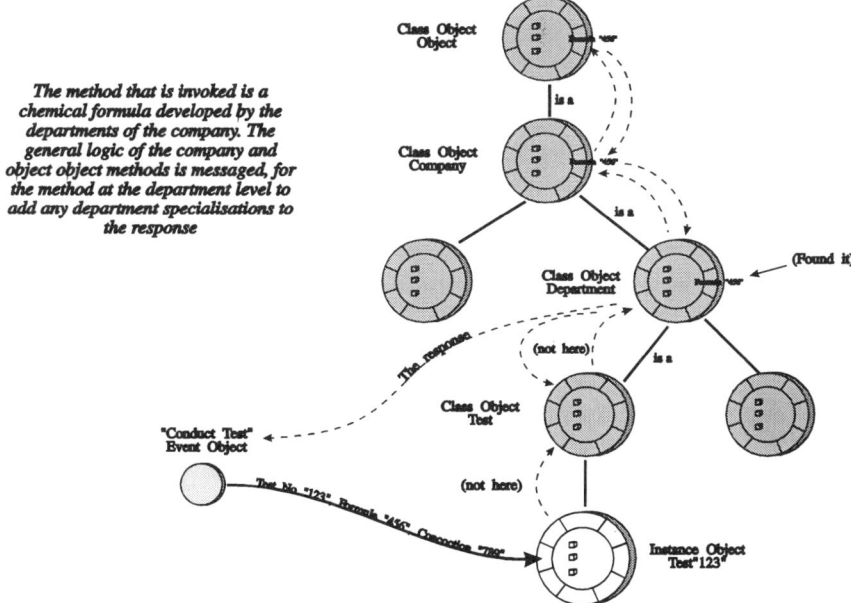

Figure 6.11 Searching for the method (specialised to general)—3

The second example is the same as the first except that there is a polymorphic modification to the formula made by the company. It is still the same named formula process but with a company variation, the variation being supported by adding a sub-class method in the Company class. The message to the instance of the Test class is therefore the same (here is an example of the benefits of information hiding and the change by addition capabilities of object orientation). The searching for the method up the class object model is undertaken and in this case stops at the Company class. The method in the Company class sends a message to the appropriate method in the Object class to obtain the response of the general case formula and then adds the changes to the general case in the invoked method in the Company class object. The method in the Company class sends the message response to the Conduct Test event class.

The third example extends the specialisation yet further. A Department within the Company has made a modification to the company's own modification of the general case formula. The message is sent as normal and the searching stops at the Department class. This sends a message to obtain the response of the Company modifications, which sends a message to obtain the response of the general case formula—additional change on additional change on the base general case. The method in the Department class sends the message response to the Conduct Test event class.

Note that the inheritance of the sub-class methods of the super-class methods is explicit through the sending of messages. This is the SMALLTALK approach. With such object oriented programming languages inheritance is obtained by the application programmers requiring to know that the method may or may not be anywhere within the class hierarchy and to send messages until it is found, with the messaging being handled explicitly and dynamically by the application programmer. With such technology property inheritance is "do it yourself" in the DML of the programming language. Note also that the event level logic of the Conduct Test abstract class is totally isolated from the changes in the inheritance class hierarchy. It continues to receive the same message response.

There is, unfortunately, no possibility of property inheritance being supported in the DDL of the database schema definition, even though the inheritance is following the fixed relationships of the sub-class object to the super-class object. It would be nice to have the DDL definition state that if there is a message to the method XYZ then send a message to method ABC in such and such a super-class object. The problem is the inability to specify when the inheritance is required in the sequence of logic in the method of the sub-class object (the messages could be sent any time between the beginning and the end of the method logic) and to the fact that the argument(s) in the message from the sub-class object method to the super-class object method from which inheritance information is sought may well vary at run time. It would be stupid to have the inheritance information "presented" to the

The Object Oriented Facilities

sub-class object method at the end of processing if the requirement in the logic is for the information at the beginning of processing. The only way to have the inherited information to be presented at the correct time is for the inheritance messaging to be DML controlled in the sub-class object method.

Another more explicit example of property inheritance can be seen in figure 6.12. An instance of a container crane is being added into the objectbase. The Date of Service and the Nickname are passed as arguments but no lift capacity. This property is instantiated at the class object level with a value of "2" tons, so that the value for the instance is inherited from the class.

Class property inheritance can be to n levels in the class object model. In the example in figure 6.7 there are four levels.

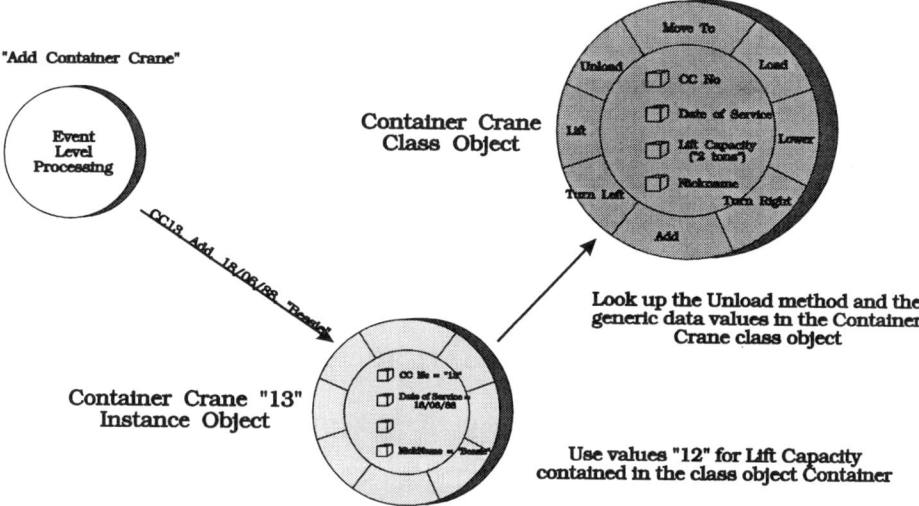

Figure 6.12 Accessing instance and class information

Aggregation property inheritance   Aggregation inheritance is obtained when their data properties and their relevant/associated methods are abstracted as generalised super-class objects. The logic properties are in third normal form to a data property of the abstracted aggregation classes. An example of this is given in section 6.6.1.2.

There can be n levels of aggregation hierarchy just as there can be for class. There could be a aggregation hierarchy of an engine has cylinders has a piston has a rod has a. . . .

6.6.1.4 Relationship semantics and semantic nets

There are three kinds of semantic descriptions of the relationships between the classes in a class model. They are illustrated in figure 6.6. The semantic types are the "is a" semantic for class property inheritance, the "has a" type for aggregation property inheritance and the "free form" semantics for the description of the business relationship between classes and no property inheritance. Each relationship between classes can be for one of these reasons, but only one. It would not be valid for a relationship between any two class objects in the class object model to have two semantic descriptions— for example, a "is a" and a "free form" semantic description at the same time.

The class model to the right of figure 2.4 demonstrates the use of the "is a" semantic. It shows that the class Star Performer "is a" Employee, who in turn "is a" Person. The purpose of the "is a" semantic is to identify a class hierarchy/common behaviour between objects, from which class property inheritance is obtained. There is a class hierarchy of common behaviour between Star Employee, Employee and Person with the sub-class objects inheriting the properties of the super-class objects.

The middle class model is of the classes Customer and Order. There is no common behaviour here—the Orders do not inherit the properties of a Customer. It would be stupid to say that an Order "is a" Customer and there are no common data properties and hence behaviour. The semantic description used is of the free form type, describing the business purpose of the relationship between the classes. In business terms customers "place" orders and orders are "received from" customers. Note that with the free form semantics there is a description at both ends of the relationship, from the master-class to the detail class and from the detail-class to the master class. This is because the business purpose of the relationship is different from the master class to the detail class. There is no super- and sub-classing of objects, only master to detail with a cardinality of $1-n$. The free form type semantic description is not specific to object orientation.

There is a school of thought that says that this class and key based approach to data modelling is too crude and does not fully model the full "richness" of the relationships between the individual data properties. The above "is a" and free form semantics are *inter-object relationship* modelling mechanisms only. They pay no attention to any relationships between the properties within an object. Yet there are such relationships.

The third type of semantic description is useful for aggregation modelling of the relationships at the data property level. What is being modelled here is *intra-object relationships*, that is, the relationship of the abstracted data properties to the prime key of the class (data third normal form) and of methods dependent on the abstracted non-key data properties (logic in third normal form). The traditional data modelling approach has been to key

The Object Oriented Facilities

based entities, where an entity is identified by a prime key data property and on which the other data properties are dependent. An entity could be a Customer with a prime key data attribute of Customer Number, and against the key would be an array of other data attributes that describe the customer, such as Customer Name and Customer Address. But the component of data modelling is the prime key of an entity and the dependent data properties.

The individual data properties can be pulled out as class objects in their own right and related to the "home" class object with the "has a" relationship, as in aggregation modelling. Thus Customer Name becomes a class object in its own right and related to the prime key of the base class with the semantic description. It follows that Customer Number "has a" Customer Name. The relationship and semantic description is of the data properties to the prime key. It would be nonsense to relate the data properties to each other, such as Customer Name "has a" Customer Address or the other way around. This data property abstraction should only be done when there is a method processing against the data property. If the data property was abstracted into an aggregation object class without an associated method it would not be object oriented design, as the aggregation object class would not contain a method. Object orientation "say" that data properties can only be accessed via a method.

One can now create a full semantic net of a class model, the semantic net showing the three types of relationship between classes, the business and abstracted class inter-object relationships and the abstracted aggregation intra-object relationships.

For the purpose of the point being made semantic nets model data at two levels, inter-object and intra-object, between classes and between the data properties of the classes. The inter-class object support is with the "is a" and the free form semantic while the intra-object relationship of the data properties within a class object is with the "has a" relationship. This can be seen in figure 6.6. The model shows the inter-object relationship between Star Performer and Employee and intra-object with Employee having a Date of Employment and a Salary.

It needs to be appreciated that there are three modelling paradigms being used in parallel in the class object model—the "is a" class and the "has a" aggregate model based model superimposed on and surrounding the key based model. The abstracted class model is an extension of the logical data model, the extension being the abstracted objects based on class and aggregation from the original entities, now also classes, based on keys. The relationship between the objects based on keys is still described by the traditional free form semantics of the logical data model, with the class behavioural and aggregation semantics for the relationship between the still key based but abstracted classes. This use of a semantic description of the relationship between objects is an excellent mechanism for creating knowledge data. Employee "is a" Person, Employee "has a" Address and Customer "places" Order is a triple, a fact, as described in section 5.2.

6.6.2 Concrete and instance objects

Objects can be at two levels—class and instance. An object can be at the class and the instance levels simultaneously. An object must be at the class level but may not have any instances. If a class object has no instances it is known as an abstract class. If it has instances it is known as a concrete object. Objects are stored at the class and instance levels. Figure 6.13 shows the concrete object Container Crane defined and stored in the objectbase schema with four instances stored in the objectbase. The instances of the class object are the individual container cranes in the container port. In a traditional database environment the class would be the record/table and the instance of the class object would be the record occurrence/table row.

Some of the objects in an class model example may be only at the class level, and are abstract. Consider figure 6.3. The class objects of Mammal and Man are at the class level only—they have no instance objects. These abstract classes contain information that is always instantiated with a value and only relevant/generic to the sub-class objects and their instances. For example, there is a data property Warm Blooded of the Man class with a value of "y", and all the sub-class object instances which can inherit from it are warm blooded.

Some of the abstract classes contain only a logic property(ies), that is, a method(s). Examples of this are the class libraries and the event classes. The event classes contain the logic of the business requirements that are not normalisable to the business objects containing user data.

The class contains a template definition of all the information properties about an object that is generic to all the instances of the object. Thus, the class Container Crane for the lifting and movement of containers onto and off vessels in a container port would contain the data property definitions and the methods appropriate to all the container cranes. This is illustrated in figure 6.12, which shows the data and logic properties of the class definition of the container crane object.

There are three data properties and seven methods/logic properties. The data properties are obvious and are the same as for current database technology. The methods are blocks of logic unique to the class, with each method describing what the container crane can do, such as load a container onto a vessel and move left. The method Set Name would update the data property Nickname. The other methods are used for instructing the container crane to do something as detailed by the descriptive title to the method (the application being a realtime system with embedded software), but would not update any of the data properties.

The Object Oriented Facilities

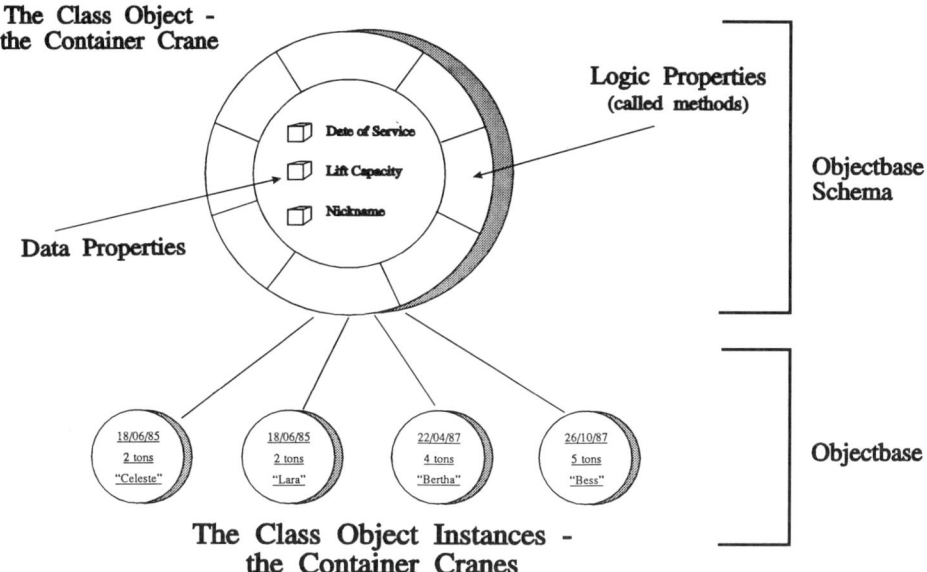

Figure 6.13 An object: its class and class instances

The class information is stored in the dictionary, instance information in the objectbase. Each class object is a "one off"—that is, there is only one instance of the class object in the dictionary schema, this being the class template definition of all the properties, such as the method name and signature (message and response formats) and the name, format and length of the data properties—for n instances of the class object, where n can be zero. The single definition of the class object Container Crane, for example, can have n instances of container cranes, 4 in figure 6.13.

If the data or logic is at the class level it is generic information, relevant to all instances of the class object. Thus, if all the container cranes had only one lift capacity the data property lift capacity would have a value, be instantiated with a value, that would define the capacity. If the capacity was 2 tons a value of 2 would be defined as an instantiated data property at the class level and all the container crane instances would be limited to 2 tons lifting capacity. Of course, some of the container cranes could have a different lift capacity, but this is incidental to the fact that most of the container cranes will have the standard lift capacity. Any data property instantiation is defined in the objectbase schema.

Some of the objects in figure 2.21 have bifurcated lines extending from the bottom of the class object. It indicates that the objects are concrete with instances of the class stored in the objectbase. Person, Employee, Broker and Star Performer objects are class objects *and* are also stored at the instance level—there are individual, real and live persons, employees, brokers and star performers. These objects contain some properties that are not instantiated with a value at the class level. For example, the Date of Birth data property in the Person class object is unique to each instance of Person. To instantiate at the class level would be meaningless, indeed it would be incorrect, as each instance of Person could have a different birthday.

The instance data properties values are not instantiated with data values in the dictionary. This is because the values to be stored in the instance data properties can vary with each instance of the class. Thus Person would have Date of Birth defined in the schema description of the class Person but stored with a potentially different value in the objectbase for each Person instance.

While the object instances are mostly used to contain unique values of the data properties for each instance of the class object the object instances can also contain logic. This is new to information modelling and is not catered for in relational technology. The employee Mary could have a unique way in which her salary is calculated and therefore requires a method to be stored at the instance level.

If the data or logic is at the instance level then it is only relevant to the single instance/occurrence of the object and none other. It could be that container crane number 12 has a lift capacity of 3 tons.

There are those of the object oriented fraternity who believe that logic at the instance level is not appropriate, but why can't Mary have a unique method for the calculation of her salary? If Mary can have unique data values why not unique logic values? After all, object orientation normalises and stores logic in the same way as data. The two are no different. One of the authors has worked on a project producing an object oriented logical design (using the techniques described in chapter 4) where some of the major parts of the business application methods were at the instance level. The application was for testing particular hardware/software configurations, each configuration being used for experiments on new types of telecommunications equipment. For example, tests of a particular nature were being frequently conducted and for each test instance there were a variety of methods. And the logic of each test instance could vary.

This latter point illustrates another use of object instances—to overrule class information. Any class values can be overridden at the instance level of the class object—the instance information is more particular than the general class information and therefore overrides it. Consider the Employee class object. A method is defined to calculate the basic salary for all employees, *unless instantiated otherwise at the instance level*. Mary's method is more particular than the classes method, so Mary gets her salary and not that of the general employee. The same is true for data. At the class level Man is instantiated with a value of having two legs, yet poor Henry has his Number of Legs property instantiated with a value of "1". This is because Henry has had a leg amputated. None of the other Employee instances contain a value in the property Number of Legs, so they can inherit the generic value of "2" from the Man class object. By contrast, with the Date of Birth property there is no need for overruling as there is no instantiation of the property at the class level.

6.6.3 Property instantiation

With relational technology the values of the data attributes are stored only at the table row level, at the object instance level. This is true even if there is a value that is common to all the instances. For example, if the attribute was number of legs then all the table rows would have a value of 2 except for those poor persons who have had one or more legs removed. If there are 1000 persons there would be 1000 values of "2" legs minus the number of amputees. This is clearly wasteful. A smart designer would create a separate table specifically for the generic value of "2" legs and use it as a table look-up for the general case. But the relational solution to the problem of generic data values is DIY.

Object orientation provides a built-in facility. It is called property instantiation, the ability to define a generic value for a data property at the class level in the schema and for the instances of the class object to inherit the generic value.

The Mammal data property with an instantiated value of "Warm Blooded" is applicable to all mammalian species and hence to all instances of the class sub-objects of Man, Person, Shareholder, Employee and Star Employee. The data property of Number of Legs with a class value of "2" is generally applicable to all instances of Man and to the sub-objects Person, Employee and Star Employee. All the sub-class objects within the class hierarchy inherit these generic values. However, the data property of Date of Birth as a property of Person is not instantiated with a value as it is not generally applicable to all instances of Persons—each Person can have a different birthday.

The data properties of abstract class objects must be instantiated with a value. It would be pointless to have any data properties that did not have a value, as there are no instance objects of the abstract class object to use the data properties with no value. The abstract class object would then serve no purpose, not even as a template description of an object. It can be seen in figure 6.4 that the Warm Blooded property of Mammal is instantiated with a value of "y" for yes and the "Number of Legs" data property for the Man class object has a value of "2". The most common abstract class objects are those that contain methods only.

6.6.4 Encapsulation

There are two parts to encapsulation—encapsulation of the data properties and encapsulation of the logic properties. As already described, objects are composed of a data component in the form of attributes and a logic component in the form of methods, and the data component can only be accessed via the logic component. The data component is thus "encapsulated" by the logic component. This is graphically illustrated in figure 6.2, with the methods surrounding the data. The great advantage of data encapsulation is that the user of the object, when sending a message to trigger one of the methods, does not need to know and does not care about the format of the data properties.

Abstract data type specifications describe class data structures not as a set of properties but as a list of services/methods available on the data structures. The only thing of interest to a user of the object is how the object appears in the interface—that is, the message and the associated response. The internal processing of the methods and the data properties they might access is irrelevant to the outside world, to the sender of the message and the receiver of the response. Therefore the methods are themselves in turn encapsulated by the interface.

The great advantage of this facility is that the user of the object, when sending a message to trigger one of the methods, does not need to know and does not care about the format of the attributes. Objects are therefore excellent implementations of the facility of abstract data types. Abstract data types specifications describe class data structures not as a set of attributes but as a list of services/methods available on the data structures. *The only thing of interest to a user of the object is how the object appears in the interface—that is, the message format—and what methods are provided and what is done by the methods to the data attributes held within the object.* The internal processing of the methods is irrelevant to the outside world.

Some examples contrasting the traditional approach to application programming and the object oriented approach can best illustrate encapsulation and the benefits thereby achieved.

Consider the following. A conventional application program requires to read in two real number variables M1 and M2, multiply them together and assign the result to a third number variable M3. Real numbers are usually represented by two integers, the mantissa and the exponent. The program reads two strings S and converts them into mantissa and exponent form. After the multiplication the result is converted back into a string and the string is written out. The code (in the style of an abstract programming language in the form of C++) for such a program could be:

```
Main ()
{
int M1, M2, M3, E1, E2, E3;
string S;
read S
convert S to M1 E1    (this is function 1)
```

The Object Oriented Facilities

```
read S
convert S to M2 E2     (this is function 1)
multiply M1 E1 and M2 E2 to get M3 E3    (this is function 2)
convert M3 E3 to S     (this is function 3)
write S
}
```

This traditional application program suffers from a number of problems. The major one is that there is no hiding of the data type from the function logic, which has to know the data type and its form. The form of the data in the initialisation and the logic in the functions are visible to each other. The "how it is to be achieved" logic of the functions is intermingled with the data. This can be seen in figure 6.14(a), where the format of the data is visible to the three functions.

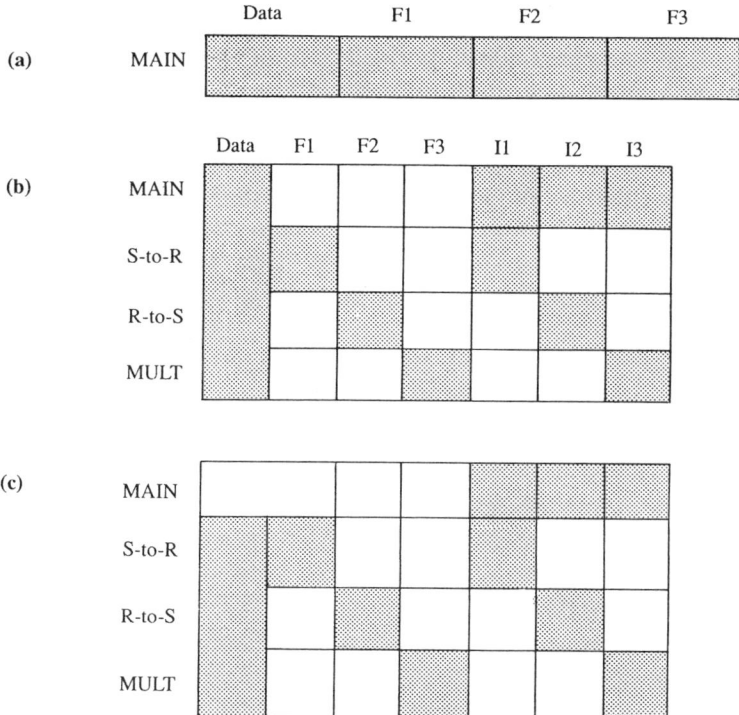

Figure 6.14   Degrees of encapsulation

A more sophisticated approach is to separate the functions, to store them as independent routines and to refer to them by name. Those references are the interfaces. Function 1 is given a reference of s_to_r (string to real), function 2 is given the reference of mult and function 3 is given the reference of r_to_s. The application program merely has to refer to the references, such as s_to_r and mult, and the functions are invoked. The code of such an approach would be:

```
Main ()
{
int M1, M2, M3, E1, E2, E3;
string S;
read S; s_to_r (S,M1,E1);
read S; s_to_r (S, M2,E2);
mult (M1, E1, M2, E2, M3 E3);
r_to_s (M3, E3, S)
write S;
}

s_to_r (S,M,E) {logic....};
r_to_s (M,E,S) {logic....};
mult (M1, E1, M2, E2, M3, E3) {logic....};
```

The advantages here is that an interface (the references to the functions) has been created for the functions, with the "how" logic of the functions separated from the "what is required" logic of the event component of the application program. However there still is no encapsulation. The interfaces can still see the data (they require to know that the number is a mantissa and an exponent), as do the functions themselves. This is illustrated in figure 6.14(b).

The third approach is to create a new abstract data type R for the mantissa and exponent of a real number. The logic of the event and the routines would be as follows:

```
Main ()
{
int R1, R2, R3;
string S;
read S; s_to_r (S,R1);
read S; s_to_r (S, R2);
mult (R1, R2, R3);
r_to_s (R3,S)
write S;
}
```

The Object Oriented Facilities

```
s_to_r (S,R) {logic....};
r_to_s (R,S) {logic....};
mult (R1, R2, R3) {logic....};
```

The interfaces cannot see the data, as far as they are concerned the data is merely a real number, and it is only the functions (which now become the methods in the objects) that contain the logic to convert the mantissa and exponent to a real number. The interfaces are only able to see the routines in that they make reference to the routines. The only link with the outside world as far as the object is concerned is the interface. Complete encapsulation has now been achieved. The object contains the interface, the data and the routines, which because the technology is object oriented, are stored as methods within the object rather then separately as in the second approach. The full encapsulation can be seen in figure 6.14(c), where the interfaces cannot see the format of the data, only the references to the functions. It is the functions, the methods, which then see and access the data. The format of the data is transparent, is abstract, as far as the interface, the message, is concerned. The three component rings of an object described in section 6.5 have been created.

Encapsulation is therefore the ultimate form of expressing stability in a computer system design. The processing logic and the data in an object can be modified to whatever degree is necessary and it matters not to the outside world. *The impact of change is restricted to the object.* There are no ripple effects to other objects.

General Format

Container Crane "12"	Unload	Vessel Cell "S17"
Object Instance key/objects	Method Name	Parameter(s)

Smalltalk Format

Container_Crane_12 Unload: Vessel_Cell_S17

C++ Format

Container_Crane_12. Unload (S17)

Note: - the object identity is the equivalent of table name + key value

Figure 6.15 The format of a message

It should be appreciated that encapsulation does not assist in software re-use, only in software maintenance. Encapsulation enables the more specialised sub-objects to be completely decoupled from the super-class object.

Not all object oriented programming languages provide the ability to enforce encapsulation. SMALLTALK hides all data values by default but C++ requires the data values to be declared "private" in order to be hidden.

6.6.5 Messages and responses

The message is the mechanism by which objects can "talk" to each other at application run time. A message is sent from a method in one class or object instance to invoke a named method in another class object or instance. The invoked method executes, accesses the required data properties of the object instances to be accessed as specified in the message and returns a response of the processed data to the source object that sent the message.

A message is therefore similar in role to a procedure call in current database technology. However, it is not a procedure call.

The format of a message (known as the signature) is shown in figure 6.15 and is different from the conventional program procedure call. There are three component parts to any message—the key/search criteria of the object instances to be accessed, the name of the method to be invoked and any parameters the invoked method is to use.

The first two parts of the message are mandatory, but the parameter/argument list are optional. The first component of the message makes the method similar in role to an SQL statement, but with the limitation that the access is record-at-a-time as in pre-relational database and not the set access of relational database. The invoked method contains the access logic to the object instances specified in the first part of the message. The name method has to be a public method—messages are not able to invoke private methods. Messages can therefore only trigger logic of importance to the business rather than the system maintenance of the objectbase data.

The first component of the message distinguishes the message from a procedure call—the procedure call does not contain a specific component for the specification of the table rows to be accessed in the database. Procedure calls, of course, contain the name of the procedure to be invoked and the argument list is the same in that it is optional.

Examples of messages to the Container Crane class are shown in figure 6.16. The first message invokes the method for unloading a container from Cell S17 of the vessel (cell number 17 on the starboard side of the vessel) the container crane is working on. The method accesses the Lift Capacity of the container crane instance number 12 to ensure that the weight of the container to be unloaded does not exceed the capacity. The second message invokes

The Object Oriented Facilities

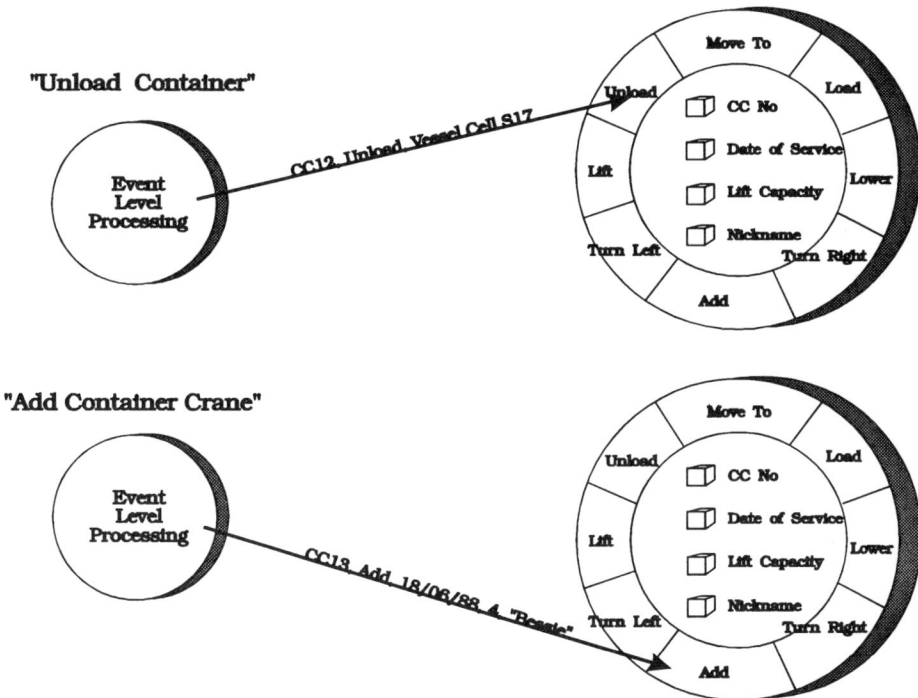

Figure 6.16 Messages to Container Crane (actually to the instances of the object class)

the method Add to insert a new instance of container crane number "12" with the data property values of a date of service of 15 May 1992, a lift capacity of 4 tons and a nickname of "Bessie".

6.6.6 Methods

Methods contain the logic of the business application and of the system processes. The way they are incorporated into the computer system design is unique to object orientation. They can either be properties of the classes containing user data (a method for Calculate Salary as a property of the class object Employee) or abstract classes with no data (a class containing nothing but a method—the Date Edit routine, a common procedure or an event class object). There are two kinds of methods, private and public. They are discussed below.

6.6.6.1 Private methods

Private methods are system triggered when an object is accessed. They are private because they are not seen and cannot be used by the users. Their prime role is threefold—to edit the data properties in the objects as they are inserted and updated into the objectbase, to prevent improper deleting from the objectbase (you cannot delete an employee if the Date Left Company property is blank) and to provide the usual objectbase referential integrity checking facilities. Private methods are therefore really concerned with data maintenance business requirements and objectbase integrity. For the first two roles many would recognise private methods as normal business processing, possibly developed as database procedures. In the latter role many would recognise private methods as database rules.

Both roles are now supported by most of the leading relational file handlers with stored procedures for public methods, triggers for private methods and rules for pre- and post-conditions.

As database procedures the private methods can be defined for inserts, updates and deletes. For obvious reasons they can only be defined at the class level. For example, it could be that all employees' salaries are within specified value ranges for a given grade. The private method would be specified at the employee class level and be automatically invoked whenever an employee is inserted into the objectbase, so as to ensure that the employee had a valid salary according to their grade.

However, a private method can also be defined at the instance level. This is only pertinent if the object is being updated or deleted, for the simple reason that the object instance must have been loaded beforehand and the private method specified subsequently for the instance. An object instance private method would be appropriate if it was necessary to edit in a unique and not a general way the data in an object instance that is being updated or when an object instance is being deleted. For example, for an update private method it could be that the employee instance Mary's salary is the exception to the general rule and that her salary is uniquely outside the general value range for her grade. A private method for editing Mary's salary would therefore be defined at the object instance level. Thus, if Mary's salary is being modified to reflect her special condition then the private method would be invoked whenever the salary update was being executed. As regards a delete it could be that at the class level employees cannot be marked as retired if their company pension information is not in order, but the employee instance of Henry has adopted a personal pension plan and opted out of the company's pension scheme. A private method for the Henry object instance would therefore be defined. Both Mary's and Henry's private methods, being instance particular, would override the class general method.

A private method can be triggered in three ways—by receiving a trigger from one of the public methods of the class to which the private methods are private (Insert Object Instance public method triggers a private method to edit some of the data being input) or automatically whenever a business or database condition is obtained. This latter form of private processing is particularly done when the method is in the form of a rule. An example of a business condition could be that "WHEN stock level is below a certain level then create an order", the conclusion being the sending of an message to the method "Create Order" in the Order class. When the given condition is true, the stock level is below that specified, the conclusion can be drawn and an order is created. A database condition is typically a referential check of the type "when new order is created check customer exists".

6.6.6.2 Public methods

Public methods are unique to object oriented systems. They contain the application logic, the logic that supports a user business requirement, but logic that is normalisable to the classes containing user data. For example, the public method Calculate Salary for the calculation of an employer's salary is defined for and stored in the Employee class.

The public methods are those that are invoked by the receipt of a message from the event class of a business requirement or from another class or object instance. The message contains the object instance value to be accessed, a method name, the name typically being a meaningful description of the function being called, such as Calculate Salary for calculating an annual salary, and relevant arguments. The method, of course, has been defined by name as a property of the object in the objectbase schema. The schema therefore points the message to the class specified in the method name, which then accesses the object instance(s) specified in the message. The public method will execute its logic, access the data properties as appropriate, process any arguments presented and return a result to the issuer of the message.

The initial message trigger for a set of public methods appropriate to a business requirement is from an application program event class. If the business requirement requires to access the data in multiple classes, then there need to be public methods appropriate to the business requirement in each class to be accessed. Each of these public methods requires to be invoked by a message.

For example, when a business requirement requires to access multiple objects, "For a specified customer display all orders for products greater than £10", then co-ordinated messages must be sent to the appropriate Customer, Order and Product classes to trigger the public methods. This can either be from the event object or from other objects previously accessed.

The data component in the objects can only be accessed by the public methods. It is the public methods that encapsulate the private data.

Public methods can contain what are called pre- and post-conditions. The prime purpose of this facility is to enable specified conditions to be true (usually in the form of data properties having a certain value that together indicate that the object instance being processed is in a given and correct state) before and after the processing of the method. It could be that the method is for the painting of a Car class object. The pre-condition is an assertion that the state of the car instance being accessed is that the colour data property must be blank. The post-condition is an assertion that the colour of the car is non-blank.

Pre- and post-conditions are an excellent facility for ensuring that logic only executes under the correct conditions and that the result of the execution is provably correct.

The EIFFEL object oriented programming language uses pre- and post-conditions for the purpose of checking that the data being input and output is valid.

6.6.7 Polymorphism

Polymorphism relates to logic and therefore to the methods within objects. It is considered in this chapter as regards its information modelling capabilities and how it affects the design of the objectbase.

Polymorphism is the ability of a function/process to have n a common business purpose as defined by its name but to have variations of its functionality/versions of its logic. Each version is defined as a method in a different class but with a common name. The version to be executed depends on the object instance being messaged.[1]

There could be a method called Calculate Salary, but within this common name there could be a version for Employee and another version for Broker, or there could be versions for Salary Grade 1 and Salary Grade 2. Calculate Salary is polymorphic. Which version is executed depends on the object instance being accessed. If the object instance being accessed is of the class Employee then the method version is for the Employee and if the object instance being accessed is of the class object Broker then the method version is for the Broker. Polymorphism therefore enables the same message to a named method to refer to many classes, each of which can produce a different response depending on the object instance being messaged. The sender of the message does not have to know that the named method is polymorphic.

---

[1] Polymorphism therefore requires what is called dynamic binding (also known as late binding). Dynamic binding binds the method to the message because the object instance to be accessed is not known until run time. Static binding (also known as early binding) compiles the names of the classes and any variable to the logic of the application program.

The Object Oriented Facilities 341

Message Calculate Salary for object instance "123" does not know whether the version of the method invoked is for an Employee or a Broker.

Polymorphism is a useful addition to the inheritance facility for aiding software reuse. Consider figure 6.17. There are two examples of polymorphic methods. The business case is the container port authority. In the upper of the examples there are three class objects—Container Crane (for the loading and unloading of containers onto and off vessels), Van Carriers (for the picking up of containers from the ground and moving them at slow speed to another location) and service trailers (for the higher speed movement of containers over longer distances). Each of the class objects have a method for checking the status of the container mover—is it free for allocating to the movement of a container? The business purpose of the three methods, one for each of the classes is essentially the same, that is, to access the status data property and ascertain the status of the container mover and if the status is free to allocate the free container mover to a container waiting to be moved. The variation of the functionality is that the instructions that go with the movement of a container differ depending on the type of container mover and how this information is to be presented to the user.

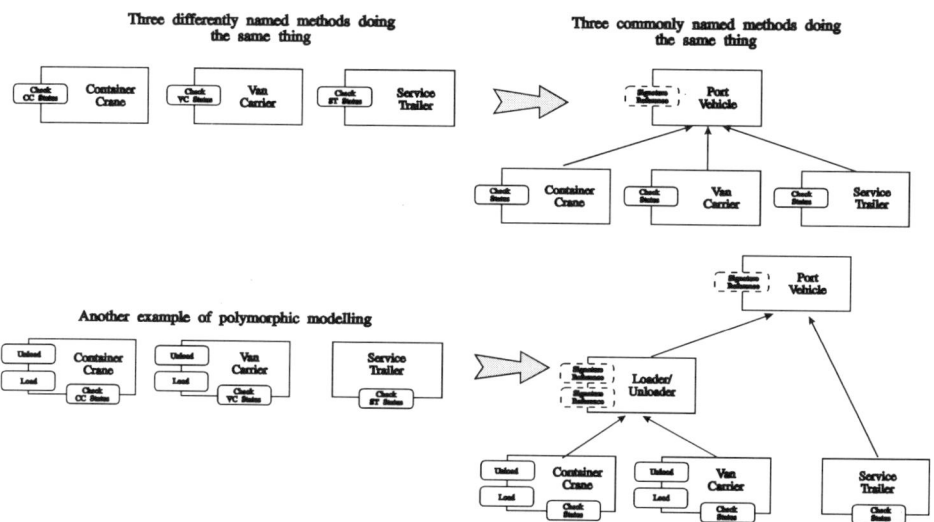

Figure 6.17 Polymorphic information modelling

The methods can be made polymorphic. Being common they can be generalised as for normal information abstraction. A super-class object of Port Vehicle is created and a polymorphic method with a suitably generalised name of Check Status is defined in the Port Vehicle class object. The three original base methods are referenced by the common name of Check Status with a common signature/message format. Each base method is still stored on disk as part of the definition of the base class object as illustrated but as a "sub-method" of the polymorphic method Check Status.

A message is sent to invoke the method Check Status, that is, the method in the Port Vehicle class. The object oriented software recognises that the name is a polymorphic signature, or interface, reference to the three sub-classes Check Status methods. If the container mover key reference in the message is a van carrier then the base Check Status method in the Van Carrier class object is "invisibly" invoked. This is not seen by the sender of the message, who merely has to say "move container 12345 with van carrier 54321" with the Check Status method. This generalisation of the Check Status method to a super-class object is seen in the top right of figure 6.17.

The second example takes polymorphic modelling of methods a stage further. The Check Status method of the first example is still present, but there is again another common method of Unload in the Container Crane and the Van Carrier class objects, but not the Service Trailer class object. Both container cranes and van carriers have the ability to load and unload/pick up and drop containers, but if this is being done by a container crane then it is in relation to a cell in the vessel being loaded or unloaded. There is no relationship to a vessel as far as the van carrier is concerned. The method Unload can be made polymorphic. But because it does not relate to all the container movers it cannot be generalised to the Port Vehicle. A new class that is a super-class to the Container Crane and Service Trailer and a sub-class of Port Vehicle requires to be created. This is illustrated in the bottom right of figure 6.17.

It has been argued that polymorphism will eliminate much of the use of the condition IF/CASE statement when used for a given entity/class in application programming.

```
If Container Crane
then do . . .;
if Van Carrier
then do . . .;
if Service Trailer
then do . . .;
type circle would disappear.
```

It is also a more powerful modelling facility for conditional logic. There could be some logic for the Employee class object that states that if the length of service is less than 10 years of service then do this, if more than 10 years of service then do that and if more than 20 years of service then do the other. There are several problems with this long practised approach—the conditionality is hard coded and potentially laborious. With polymorphism there could be three separate methods for each of the above two sets of conditions, each condition given a suitable polymorphic name and treated

The Class Model

as a polymorphic method with a signature reference in a super-class object. Problem solved—and one could add new methods to new conditions without affecting the existing methods.

6.7 THE CLASS MODEL

The OO class model is to object orientation what the logical data model/entity model is to relational technology. The basic structure of an object class model and the different types of class objects are illustrated in figure 6.18.

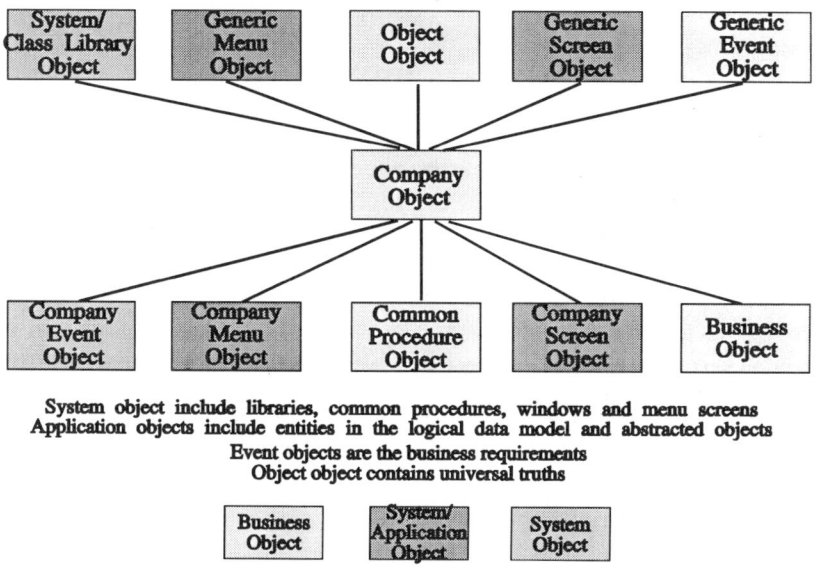

Figure 6.18 Basic structure of an object oriented information model

There are two basic groups of object class—the business objects that support the application being designed and developed and the system objects that support the computer system on which the application will run.

6.7.1 The business object classes

The origins of many of the business objects in a class model are as entities in a traditional logical data model. The entity Employee becomes the class object Employee. The business classes are therefore based on the entities and are therefore classes containing user data. But there may well be other derived business classes abstracted as super-classes through generalisation or as sub-classes through specialisation of the base entities according to object oriented abstraction rules. These business objects of the base entities and their class and aggregation abstractions contain both user data and methods.

The other business object classes contain a method property of the event level logic of the business requirement. Each business requirement is a "Do something" event, such as "Calculate Salary" and "Make Payment". As already explained, some 80% of the logic of the event can be normalised to the classes containing user data. Part of the logic of Calculate Salary relates to Employee and is therefore defined as a method for the Employee class. The residue of the logic is not normalisable to the classes containing user data and remains at the event level. This logic becomes a method-only event abstract class.

Given that the base entities of a logical data model continue to be preserved in a class model but can be abstracted into further super- and sub-classes, plus the addition of the event and system class objects it follows that a class model is larger, usually much larger, than the equivalent logical data model.

The base entities of a logical data model are shown in figure 6.19 and the resultant class model is shown in figure 6.20. Each rectangle in the logical data model represents an entity, a "thing of interest", to the application system about which data can be recorded, and the line between the entities with an arrowhead at one end represents a one-to-many relationship between two objects, the arrowhead indicating the direction of detail class to the master class. Commission is a detail of Broker. Each rectangle in the class model represents an object class and the line between the classes with an arrowhead at one end represents a relationship between two classes. When the arrowhead points downwards the relationship is between the business classes remaining from the source logical data model entities from which the class model has been abstracted. This can be seen with the Broker and Share Allocation classes. Where the relationship arrowhead points upwards this is because of a class or aggregation generalisation abstraction of super-classes or a class specialisation abstraction of sub-classes. The arrowhead indicates the direction of property inheritance and pointing to the super-class object, with the arrowbase pointing from the sub-class object. Employee is a sub-class of Person in a class hierarchy. Address is a super-class of Employee in an aggregation hierarchy.

The generalisation from the base entities in the class object model are Man and Person and the specialisations are Star Performer and Star Bonus. There has also been sub-typing in that the Company entity has been split into two sub-types. Some of the data properties are relevant only to the Employees of the company and others to the Valuation of the company. The data about the internal organisation of the company is relevant only to the Employees and the data about the financial value is only relevant to the Share Allocations

The Class Model 345

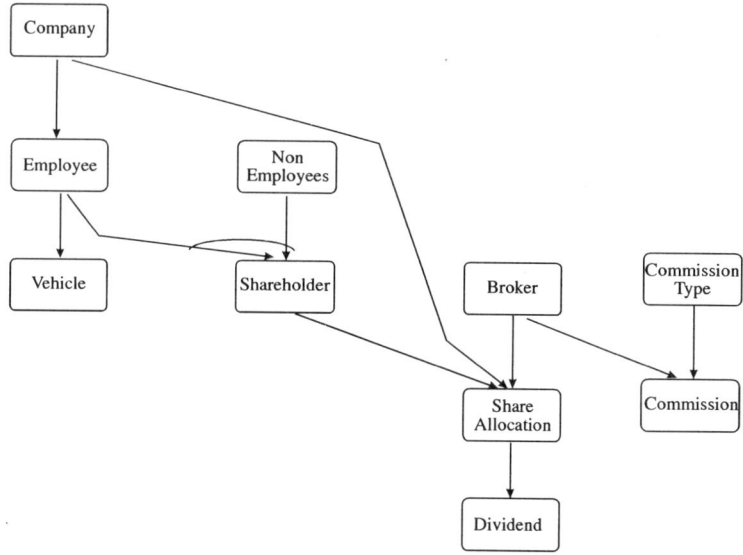

Figure 6.19 A logical data model

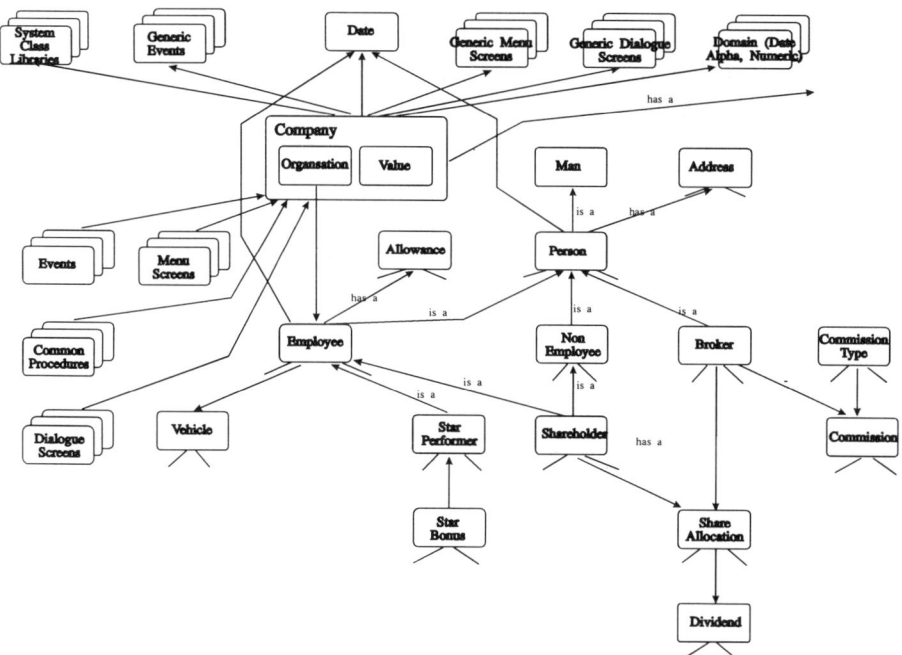

Figure 6.20 An object oriented class object model

of the Company. The Company class object has therefore been split into Company/Address and Company/Valuation.

There are six more business class objects in the class object model than entities in the logical data model—two super-class objects, two sub-class objects and two sub-type class objects.

6.7.1.1 The Object class[1]

Object orientation is able to model information to its appropriate level of abstraction. As described in chapter 1 this is not possible with logical data modelling. The result is that objects "higher up" the class object model are more general than the objects "lower down" the class object model. The more that one goes up the class object model, the more general the information and the more one goes down the class object model, the more specialised the information. It therefore follows that information that is universally true must be at the top of the class model.

The Object object is the class that contains information that is recognised to be universally true (such as the world is round and that if you are in location A you cannot be in location B) and of relevance to the business application. The author worked on a project concerned with chemical formulas. The company worked within the universally accepted laws of chemistry but mixed the various chemicals in such as way as to produce company- or department-specific products. The raw chemical formulae were defined as method properties of the Object object and the results of the mixing were defined as specified methods in class objects lower in the class object model, such as the Company and Department class objects. This example was used in figures 6.9–6.11 to show the inheritance of methods.

6.7.1.2 The Company class

This class contains the information that is more specific, more specialised, than universal truths, but is information that is applied corporately wide. Such company information could be a business rule that states that overtime for all employees is standard pay × 1.25. This would be stored as a method within the company class as the method logic is applicable to all employees within the company.

---

[1] This is sometimes called the base class. In SMALLTALK this class object is called Object and in Object Pascal TObject. For C++ the top class object is anonymous.

6.7.1.3 The Business class

The most common classes in the class model are the Business objects, all based on the entities in a logical data model. A typical Business object would be Employee, which contains the data and logic that is pertinent/normalisable to employees. Examples of this could include the Telephone Number of the employee for the data component and the Calculate Annual Salary of the employee for the logic component.

Users of relational databases would recognise the Business object as a standard entity/table of information. The Business object almost always contains user data, as in a conventional logical data model. The only Business objects with no user data are those where all the data properties have been abstracted.

All the entities in the figure 6.19 logical data model are in the OO model, with class and aggregation generalisation abstraction and class specialisation of the base entities creating additional object classes. Examples of class generalisation from the logical data model are the class objects of Person and Man and of class specialisation are the class objects Star Performer and Star Bonus. There are also two aggregation generalisations, Address and Allowance.

Large parts of the logical data model remain unchanged, such as the Broker, Share Allocation and Commission and Shareholding entities. These objects are classes in that they all have a template objectbase definition, but none of them belong to a class hierarchy because there is no behavioural abstraction of factored out common information.

It need not be assumed that information abstraction is appropriate to all parts of a logical data model. Much can remain unchanged. *Not all object oriented facilities are universally applicable all of the time.*

6.7.1.4 The Event class

The event class contains the logic that is relevant to the event/business requirements of the application and not to the classes containing the user data. Typically they contain the 20% of logic that is not normalisable to the user data class objects. They become class objects in their own right but containing a method only. *But being specific to a business requirement event classes are the only portion of a class model that is not application independent.*

The Event class object generally contains the logic for the beginning and the end of the application program. The beginning logic would typically be for the receipt of the input screen information, the initialisation of the variables and the sending of the messages to one or more objects in the class model. The end logic would typically be any final processing of the response information and the presentation of the output screen. The logic in between would be for the sending of messages to the class objects containing the user data the business requirement requires to access and the synchronisation of the responses before final processing. Since each event is unique and distinct in its own right it cannot be part of a class hierarchy.

The placement of the Event class in the class model has been a matter of some debate. Perhaps the best place to put it from the point of view of information abstraction is as a sub-class of the Company object, as the events are obviously relevant to the company. Such a company event could "Receive Customer Order". These events are triggered by the company's internal operations. But there are events that are of interest to the company that are not company-specific in that they are not triggered by the company— they are events that are external and also affect other companies, but nevertheless the company is interested in the events as it requires to react to them. Such events can be classified as generic as opposed to company-specific. A generic event could be a government's decision to raise interest rates. Another example. The company could be a chemicals manufacturer and the events regarding the manufacture of generally available chemicals are not specific to the one company, but to any company manufacturing chemicals.

From an object oriented information modelling point of view there therefore needs to be a distinction between the generic and the company events. Being generic to n companies generic events need to be modelled as super-class events to the Company class. Company events are specific to the company and need to be modelled as sub-class events to the Company class. This can be seen in figure 6.20 where some of the events are identified as being above the Company class and some being below the Company class.

Event class objects have most important roles to play within object orientation, specifically:

- being the main procedure for an object oriented application program and acting as the controller of the "central policeman" messaging strategy (see section 6.8.7);

- acting as the "glue" between the business and the system class objects in that messages are sent from the business classes to invoke appropriate methods in the system class objects.

6.7.1.5 The menu screen class

Menu classes can be used at both the business and system levels. Those for the business of the company are designed for the user to select business requirements. Usually there is a main menu for each user role for the appropriate user to select a high level business function from an array of

The Class Model 349

functions relevant to the user role. There is then a cascade of subordinate screens for the progressively more detailed selection of individual business requirements. These business menu screens are sub-classes to the Company class. The system level menu screens are discussed in section 6.7.2.2.

6.7.1.6 The dialogue screen class

The dialogue screens support the selected business requirement from the business level menu screens. If the format of the screens is unique to the company then this class is a sub-class object to the Company class. If the screen is not unique to the company, typically because it has been bought in with application package software and therefore not only relevant to the company but also to other companies, then it is a super-class to the Company class.

6.7.1.7 The repeating value object

This type of class object is not modelled in figure 6.19 but illustrated as the Address class object in figure 6.20. Relational data analysis works on the principle that the data is dependent on the prime key and that there is only one value for the data item for a given value of the prime key. Thus there could be n customers each with an address and all living at the same address. The address value would be repeated for each of the customers without any breaking of the rules of data normalisation. Relational data analysis would not spot the wastage of disk space, which could be considerable (number of the repeated addresses × address length). An example of this is illustrated in figure 6.21. Of the five employees in the table three live at the same address of 12 Chase Side. This is a 60% duplication of this representative table. The class model shows that this address has been generalised and made a super-class object to all the class objects with addresses, and supported by aggregation inheritance.

Object orientation is able to cater for this situation in that the address data property can be abstracted as a super-class object to those business classes with an address data property. In the OO information model this has been done with the address class.

A relational solution to this design problem is to create a separate table of the addresses and use the table as a "table look-up". The separate table would fulfil the role of the abstracted super-class.

6.7.2 The system classes

The system classes are new to information modelling. These non-business objects are unfamiliar to those from a traditional data processing background and

Employee Table					
Employee No	Name	Address	Date Of Employment	Salary	Retirement Date
12,345	Henry Smith	12 Chase Side Slough Berkshire	15/01/86	17,000	31/12/1999
12,346	Mary Bloggs	12 Chase Side Slough Berkshire	17/06/82	42,000	12/11/1995
12,347	John Brookes	The Form Field House Close Ascot Berkshire	25/11/80	17,500	12/07/2005
12,348	Susan Parker	17 High Street Edinburgh Scotland	12/12/78	25,750	15/09/2001
12,349	Andrew Eagle	12 Chase Side Slough Berkshire	15/02/77	35,000	13/02/2010

Relational data analysis fails to spot an address values repeats, so wasted data

Figure 6.21 A relational table in third normal form

are not found in a logical data model. They are unique to object orientation. They are classes that contain information for the running of the application on the hardware/software configuration. The system classes contain the logic supporting system functions, such as general data edit routines, print routines and technical processes such as windowing and file handling. Many of these system classes can be purchased from specialist vendors of classes.

All system objects, being of generic functions, are above the Company class in the class model. They are almost all abstract, containing a method of the functional purpose of the system process.

6.7.2.1 The domain class

The concept of the domain is the same for object orientation as it is for a relational file handler. There is no difference. Domains, therefore, also contain information that is universally generic, such as date, or generic to a company, such as the salary range being between £10 000 and £20 000. All the classes that contain the Date and Salary domain classes will inherit all data and logic pertinent to valid values of date and salary. Date would be a super-class to Company and Salary would be a sub-class.

Enhancements to SSADM

6.7.2.2 The Common Procedure class

The Common Procedure class is a class that contains only a logic component, the logic being a common procedure that is application generic and relevant to the event classes as well as the methods in the business classes. The methods in the Common Procedure class or object instances are therefore similar to the classical common procedure of conventional application programming. In the class model the Common Procedure object is a sub-class of the Company class as it is relevant to all the business objects in the class model.

6.7.2.3 The class library class

The instances of this class are usually supplied as what are called class libraries. Typical class libraries are objects for system software support, such as bit string handling, array handling, inserting an object, deleting an object, character editing, numeric editing and printing an object. There are a number of vendors of class libraries, both independent third party, such as Glockenspiel, and suppliers of object oriented programming languages, such as AT&T for C++, Interactive Software Environment Inc for EIFELL and Xerox Parc for SMALLTALK.

These class libraries objects are of a very generic nature, to the extent that they are universal type functions for system processing, and could therefore be stored as method-only super-classes to the Company class, as shown in figure 6.21.

6.8 ENHANCEMENTS TO SSADM

6.8.1 The class model

The class model is to object orientation what the logical data model is to current database technology. But the class model is more than this, as it also models the structure of the application and systems logic. The processing logic of the application is normalised as appropriate to the classes containing user data and the system processes are also included in the building of the class model.

The basic approach of this technique is to build a logical data model as done with the current structured design methods and from this extend the model through abstraction to include object oriented features. The merits of this approach are its objective starting point and mechanistically produced, and hence objectively based, finishing point, the class model. The objective starting points are the key to the entities, to which are applied the mechanistic

352   *Additional Data Processing Environments for SSADM—Object Orientation*

rules for class and aggregation abstraction. The result is a correctly designed class model, not like the guideline based approaches of other object oriented structured methods. Nothing "wishy-washy" here. The approach is data driven so as to conform with the requirements of the object oriented concept of information normalisation. The tasks for this technique are:

- build the logical data model of the key based entities;
- identify class and from this any class objects with common behaviour of the data properties to be abstracted from the key based entities of the logical data model. Identify any additional class objects containing only methods;
- abstract the data properties that need generalisation for the aggregation modelling of complex objects;
- define the class aggregation and free form semantics as appropriate to the relationships between the classes;
- define the data properties for each of the classes and instantiate those with a value if the value is applicable to all instances of the class object;
- add the event/business requirement classes;
- add common procedures and the human/computer interface classes;
- add the Class Libraries.

There is a school of thought that believes that the first step is not required and that one should build a class model from the beginning. The problem with this line of reasoning is that it is not possible to apply the rules of abstraction unless one has key based entities of a logical data model from which to identify the need for class and aggregation abstraction. If there is no logical data model then the construction of the class model has to be based on the flair and experience of the practitioner. Flair and experience are not the hallmark of a structured method.

The best way to illustrate the application of the above steps is to build an example object class model.

6.8.1.1 *Build the logical data model*

The first step is to build a logical data model. The model is shown in figure 6.22, along with a set of data attributes and volumes to some of the entities. All the entities are based on keys, such as the key of Employee Number to identify the Employee entity and Broker Number to identify the Broker entity.

The technique for the building of the model is not described here as it is well known and covered in numerous manuals of many structured design methods. The important point to realise is that the initial identification of classes is based on keys, as with the traditional structured methods. The

Enhancements to SSADM

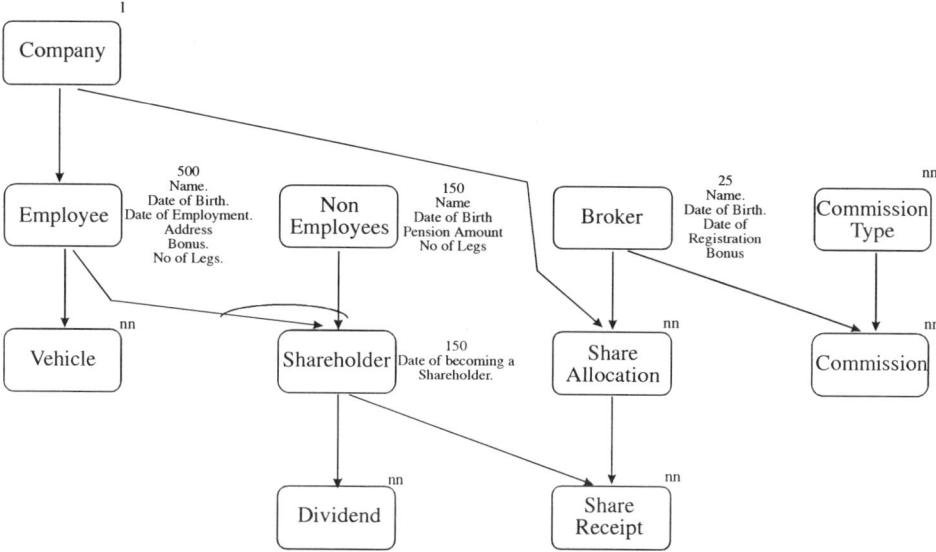

Figure 6.22 The logical data model

logical data model and its key based entities is the objectively based starting point for the building of an object oriented class model.

Many of the books that have been written on the techniques for the identification of objects have stated that the identification of objects is an art form flair of the skilled practitioner. The author very much disagrees with this notion. He believes that the application of a set of rules should be used for the mechanistic identification of objects and from this the construction of the class model. These rules are described below. The logical data model is now to be converted into a class object model.

6.8.1.2 Identify and model class

Class is the facility that models common behaviour between objects. Sets of objects with common behaviour have a common class. Common behaviour is identified as data or logic properties that are common to more than one class. In figure 6.3 the class objects of Employee and Broker contain many properties that are common, such as Date of Birth and Warm Blooded. The classes therefore have common behaviour. This requires to be modelled through abstraction.

Rules have been developed by the author which will enable any common behaviour of the data properties to be identified and to ascertain the degree of abstraction. The basic approach is that the data properties are tested to identify and model class abstractions. To this are then tested the logic properties, which when in third normal form, will create abstractions of aggregation classes. *Data abstraction produces class hierarchies; logic abstraction produces logic abstractions.* The data abstraction rules are:

- The Behaviour Rule.
  Is the data property common to more than one class object?
  If it is then there is common behaviour between the class objects and abstraction of the common data property is required, the abstraction being either generalisation or specialisation.
- The Data Generalisation Rule.
  Is the property common to all instances of the class object?
  If the answer is yes then the data property is to be generalised to a superclass object of the source class. An example of this is the property of Date of Birth—it is common to both Employee and Broker and all the 100 instances of employees and the 20 instances of brokers have birthdays. Date of Birth therefore requires to be generalised to a super class. And the same is true of Favourite Food, all 120 instances of Persons having favourite foods.
- The Data Specialisation Rule.
  Is the property common to only some of the instances of the class object?
  If the answer is yes then the property is to be specialised to a sub class of the source class. The property of bonus is common to Employee and Broker but not all employees and brokers get bonuses, only 12 employees and 3 brokers being so entitled. Bonus therefore requires to be specialised to a sub class.
- The Information Role Rule.
  This rule relates to the question, how far should the data properties be generalised and specialised? The rule is that a property is abstracted to the level in the class model appropriate to its role.
  The role rule is that of the practised eye. [1]
  The best way to explain this is by example. The data properties common to all instances of both Employee and Broker of Date of Birth "sensibly" belong to a super class of Person, No of Legs "sensibly" belongs to a super class of Man and Warm Blooded "sensibly" belongs to a super class of Mammal. This "sensibleness" in the placement of properties to the classes matches the "role" of the data property to the role of the class. Date of Employment is not a data property common to many of the classes and is not abstracted. It is relevant to the role of employment and remains as a data property of Employee. Being warm blooded matches the role of being mammalian.

---

[1] To describe role as being based on a rule is perhaps an exaggeration. Rules require a mechanism to apply mechanistically, the practised eye requires art, skill and intelligence.

- The Logic Abstraction Rule.
  The rule is the normalisation of logic to the abstracted class objects containing user data.
  This rule can only be applied after the data properties have been abstracted into class hierarchies. What this means is that, if you accept the argument about object orientation requiring the normalisation of logic, then object oriented logical design *has to be data driven*—data normalisation precedes data abstraction precedes logic abstraction through logic normalisations. The position for the logic properties is different from the data properties. Almost without exception the logic properties, the methods, are defined at the class level, there being only occasionally logic at the instances of the class objects. This latter point is certainly true when the application system is being analysed and designed, as the object instances which could contain methods have not been created and entered in the objectbase. The instance objects of the application are not yet known, so it would be impossible to have different versions of logic at the instance level. The data abstraction questions "Is the property common to all instances of the class object?" and "Is the property common to only some of the instances of the class object?" are therefore not relevant to the logic properties.

  The mechanism for the abstraction of the logic into methods and the placement of the methods to the classes is based on the rules for the normalisation of logic. What the rules are and how to apply them is described in section 6.8.2.2. But the logic component also follows the same principle of abstraction/specialisation as the data.

  The public method Calculate Salary is relevant at the Employee class level, as all instances of Employees require to have their basic salary calculated. This was the original event based logic called Calculate Salary. It contained all the logic for the calculation of all forms of employee remuneration. But the application of the rules for logic normalisation shows that the logic of Calculate Salary requires to be split, as some of the logic has nothing to do with Employees but with the data abstracted super classes and sub classes. There is a specialised method of Calculate Bonus for the sub-class Star Performer. The logic in this method follows the principle that it is more specialised than the more general logic in the method in the superclass Employee object—it is an extension of, an addition to, the general Calculate Salary logic. It is a specialised abstraction from the base logic of Calculate Salary, the residue of the base logic being Calculate Salary. The more specialised logic would be for the calculation of the Star Performer's bonus, whereas the more general logic would be for the calculation of the Employee's basic salary. And the logic for compensating for the loss of a leg is even more general than the original base logic and is a generalised method to the Man class object. *The normalisation of logic therefore produces the same results as the abstraction rules for data in first and second normal form. The rule of third normal form produces aggregation classes.*

356 Additional Data Processing Environments for SSADM—Object Orientation

The application of the practised eye is useful to confirm the correctness of the normalisation of logic. The process Calculate Lost Leg Compensation is appropriate to the role of the class Man (it accesses the object instances to see if the value is less than 2 and if it is calculates the appropriate compensation), Calculate Salary is appropriate to Employee and Calculate Bonus is appropriate to Star Performer. And a process called Calculate Commission would be appropriate to Broker.

If Calculate Salary was also appropriate to Broker then it would be good design to create a sub-class object to Person of Employed Person.

The generalisations and specialisations of the logical data model are shown in figure 6.23. The arrowhead on the lines of the relationships point from the master classes to the detail classes and from the sub class to the super class where there has been class and aggregation abstraction.

Such classes are Share Allocation and Dividend. The relationship to the abstracted class and aggregation objects depends on whether the abstraction is a generalisation or a specialisation. The logical data model is extended upwards through generalisation and downwards through specialisation. The resultant structure of the final class model from the application of the data and logic abstraction rules is illustrated in figure 6.24. This shows that the core of the class model is the logical data model and is the source of all class and aggregation abstractions, with the class and aggregations creating super classes and class also creating sub classes.

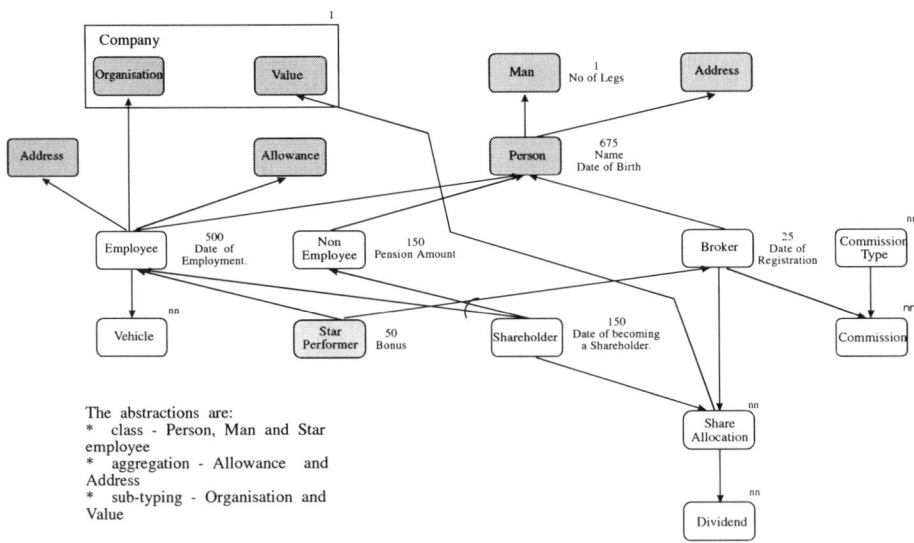

Figure 6.23 Add class and aggregation abstractions, sub-types and build appropriate hierarchies

Enhancements to SSADM

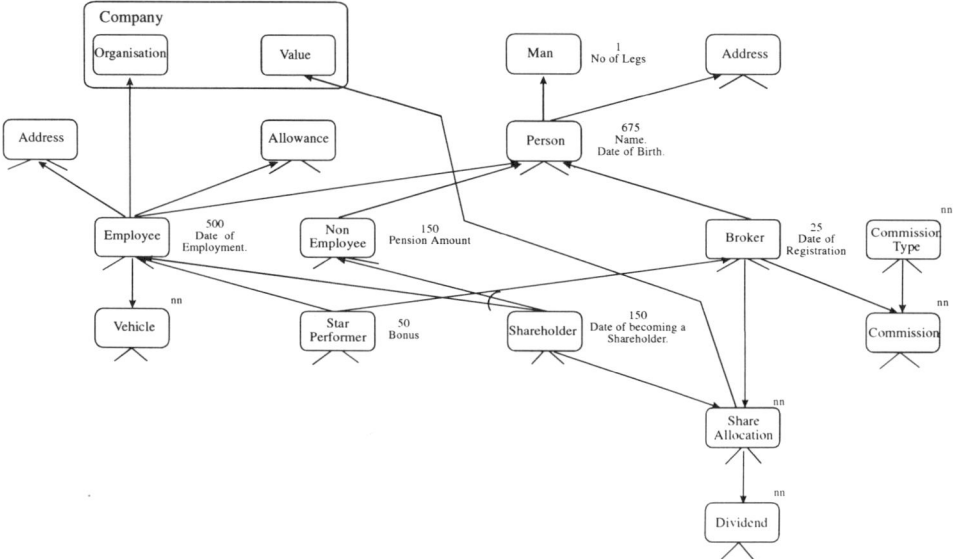

Figure 6.24 Identity abstract and concrete classes

As part of the class modelling it is necessary to identify which of the class objects are abstract in that they support no instances of the class. In the class object model in figure 5.6 Man is an abstract class object, there being only a definition of the class object in the objectbase schema.

6.8.1.3 Create sub-type class objects

Consider the Company class object in figure 6.22. Some of the data properties are not relevant to all the detail classes. The data about the Company Organisation is relevant to the company's Employees but this is of no concern to the other detail entities of the Company entity. Likewise the value of the company has nothing to do with the Employees but is decidedly of relevance to the Share Allocation as the value of the Company affects the value of the Share Allocation.

This business context-sensitive data needs to be specialised as sub-types of the base class object and the correct relationships to the sub-class objects of Company established. This is illustrated in figure 6.23, so that the Employees relate to the Company/Organisation and the Share Allocation to the Company/Value.

6.8.1.4 Identify abstract and concrete classes

Mark the classes that define no instances. These become abstract classes. Any data properties they contain must be instantiated with a value. The property Number of Legs for the Man abstract class object has a value of "2". This is illustrated in figure 6.24. The other classes are concrete objects. The abstract classes are identified by the absence of the bifurcated lines at the bottom of the class rectangles.

6.8.1.5 Add the relationship semantics

This is a simple matter of defining the appropriate semantics to the relationships of the class model so far constructed, as it is the semantics that define class and aggregation hierarchies. This can be seen in figure 6.25 where the "is a" semantic description models class hierarchies, the "has a" semantic the aggregation hierarchies and the free form the relationships that are not class or aggregation based but business based.

The top super-class object to a class hierarchy is the Man class and from this the sub classes of the Man class hierarchy can be seen, bifurcating in the middle to include Employee, Non-Employee and Broker and coming together in the Star Performer. The aggregation semantic to the Address abstracted data property for the Employee class object can be seen. An Employee "has a" Address.

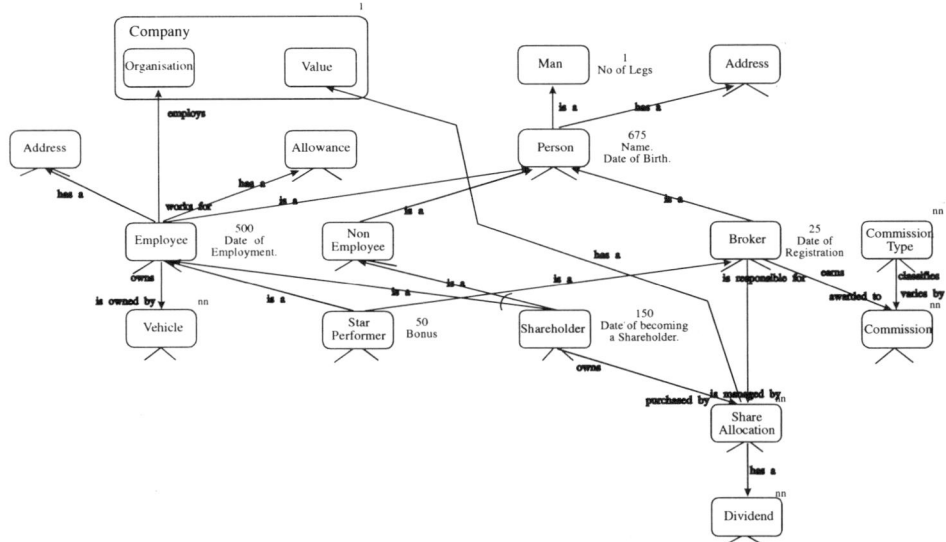

Figure 6.25 Specify class, aggregation and business semantics

Enhancements to SSADM 359

The remaining semantics are free form, describing the business purpose of the relationships between the classes that are the discrete descendants of the entities in the logical data model. An example is that an Employee "works for" a Company and a Company "employs" Employees.

6.8.1.6 Define and instantiate the data properties

This is a simple matter of defining the name, format and length of the data properties and any instantiation if a property has a value that is generic to all instances of the class object. An example of instantiation could be a value of "2" for the data property of Number of Legs for the class object Person.

6.8.1.7 Add the event classes

There are, from a modelling point of view, two types of events, those that are internal to the company and are triggered by users in the company and those that are external but are of interest to the company and triggered by persons not in the company's employment. The need for this distinction is that the information in the class model is more general as one goes up the class model and more particular as one goes down the class model. Events that are external to the company of their nature are more general than those that are triggered within the company. Events that are external to the company are generic to n companies and therefore more general than the company class. They therefore need to be modelled as super classes to the company.[1] An external event could be the raising of interest rates. The company is interested in that it will now have to pay more on its borrowings.

Internal events are specific to the company and are therefore modelled as sub class to the company class object. An example of such an event is Calculate Dividend. Others are Raise Invoice, Make Payment and Create Customer. Both kinds of events are illustrated in figure 6.26.

6.8.1.8 Add common procedures and the human/computer interface

Any logic that is found to be common to n methods can be abstracted and defined as a more general class. The placement of the class Common Procedure follows the same principle as the event class. This is also the same for the classes that support the human/computer interface—dialogue and menu screens can be specific to a company or generic to the computer industry. There are two kinds of menu screens—those for windows technology with their slide-off and pull-down menus and those

---

[1] From a physical development point of view there is no difference in the treatment of the external and the internal events—both are developed into application programs.

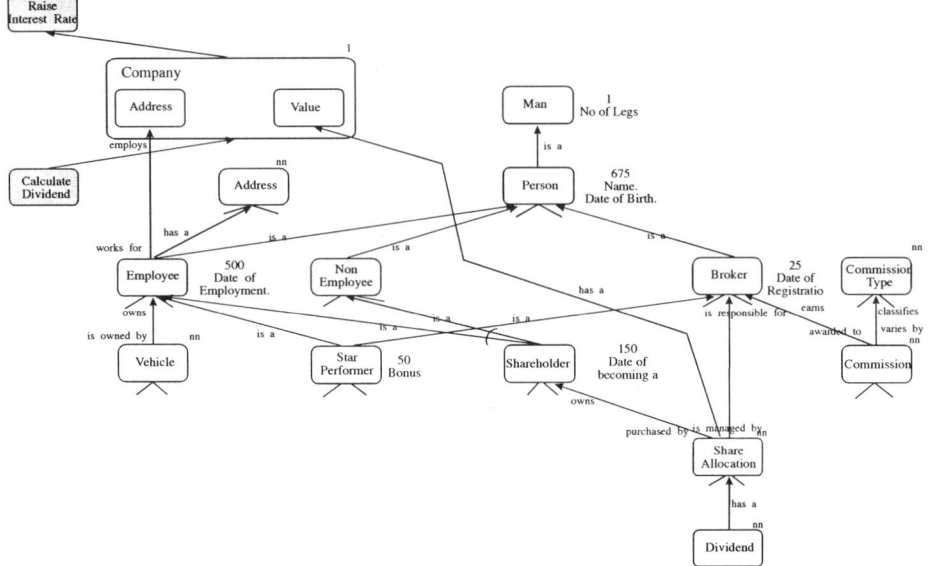

Figure 6.26 Add the event classes

for the selection of business requirements. The former are super-class objects to the Company class object if they are screens from windows packages, such as Hewlett-Packard's New Wave and Microsoft's ® Windows™, but would be sub classes if developed in-house. And the same would be for the dialogue screens, screens for the display of information pertinent to a business requirement once selected from the menu screens. If the screen was unique to the company it would be a sub class of the Company class—if it was generic/bought in then it would be a super class to the Company class. This is illustrated in figure 6.27.

6.8.1.9 Add the class libraries

These are the libraries of generalised system classes that are obtained with many of the object oriented programming languages and from specialist software vendors. They can, of course, also be developed in-house. Being of a general functional nature, such as print and sort objects, they are modelled as super classes to the Company class. This is illustrated in figure 6.28.

Enhancements to SSADM

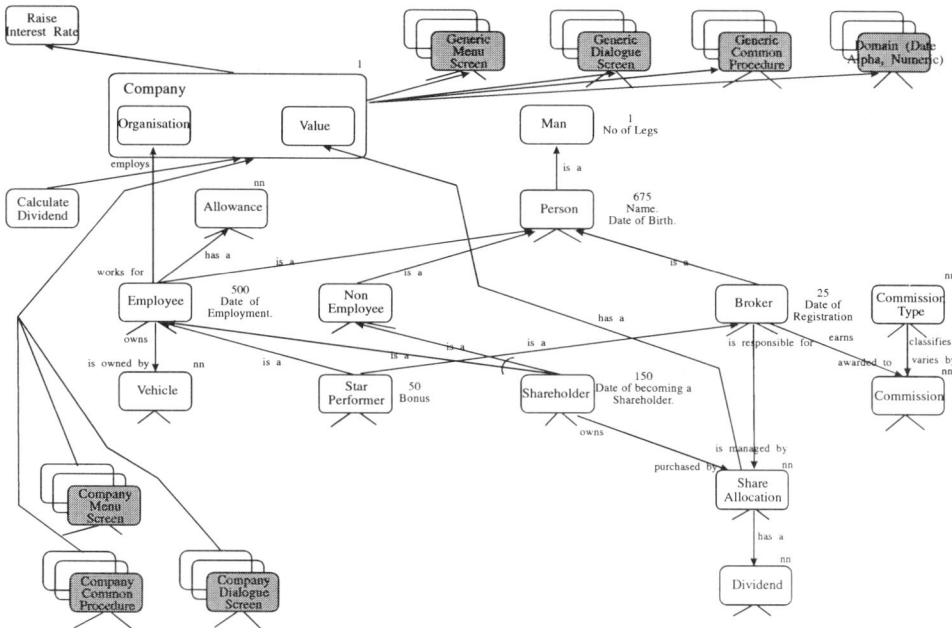

Figure 6.27 Add common procedure domain, menu and dialogue classes

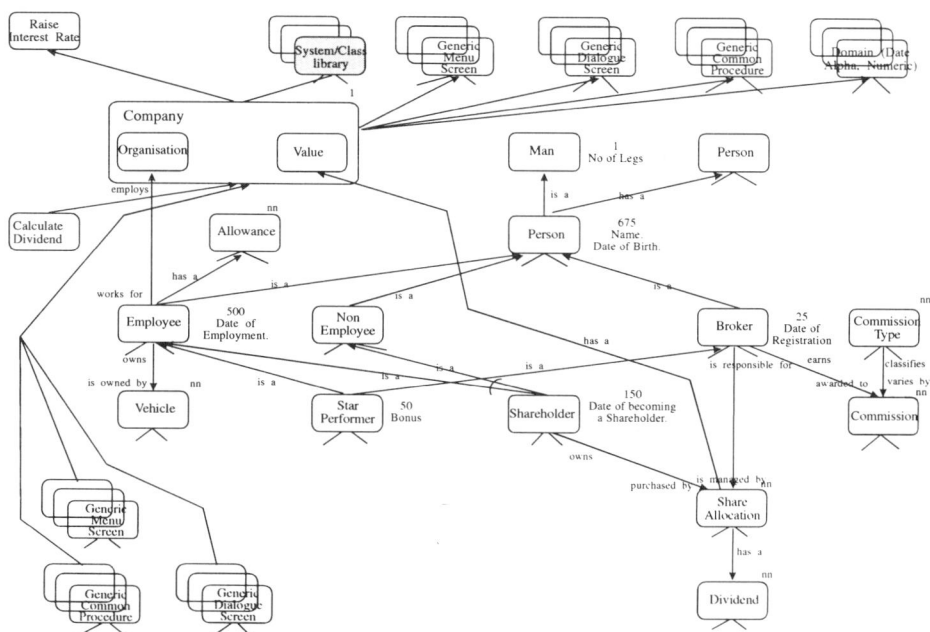

Figure 6.28 Add system/class libraries

6.8.2 Information normalisation (formerly relational data analysis)

The term relational data analysis is no longer adequate because it no longer is appropriate to relational technology or to data. It should be replaced by the term information normalisation, because it is now realised that the rules of normalisation that have been for so long applicable only to data are equally applicable to logic. *The result is that all information can be normalised.*

This section will not repeat the technique of the normalisation of data. Many textbooks are available on the subject. All that will be described is the definition of the data normalisation rules and their redefinition, with examples, for the normalisation of logic, both declarative logic and procedural logic. The rules of data normalisation are:

- First Normal Form—"Take out Repeating Groups", that is, any data property that has many values for a given value of the key. Remove them and create a new relation of the repeating group.
- Second Normal Form—"Test for Part Key Dependence", that is, if a data property is dependent on only part of a multi-part key relation remove it and create a new relation of the part key dependence.
- Third Normal Form—"Test for Inter-Data Dependence", that is, if any of the data properties in the second normal form relation are dependant on any of the other data properties. If they are remove them and create a new relation of the dependence.

There are two aspects to consider for the normalisation of information, both data and logic, in an object oriented environment:

- The introduction of the concept of the role of the information that is being normalised.
- The normalisation of logic.

6.8.2.1 The normalisation of object oriented data

Given that a logical data model requires to be built as the beginning of a class model, the rules for data normalisation to third normal form developed by Dr Codd require to be applied but unchanged. The rules for data normalisation do not, however, support the concept of the role of the information to be normalised. This is an omission that needs to be rectified for object orientation.

Consider the class Man, Person and Employee and the data property of Date of Birth in figure 6.3. With a relational approach to data modelling the Date of Birth could happily reside in the Employee entity and not break the

Enhancements to SSADM 363

rules of data normalisation—is the data dependent on the key, the whole and nothing but the key? Yes. But with object orientation this is not adequate. Is Date of Birth relevant to the role of Employee? No. Date of Birth has nothing to do with employment. But the Employees are Persons and therefore the class of Person is abstracted as a more general super class to Employee. Should the Date of Birth be defined as a data property of Employee, Person or Man? From the role point of view the data property is relevant to Person and not to Employee.

But what about putting Date of Birth as a property of Man? Notice that Man is a class object with no instances—it is an abstract class object. Man is a generic "thing" with no instances, but containing information that is generic to any sub classes, to all instances of Person and Employee. This means that any data properties are instantiated with a generic value relevant to all the instances of the sub-classes. To instantiate Date of Birth with a generic value would clearly be nonsensical, as each instance of a Person could have a birthday on a different date. Date of Birth is a property that describes individual instances of Person and thereby, through property inheritance, Employee. Notice that Person is a class object with instances, because there are individual Persons who might also be Employees. Date of Birth "sensibly" belongs to the role of a Person, not with the role of Man, to a concrete and not an abstract class. Thus all the instances of Person and Employee can have individual birthdays. The concept of the role of data needs to be added to the rules of data normalisation.

Consideration of the role of the data needs to be applied after the standard normalisation rules have been used to build the standard logical data model and the rules of class abstraction to add the class hierarchies.

6.8.2.2 The normalisation of object oriented logic (procedural)

The work of Dr Codd defined the rules for the normalisation of data, and the rules have been widely accepted and applied. *The author believes that the rules for data normalisation are as relevant to logic as they are to data.* Why not—is it just information? All that needs to be done is to alter the phraseology of the rules to cater for logic.

An example of the normalisation of logic up to first normal form is given in figure 6.29. The rule of first normal for data is "Take out repeating groups". The test for repetition is "For a given value of the key is there more than one possible value of the data?" Suitably rephrased for logic the wording becomes "For a given class is there more than one version of the logic?"

Figure 6.29 shows the normalisation of the logic for the business requirement of calculating salaries, the logic of which varies, is relevant to different objects. The logic varies depending on whether the class is an Employee, Star Employee, Star Employee Bonus or Employee Bonus. The base logic for the event/business requirement is in the event class

Logic First Normal Form Rule
"For a given class is there more than possible version of the logic"?

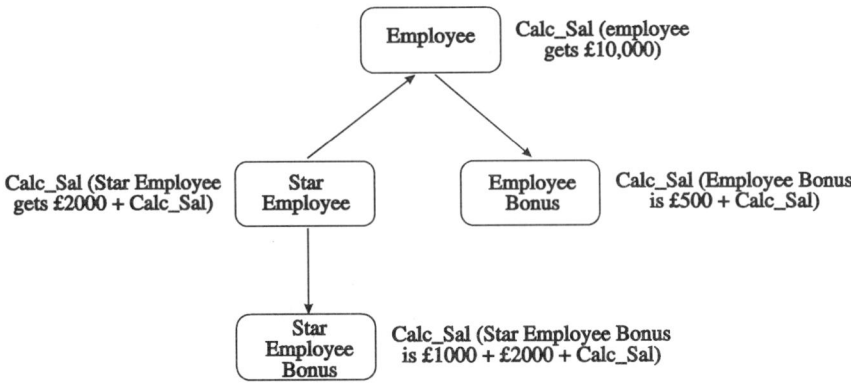

Figure 6.29 The normalisation of logic (1NF at the class level)

object Calculate Salary, with the remaining logic "normalised" to the classes containing user data the logic is pertinent to. All Employees receive a salary of £10 000; the Star Employees receive an additional £2000; the Star Employee Bonus is a further £1000 and the Employee Bonus is £500 on the basic salary. There are therefore n versions of the logic in the original Employee object class.

The logic for calculating the bonus of £1000 is appropriate to Star Employee Bonus and is therefore normalised to the Star Employee Bonus class object. *It can be seen that the sub-class object contains the logic that is additional to the logic of the method of the super-class.* Thus the calculation of the Star Employee Bonus is the combined logic of the Star Employee Bonus (£1000), Star Employee (£2000) and Employee (£10 000). The sub class methods inherit the logic of the more general case super-class methods.

Figure 6.30 shows an example of logic that requires to be normalised to second normal form. The rule of second normal form for data is "Test for part key dependence". The test is "Is the data dependent on the whole key or part of the key?" Again suitably rephrased the test would be "Is the logic dependent on the whole key or part of the key?"

The key of the stock is a compound key of Product Code and Depot Code, as it contains the stock of a particular product at a particular depot. The method Calculate Stock Value is in second normal form in that it is calculating the value of the stock of a product at a depot. The method Increase Product Price is not in second normal form in that it is only relevant to product—it is calculating the increase in the price of a product and has nothing to do with the stock at a depot. The method is in the wrong class and requires to

Logic Second Normal Form Rule
"Is the logic depedent on the whole key?"

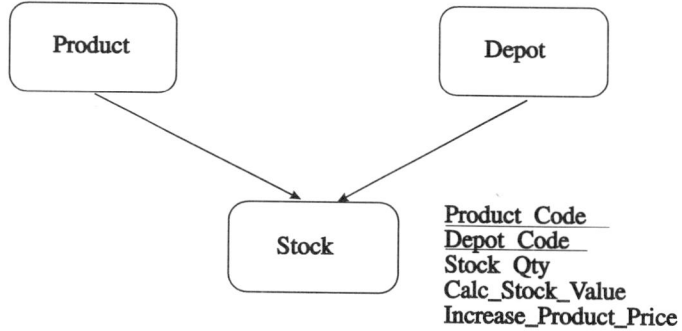

Figure 6.30   The normalisation of logic (2NF at the class level)

Logic Third Normal Form Rule
"Is the logic dependant on a data property that is not the prime key?"

Figure 6.31   The normalisation of logic (3NF at the class level)

be moved to Product, so that the logic of the method is now appropriately related to the class to which it is relevant.

Figure 6.31 shows an example of logic that requires to be normalised to third normal form. The third normal form data normalisation rule is "Test for inter-data dependence", that is, are some of the data properties dependent on each other rather than the prime key? Suitably rephrased the rule becomes "Is the logic dependent on a data property that is not the prime key?"

The method Check Name Range is a logic property of the class object Customer. The logic requires to access the Customer Name and retrieve customer details if the name of the customer is within the specified range. The company markets shirts emblazoned with personalised names and wishes to supply a free set for a given set of customers with specific surnames. The logic is dependent on a data property of the class object that is not the prime key, in this case the Customer Name. In the example of Calculate Stock Value the logic was dependent on the prime key of Product. But Customer Name is not the prime key, Customer Number is. The method needs to be abstracted with the Customer Name data property and generalised as a super class. It is this rule of logic third normal form that creates the aggregation classes, classes that therefore leave one data property only and one or more methods.

The examples given so far are to business objects that one would see in a traditional logical data model. The normalisation of logic is also appropriate to the system objects and other objects not normally found in a traditional logical data model as well:

- The Object class. The Object object contains logic that is universally applicable. For a chemicals application the chemical formulae convert into methods that contain logic that are universally recognised to be laws of chemistry and therefore universally true. The logic for these chemical formulae would become method properties in the Object object.
- The logic contained in the system classes, typically for system type procedures, such as numeric editing, storing an object and string handling. Being generally applicable the objects would be defined as sub classes to the Object.
- The logic contained in the event class, that is, the logic that is specific to the business requirement. This event level logic by its nature cannot be normalised to a business object class, as the logic is not relevant to any key of a business object. Typically the logic would be the issuing of the messages to the other classes, the synchronisation of the message responses received from the classes and any final processing before information output and screen formatting.

6.8.2.3 The normalisation of object oriented logic (procedural) for abstract classes

Object orientation is able to model abstract class, that is classes without keys, or more accurately user defined keys. Dr Codd's rules for data normalisation, being based on user defined keys, are therefore not suitable for abstract classes. Another new rule requires to be defined.

Since keys cannot be used for information normalisation it is necessary to relate the data and logic properties to the role of the class, the role of the properties matching the role of the abstract class. The rule is the very simple one of "Does the role of the property match the role of the class?" Using the examples in figure 1.7 with the abstract object of Man and data property of No of Legs and the logic property of Calculate Lost Leg Compensation (to be invoked if one or more legs have been amputated) both properties are relevant to the role of being a man, a homo sapiens, and are therefore corrected related to the class Man. If the data property was Date of Employment this has nothing to do with being a man but an employee, and the same would be true for the logic property of Calculate Salary. To place these properties within the class Man would be wrong, they relate to the role of being an Employee and would therefore be correctly placed within the class Employee.

6.8.2.4 The normalisation of object oriented logic (declarative)

The form and nature of declarative logic was discussed in chapter 5.

The issue of the normalisation of declarative logic has caused the author much thought and the answer is still not certain. There is no reason that declarative logic cannot be normalised like procedural logic—it is nothing more than another way of representing logic. But that reasoning is the simple part. The dilemma is this. Given that each rule is a statement of some truth and is therefore a piece of logic that can be stand alone, does one normalise each rule on a standalone basis or as a set of rules in the ruleset (for the purposes of this section the nearest equivalent of an application program in declarative logic) and, if the normalisation is to a single rule, does one normalise to the condition of the rule or to the conclusion if the two relate to different "things"/objects?

Consider the problem of normalising a rule on a standalone basis. The situation is all too often that the condition is usually about a "thing" of A and the conclusion about a "thing" of B. To which object, A or B, should the rule go? And what if there are n conditions about A, B, C and D and there are n conclusions about E, F and G? An extreme example could be "If the world is round and Joe Bloggs is big and the President of the United States is on the campaign trail then it is raining and necessary to get your car repaired". N conditions and n conclusions each relating to different classes and instances of the class. Where does one start with that rule?

How about normalising the rules to the query that they are providing advice to—that is, at the ruleset level? This seems to be the most plausible approach. The approach of normalising rules on a standalone basis does not seem to be viable, even though rules have value in their own right. Thus the ruleset rules appropriate to changing a tyre would normalise to the tyre class object and the rules in the ruleset for maintaining a car's electrics would normalise to the electric sub class of the super class of car. This is normalising declarative logic in the same way as procedural logic—and this approach is therefore the more attractive because it is common.

But what if the ruleset has rules that are to be found in multiple rulesets? There could be a ruleset of rules about purchasing a computer with a printer. There could well be another for selling a computer with a printer and some of the rules in the purchasing ruleset are also in the ruleset for selling. No problem. They are common functionality and should be generalised as a super class containing part of the ruleset for both purchasing and selling.

If declarative logic is combined with procedural logic then both require to be normalised as standard. This most frequently occurs with the use of rules as the mechanism for defining the pre- and post-conditions for method processing of the application class objects. "If colour of car is non-blank then print error message 55" could be a pre-condition rule for a method Paint Car. The rules for information abstraction can be applied here. If the rule is specific to painting a car then it is not abstracted. If the rule can be used elsewhere then it is abstracted. There is no problem with intermingling procedural and declarative logic.

6.8.3 Dataflow diagrams

The technique is unchanged, the only addition to consider being the decomposition of the processes. Advice to both of these questions is provided in section 3.2.1.

Object oriented processing provides the answer to one of these questions—the level to which you decompose. *What the author has found most practical is to decompose the processes to the event/business requirement level, and then to normalise the logic to the classes to which they appropriately belong.*

Although logic can be appropriate to object instances this cannot be identified during the logical design of the application for the simple reason that the instances of the classes cannot at this stage be identified. This has the advantage that the traditional approach to process decomposition is preserved as far as possible. There is therefore no point in decomposing processes below the event level for the object oriented use of the dataflow modelling technique.

The event level logic is normalised to the event class and the remaining logic, some 80%, to the business classes as appropriate.

6.8.4 The I/O structures

This product is unaffected by the introduction of object orientation.

6.8.5 Modelling object oriented application programs

Object oriented processing is still event level triggered. The sending of messages between the objects to be accessed for the event are all initially triggered when the event/business requirement occurs. As we have seen in non-object oriented systems, application programs should be pitched at the event level and the constituent modules at the problem-to-solve level. What is true of traditional batch and online centralised processing is also true of object oriented processing. Once again it is a case of the facilities of centralised processing being generic, with the facilities of object orientation being add-ons.

Enhancements to SSADM 369

Application programs in object oriented systems are also event based but, instead of being composed of problem-to-solve modules, are composed of a set of normalised class based methods. The "program" is triggered from the event class object for the business requirement. The succeeding application objects are "chained" in the sequence required to be accessed for the processing to proceed, as in conventional processing. The other classes could be system class objects for common routines, menu and dialogue screen processing and common procedures.

There is a "battle" going on over the best way in which to model the structure of object oriented application programs. The two leading techniques are the Jackson approach and Action Diagrams. Within the Jackson approach there is another battle as to whether to use the object state or event process models as the basis for designing object oriented application programs. Both techniques are widely used. Both can be easily tailored to support object oriented logical design of event processing.

While both methods are equally valid, the author believes that the Jackson approach will "win the day" because it is very diagrammatic, with the diagram beautifully modelling the structure of object oriented processing. The technique is also already object based.

6.8.5.1 Enquiry Access Paths and Effect Correspondence Diagrams

These merely need extending from the access path to the logical data model to reflect any abstraction of super and sub classes in the class model. For example, if the access was to all the data attributes in the Employee entity in figure 1.7 then the object oriented Enquiry Access Paths and the Effect Correspondence Diagrams would require extending to the abstracted classes Person, Man, Mammal and Star Performer to obtain all the data properties. The access to the abstracted classes would be one-for-one access.

6.8.5.2 Access path analysis

The object instances are accessed and generate disk I/O just as the table rows of relational database. The objectbase design can be optimised (see chapter 3) so as to minimise disk I/O if the accesses to the object instances are modelled in the object oriented logical design of the application. The logical access paths therefore need to be modelled just as with the existing structured design methods.

This aspect of access path analysis is part of the logical design process that, almost without exception, the existing structured design methods are poor at modelling. SSADM version 4 models the accesses to the entities in the logical data model with the Enquiry Access Paths and the Effect Correspondence

Diagrams, but does not count the number of accesses!!!! And it used to count the access in version 3!!! So the method cannot ascertain the accesses when producing the initial database design! Information Engineering is much better and counts the accesses and produces summary access statistics, thus giving an application overview of all the accesses to the logical data model.

It needs to be appreciated that if the object oriented programming languages compile the methods then the logical accesses to the method and the potential for disk I/O accessing the method-only classes are eliminated. If the language is dynamically bound to the methods, as in SMALLTALK, then the accesses to the appropriate method-only classes will generate disk I/O in the physically developed system.

6.8.5.3 The Object State Models/Entity Life Histories

This technique is based directly on the Entity Life History technique and is suitably renamed the Object State Model technique for object orientation. It is ideal for modelling the state of the objects in an objectbase and is only to be used against the business concrete objects containing user data in the class model. An object state model is produced for each such class object. There is no need to model the life history of abstract classes. They are created with substantiated data properties and that's it—or should be.

The entities/objects now contain logic properties as well as data properties. Both property types can be inserted, updated and deleted and therefore both property types have lives. The approach adopted by the author, however, has been to keep the two property types separate—they have very different and separate lives, even though they belong to a single class.

The data properties can have many different events affecting them, they can occur in sequence, can iterate and can be selective, can incur quits and resumes and can occur in parallel, all in a potentially highly complicated manner. The logic properties, in contrast, have simple lives—they are entered, potentially undergo n iterations of code modification and are deleted. Each modification of code in a logic property is the same event, although clearly different lines of code can be affected. The life of the logic properties is therefore in a standard ELH form—an insert, followed by n iterations of updates, followed by a delete.

To be honest the author has not bothered with constructing an entity life history of the logic properties in an object.

The final reason for separating the lives of the logic properties from the data properties is that their lives are also independent of each other. The updates to the logic in no way affect the lives of the data properties.

The ELH technique does not require change to become the Object State Model techique. All that is required is that the operations must be interpreted in an object oriented way, so as to identify the methods and the sequence in

Enhancements to SSADM

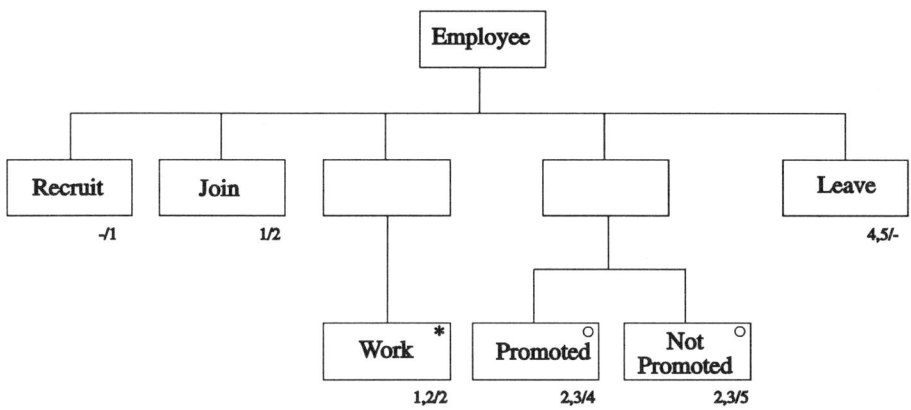

Figure 6.32 An entity life history/object state model example

which the methods are to be messaged during the life of the class object instances.

A worked example is illustrated in figure 6.32, which shows that an Employee is recruited into the objectbase. Some time later the employee joins the company and undertakes many iterations of work. Some time later there is or is not promotion and after that the employee leaves the company and the objectbase. The fact that the employee does not do any work after being promoted is neither here nor there.

In the Object State Model the processes are the normalised methods that affect the life of the class object being modelled. The object state model process boxes are therefore at a lower level of detail than the event level processes of the Entity Life History.

Methods that merely retrieve the data properties are not modelled as they do not change the state of the object instance and can therefore be invoked in any order. Such methods are often called observers.

What is the purpose of this standard technique for traditional entity state modelling in an object oriented environment? The usefulness of the object state model to object orientation is that it models:

- The methods that affect a class. In figure 6.33 there are five methods for both classes 1 and 2.

- The operations of the methods. From this the commonality of operations across the methods in different object state models shows those which can be generalised and made common procedures. The residue of the operations become methods in the class.

Figure 6.33 requires detailed interpretation. Operation 1 is common to method 1 for class 1 and to method 6 for class 2. It is an object instance insert

Additional Data Processing Environments for SSADM—Object Orientation

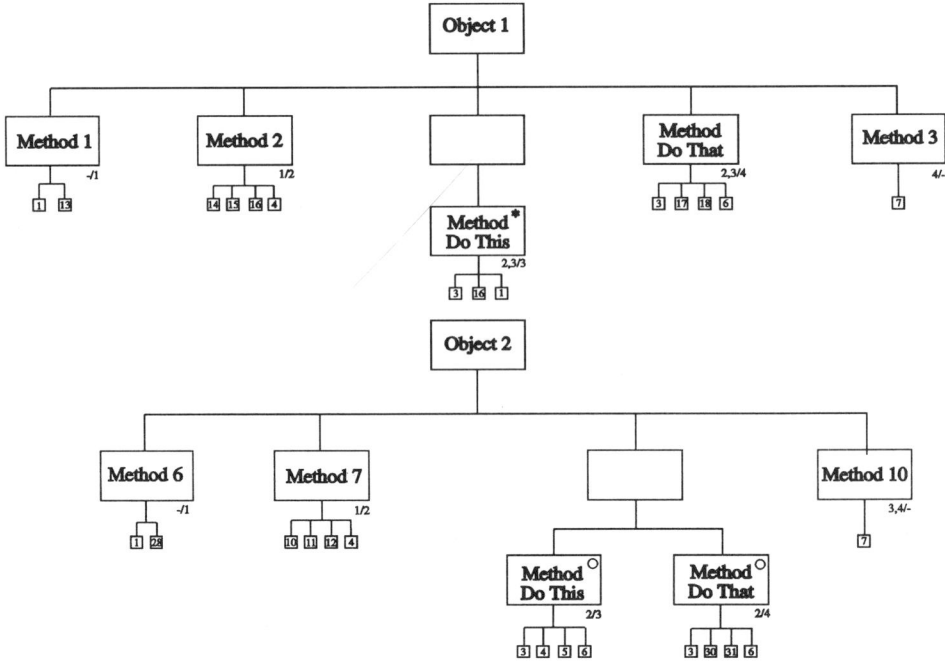

Figure 6.33 Object state models

operation. Being common it represents common behaviour of some logic across more than one class method and so can be generalised. It becomes a common procedure/class library class. And operation 7 is a common operation for methods 5 and 10 being a delete object instance process. It can be likewise generalised. Operation 6 is common to methods Do That and Do This. Operations common to several methods include 4 and 16 (common to two methods for the same class). Such common operations can also be generalised as common procedures.

The residue of the operations not generalised become the source method attached to the class being modelled. Thus method 2 for class 1 has the operations 14 and 15 and method Event Do That for class 2 has the operations 30 and 31.

- The sequence in which the class methods that affect the object instances must be invoked by the receipt of messages from the event class is based on the sequence of the methods in the Object State Model. For class object 1 in figure 6.33 the sequence of the methods that affect the object instances is method 1, 2, Do This, Do That and 5.

- The pre- and post-conditions that must exist prior to and post the execution of the method affecting the object instance. It is a most useful mechanism

Enhancements to SSADM

for ensuring that the logic of the method executes correctly. It is thus the basis of a "contract" between the class sending the message and the class receiving the message.

This can be done either crudely or in detail. The crude way is to check the value of the state indicator in the logic of each method. For method Do This for class 1 the method logic would start as a pre-condition by checking that the value of the state indicator was either 2 or 3 and if not valid send an error message and stop processing. If the state indicator was OK then to accept the message and invoke the method. At the end of processing there would be logic to set the state indicator value to 3. The post condition would check the state indicator was 3 and if OK send the response.

The more detailed approach is to check the values of the data properties that are pertinent to the business function of the method. Assume that the object is a car and that the method Do This is to paint the car. The data properties of car colour would need to be checked as a pre-condition to ensure that the colour is blank. As a post-condition the logic would be to check that the colour is non-blank. There could, of course, be a range of data properties that require to be checked in the pre- and post-condition testing.

The operations added to the ELH are concerned only with the keys and relationships of the entities. The full specification of access and process logic is recorded in the Process Models.

The need for parallel lines is no longer required in object orientation. Where there is, with "traditional" state modelling a need for a parallel life, an Employee being simultaneously a gardener, a parent and a sports person for example, object orientation creates sub classes. Create an Eiffel/object state model for each sub class.

6.8.5.4 The Event Process Model

One of the beauties of SSADM version 4 is that it is already object based in the Process Modelling technique. The operations are defined against the affected objects/entities for the event. The Process Models already reflect the structure of an object oriented application program. The operations at the event level are defined against the "root" box for the event the Process Model is modelling, and the operations for the objects/entities being accessed for the event are recorded against the entities.

The event process model models:

- the structure of an object oriented application program;
- the methods that require to be messaged for an event/business requirement;

374 *Additional Data Processing Environments for SSADM—Object Orientation*

- the logic that is to be executed against the data in the objectbase as required by an event/business requirement.[1]

An example of the business requirements/events Do This and Do That process model is illustrated in figure 6.34. The model shows the sequence, selection and iteration of the methods the events require to send messages to invoke the processing the events require.

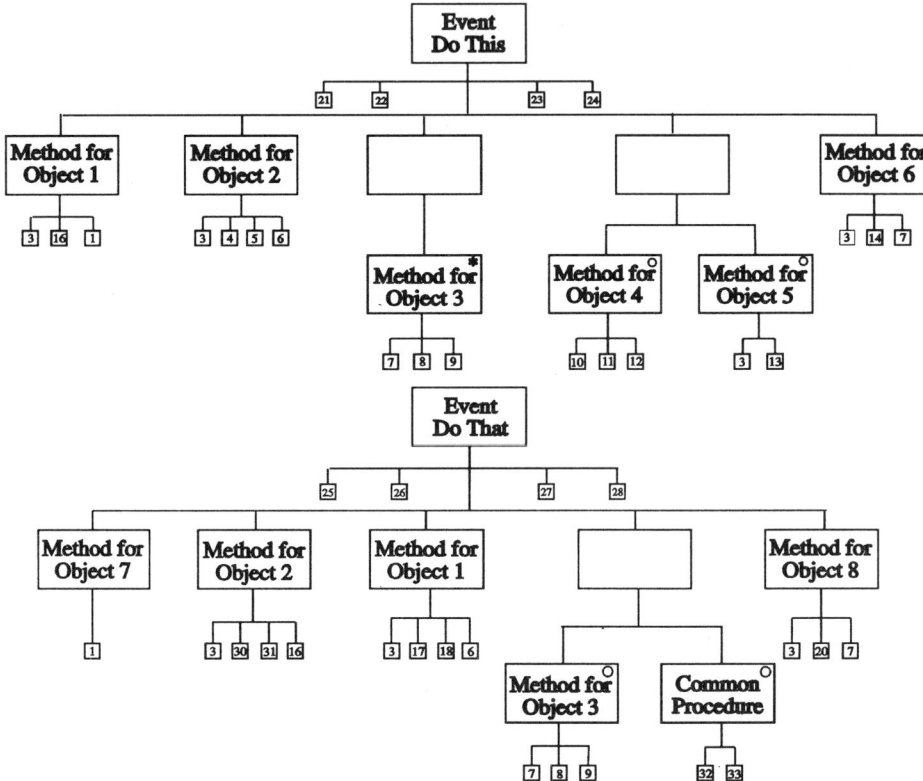

Figure 6.34   Event process models

---

[1] G. Booch (*op. cit.*) argues that the finding of the right class objects and then organising them into separate modules are entirely independent design decisions. The author does not accept this argument. The identification of classes, their abstraction and placement in the class model, and the normalisation of logic into methods are based on the rules defined in this chapter. The technique of Event Process Modelling is then based on the classes and the normalised logic and the model is then used *directly* on a one-for-one basis as the structure of an object oriented application program. There is a continuous trace of the techniques and their products from the beginning of the identification of classes and ultimately the modelling of the object oriented application programs.

The rectangle at the top of the model is the event/business requirement. Operations are also detailed at the event level. This event level process is, in effect, the main procedure of an application program, with the messages to the methods fulfilling a similar role to program procedure calls. This part is the 20% of the business logic that is application dependent.

The rectangles at the bottom of the model represent the methods in the classes containing user data or classes containing only a method (typically a class library or common procedure) to be invoked. Operations are listed against each of the methods to detail the logic that is to be executed. The methods at the bottom of the event process models are application independent.

The usefulness of the event process model to object orientation is that it models:

- the methods to be invoked by the event/business requirement class objects;

- the sequence, iteration and selection of the methods to be invoked for the event/business requirement.

    In effect what one is modelling is the access path of the business requirement to the object instances in the objectbase. In figure 6.18 the access path for Do This is object 1 (a specified Customer), then object 2 (the latest Order for the Customer), object 3 (all the Order Lines for the Order), objects 4 or 5 (either an Invoice or a Payment Request) and object 6 (a specified Product).

    The method does not have to be defined for an application class containing user data, as implied by the above example. For the event Do That there is a Common Procedure method that is invoked and is defined in an abstract class that contains no user data, just the method.

- The specification of the logic in the methods, both at the event level and for the methods defined for the application classes with user data and method-only classes.

    The operations can be interpreted in exactly the same manner as for the object state model. Operation 3 is a read for update and is common to several methods. It can be generalised as a common procedure. Operations 1 and 7 are the same, being inserts and deletes of object instances and can likewise be generalised. And so on. The residue operations become the methods of the classes the methods are operating against. Thus 30–31 are the operations of the method for class 2.

- The structure of an object oriented application program. The object state model remains useful for just one thing—object state modelling.

    The "program" for the event/business requirement Do That is composed of:

- the event level processing of operations 25–28, these being concerned with the sending of messages to the methods modelled in figure 6.34 and processing the replies and any other needs as appropriate;
- five methods defined for the application classes containing user data and one common procedure method.

• This structure may require finessing due to the messaging strategy—see section 6.8.7.

6.8.5.5 Modelling polymorphic processes

There is a good way to ascertain if a process is polymorphic. If the logic is of the kind that states:

> If object instance is X then do ...;
> If object instance is Y then do ....;

this is a clear indication that the logic is dependent on which instance of a class object is occurring. One then has to name a common polymorphic name for the processes with the instance variations on the lines of figure 6.17. The recognition of a polymorphic process is best achieved by studying the Object State Diagrams. If there is a commonly named process that updates several classes with many but not all the operations being the same then the process is a candidate for polymorphism. A model of a polymorphic process is illustrated in figure 6.35.

There is the method, Method 2, that is common to two classes, Container Crane and Van Carrier, but as can be seen from their operations they have a common set of operations 14 and 15 but thereafter differ. There is therefore some common processing with some different processing in each of the similarly named processes. Very much polymorphic processing.

6.8.6 Human/computer interface design

There are two types of screens for the Human/Computer Interface—menu screens and dialogue screens. The menu screens support the man/machine interface for the selection of a business requirement and the dialogue screens the man/machine interface of the selected business requirement.

6.8.6.1 The menu screens

The menu screens are not pitched at the event/business requirement level. They contain options for/are generic to the selection of many business requirements. The menu selection logic is therefore not normalisable to a

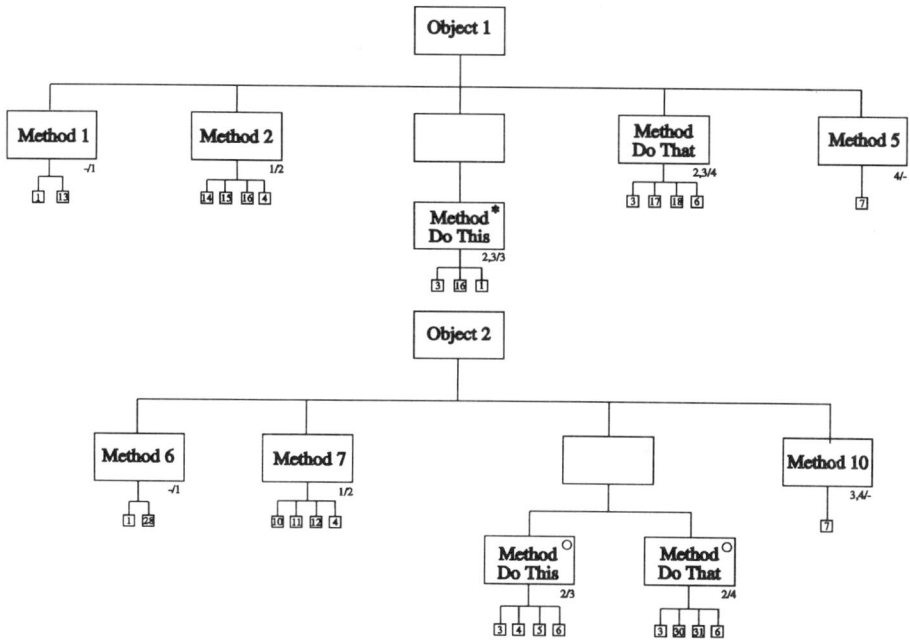

Figure 6.35 Polymorphic process modelling

business object in the class model. They should be normalised to their own classes. If the menu screens are bought in software, such as pull-down and slide-off menu construction facilities or whole screen formats, then the classes are super classes to the Company class. If the menu screen designs are specific to the company then the classes are sub classes to the Company class.

Most of the structured methods rightly pitched the menu screens at the user role level. One could design a user role class and put the menu selection logic for the user role into the appropriate user role class. The author has tended to create a separate menu class that fulfils a similar function. But a more accurate approach is to build a structure diagram as illustrated in figure 6.36. The diagram is *à la* Jackson but without iteration or selection. The logic/operations for each screen is defined in the same way as for the event process models and become method-only classes. There would be a method-only class for each of the option screens and the main menu screen of figure 6.36.

6.8.6.2 *The dialogue screens*

The same Jackson-like structure diagram can be used for modeling the dialogue screens for each event/business requirement of the application.

378 *Additional Data Processing Environments for SSADM—Object Orientation*

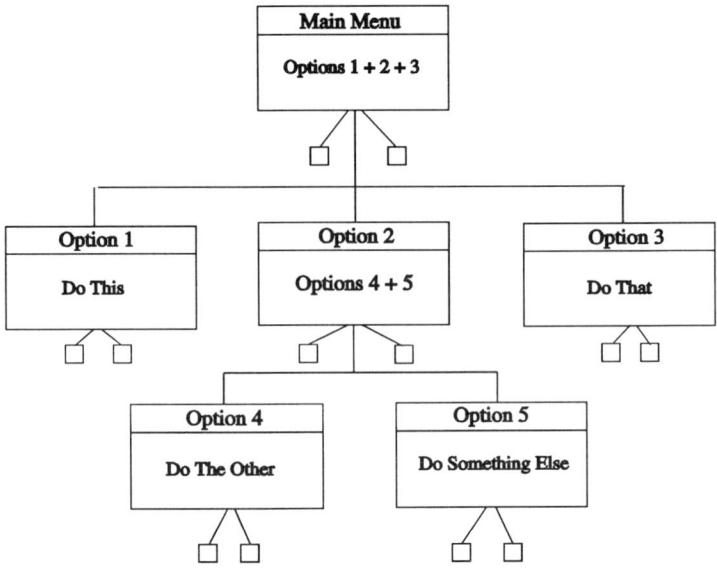

Figure 6.36   Menu screen structure model

An example is illustrated in figure 6.37, where the dialogue screen for the event/business requirement Do This has five screens in sequence, with iteration and selection. The operations for the screen processing (editing an input screen attributes, processing the values in the screen input) are specified as in the event process model and are interpreted for common processing in like manner. Any common operations can be generalised as super-class method-only classes. The residue operations become the methods for the dialogue screens.

If the logic for dialogue screen handling is generic to many business requirements then the logic should be normalisable to a common procedure class.

A failure of some structured methods, such as SSADM, is to recognise that with the advent of client-server hardware architecture the processing of the man/machine operations can be done on the front-end client processor, with the database processing operations of the application classes being done on the back-end mainframe type server processor. No operations are currently specified on the dialogue or menu structure diagrams. There is a simple solution. The database operations on the server are defined in the event process models and the client processing is defined in the Dialogue Structure Models.

Enhancements to SSADM

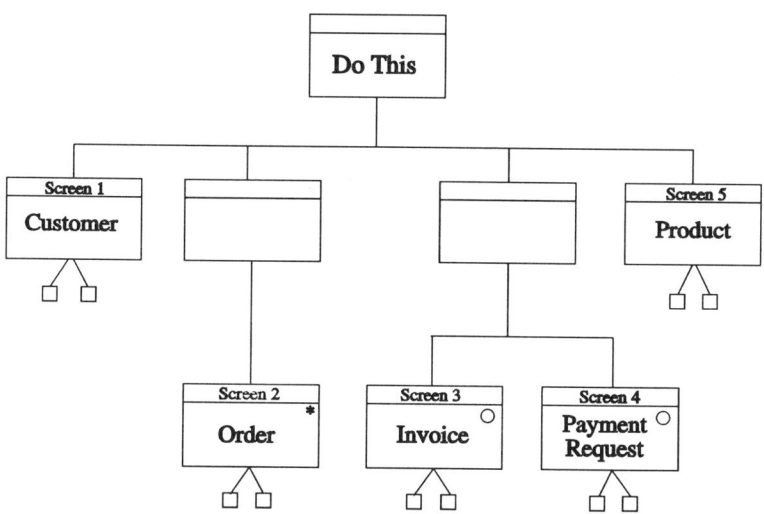

Figure 6.37 Dialogue structure model

6.8.7 Message strategies

There is a serious issue that fundamentally affects the correctness of the normalisation of the logic to the methods. It is the strategy to be used for the sending of messages between the classes. There are two basic strategies that can be used, Central Policeman and Round Robin. The strategies are illustrated in figure 6.38.

One of the central concepts of object orientation is the normalisation of logic. The benefit is that all logic that can be normalised to the business classes in the class model are application independent and therefore much more stable than the residue of application dependent logic at the event level.

The central Policeman messaging strategy is based on the principle that all messaging to the business classes is done from the event class. The messages are, in fact, data access calls to the object instances that contain the data pertinent to the event/business requirement, but, because of encapsulation, the calls have to be in the form of messages to methods in the classes to be accessed. The messages are therefore logic that is relevant to and conditional on the business requirement, the event. Thus, it is correct from a normalisation point of view to place all messages to the business classes at the event level. The event knows what classes it needs to message. Each of the accessed instances of the classes containing the relevant data is completely unaware of the other instances to be accessed for the event/business requirement. This enforces encapsulation. All business access logic to the business classes is at the event level with the Central Policeman messaging strategy.

380 *Additional Data Processing Environments for SSADM—Object Orientation*

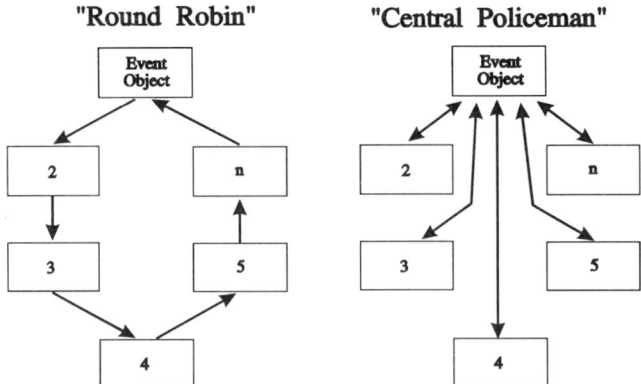

* Central policeman strategy produces code that fully normalises logic correctly
* Round Robin includes event level access logic to the next object in the methods in the business class objects
* Use the Round Robin strategy for the "has a" relationships on the class object model

Figure 6.38 The messaging strategies

The Round Robin messaging strategy is based on the principle that the business classes containing user data can send a message to the next business class to be accessed for the event/business requirement the method is supporting. Thus, using the above example, the method in the class number 2 would send a message to class 3 for the event. The business classes being messaged require to know which is the next class to be messaged for the event/business requirement and therefore are not independent of the other class. The methods in the business classes require to contain access logic pertinent to the event/business requirement regarding access to the next business class. The business classes therefore contain logic in the methods that is pertinent to the event/business requirement and not just the class. The rules of the normalisation of logic are therefore being broken. Notwithstanding this, this practice has been followed on many object oriented design projects.

The practice of object orientation is indicating that the round robin messaging strategy is wrong under all circumstances, that the central policeman access strategy should always be used and that object messaging should be centred on the event class.

It is also being realised that there is a need to base object messaging on the base object of an aggregation, the source object class of a set of "has a" relationships to its derived aggregation classes. The need for this additional central policeman base for messages is that the user of the system should not require to know that the designer of the system, because of the rules of some esoteric rule called third normal form of logic normalisation, has made a class

Enhancements to SSADM 381

understood by the user, a car for example, as the basis of decomposition into an aggregate structure of a set of classes, that the Car in actuality is made up of this, that and the other. All the user knows is that there is a car which includes all sorts of facilities, such as a carburettor and an engine. Look, you can see that it has—what's all this nonsense about logic normalisation? Thus, the user would send a message to the car with all the details of the car, including in the message argument list details of such things as the carburettor and engine or whatever, if that is required to be accessed. The invoked method in the Car object class would be the send messages, unseen by the user, to the Carburettor and the Engine classes as appropriate. Thus basing messages on the base class of an aggregation enforces encapsulation of the complex multi-class structure of what is a car as far as the user is concerned.

In figure 6.39 the messages from Customer Name to Customer Address reflect the need to go from Customer to Customer Name and to Customer Address for a business requirement that wants to access some data properties from the Customer class and the aggregation classes related to Customer. The classes Customer Name and Customer Address have been abstracted because of there being methods that process Customer Name and Address. Apart from the fact that the logic is not in third normal form the trouble with this round robin approach is that, if the business requirement changes and there is no need to access Customer Name, then three methods require to change their code with the round robin approach, the method in Customer not to send a message to a method in Customer Name, the Customer Name method to disappear, and the Customer Address method would require to be messaged from Customer. What should be done is to have a central policeman access strategy from Customer to its composition/aggregate objects, as illustrated in figure 6.40.

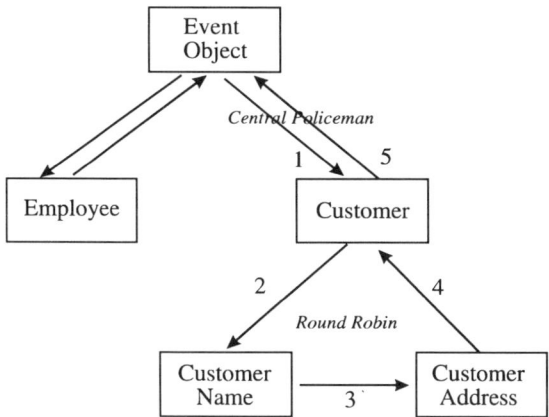

Figure 6.39 Message strategy example

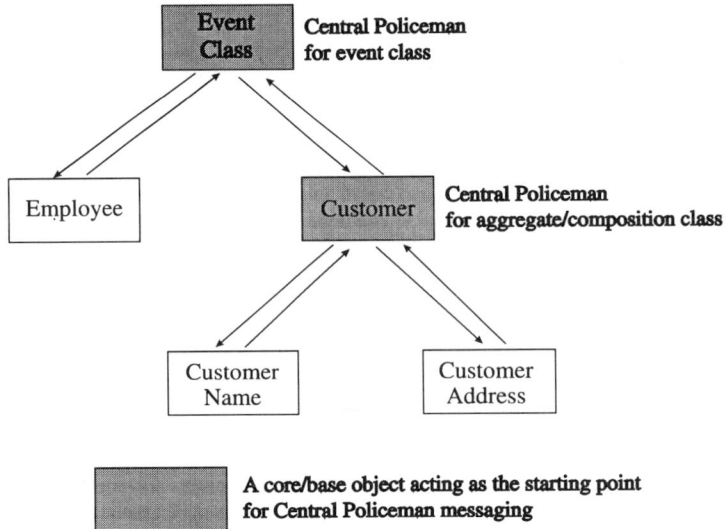

Figure 6.40 The "Central Policeman" messaging

The methods to be invoked to access the aggregate classes would be only those appropriate to the business requirement. In the example in figure 6.40 there is a need to access the aggregation classes of Customer Name and Address and no other aggregation classes. The access requirements of the business requirement to the aggregation classes would be defined in the message to the Customer class. The appropriate aggregation classes would then be messaged as defined in the message to the Customer class.

Should there be the same change in a business requirement as detailed above the messaging approach of central policeman would be to alter the appropriate methods in the two object classes only, the Customer, the central point for the sending messages to the abstracted aggregation objects and the affected aggregation classes. This is a 30% improvement to the above example.

6.8.8 Modelling pre and post conditions

This is becoming a much more commonly used facility in object orientation, particularly with the Eiffel programming language, which provides facilities for it. It is a most useful facility, as it enables the state of the object instance being accessed to be correct before a method is triggered, the execution of the named method to be guaranteed as being correct and the state of the object instance to be correct when the method has completed its processing. The changes to the object instances are thus provable correct. The facility thus enables a "contract" to be obtained between the method that sends a message

Enhancements to SSADM

and receives a response and the method that receives the message and sends the response. The sending method can send the message knowing that the named method will not be triggered if there is anything "wrong" with the object instance being accessed by the named method and knowing that the response contains valid information because the named method has executed successfully and done what it is functionally supposed to do, with the object instance being in the correct new state.

There are two ways in which the pre and post conditions of an object instance can be tested—by checking the values of the data properties of the object instance as set by the last update process or by checking the state indicator of the object instance as set by the last update process. The former approach is required for realtime applications and the latter is often used for batch and online processing, using the object state modelling approach illustrated in section 6.4.6.

Either way the use of declarative logic is being increasingly used for pre and post condition testing. Declarative logic is quite different from the long practised procedural logic. Procedural logic is positional logic with each line of code only relevant in the context of the line of code that precedes it and succeeds it. A line of procedural code has no value in its own right. There are also many constructs for procedural code, the main ones being sequence, selection, iteration and branching for data processing and read, write, update and delete for data access.

Declarative code is quite different:

- It is not positional, as the lines of code are symbolically related.
  Declarative logic is in the form of rules. The rules are in the form of "If condition A then conclusion B". There could be rules "If condition A then conclusion B" and "If condition B then conclusion C". There two rules relate in that the conclusion of one rule matches the condition of another rule. The rules chain together on the basis of the value of the conditions and the conclusions. This means that the rules do not have to be positional and can be written in any order.
  The great benefit of this freedom is that pre and post conditions can be added to an object method at any time without affecting the condition testing already in place. This greatly aids maintainability.
- There is only one construct—"If condition then conclusion".
  This single construct contains both process and access logic. The one construct of such functional power substantially reduces learning curves.
- Each rule is a statement of some truth so it has value in its own right.
  The rule could be "If today is Monday (fact) then tomorrow is Tuesday (fact)" or "If today is sunny (a fact) then go sunbathing (a command)", the meaningful association between the facts and command being the "if...then...clause". Both of these statements are meaningful as standalone rules.

So we have a situation where a rule can be:
—if a fact(s) is (are) true then some more fact(s);
—if a fact(s) is (are) true then do something;

Both forms of rules are useful for condition testing but particularly the latter. The condition of the rule can be used to test the state of the object instance and if it is OK the conclusion can be drawn and the class method invoked, with a command being the named method as a procedure call. The pre-condition test could be "If object instance "123452.State_Indicator = "3" then call method "Do This", this being the named method in the newly received message. The post condition test could be similar to the pre condition test but would test the updated state indicator updated by the invoked method. The post condition test would be "If object instance "123452.State_Indicator = "4" then send response with value "whatever". The great benefit of this for condition testing is that it is conducted only against known positions of truth/validity, this being an excellent basis of a contract between the sender and the receiver of a message. The pre condition is the contract as far the named method is concerned—it only operates when the object instance it is processing is in the correct state. The post condition is the contract as far as the sender of the message is concerned—it only receives a valid response because it can be proved that the named method has executed successfully and processed correctly.

There is also a further benefit of this approach. The command based conclusion is an excellent way in which declarative logic can be linked with procedural logic, thus combining the best of both forms of logic. Procedural logic has the advantage of efficiency and declarative logic has the advantage of flexibility. The conclusion command in the declarative statement invokes the procedural code of the named method in the command.

An example of the structure of pre and post condition testing is shown in figure 6.41 with an example of the way that the object state model can be used for modelling the test conditions in figure 6.42.

An example of the alternative approach to the testing of the state indicator of an object instance is given below, this approach being the testing of the values of one or more of the data properties.

Method Fly_Aircraft
Check pre-condition such as:
  "Select Engine.State, Fuel.State, Aircraft_Height
  From Aircraft
  Where Engine.State = "Full blast", Fuel.State = "Full" and Aircraft_Height = "100"
  If OK then Execute Fly_Aircraft
  Do whatever...

Enhancements to SSADM 385

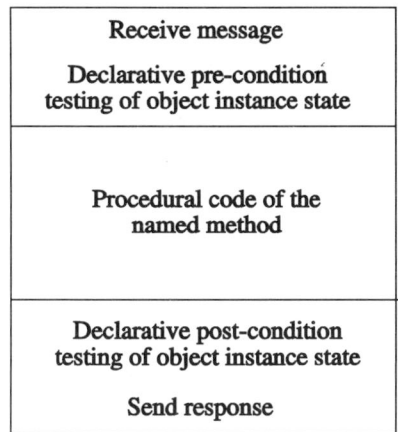

Figure 6.41 Pre and post condition testing (1)

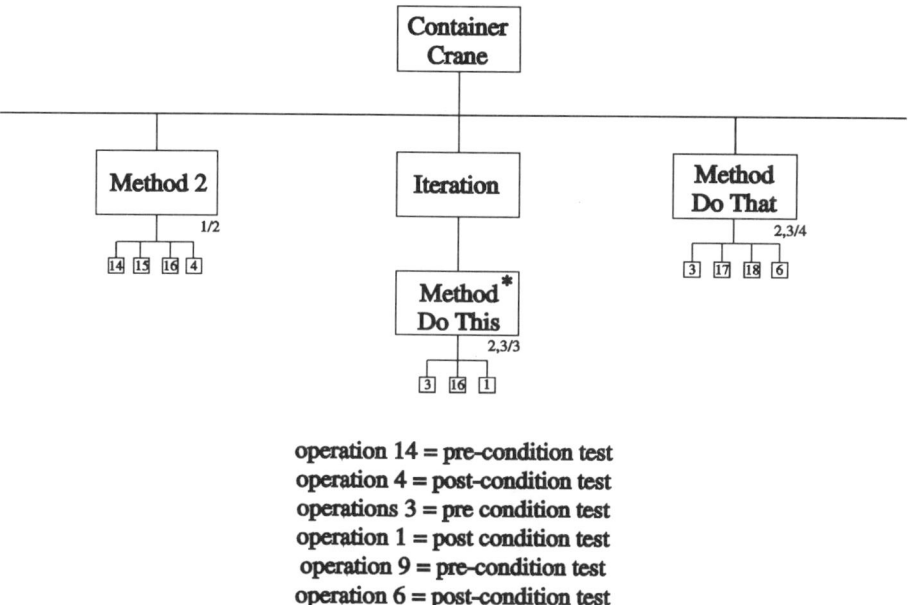

Figure 6.42 Pre and post condition testing (2)

Check post_conditions such as:
If Engine.State = "Idle", Fuel.State = "Empty" and Aircraft_Height = "0"
Then send response "ABC, EFD.HIJ, etc
End Fly_Aircraft;"

6.8.9 Private methods

If the public method being modelled in the event process model requires to invoke a private method in the class being accessed then this has to be modelled separately, as illustrated in figure 6.43. This shows that the public methods send messages to and expect responses from the private methods in exactly the same way as the event level object communicates with the public methods.

The private method can also be invoked by the:

- triggering of a system event such as the need to check for database referential integrity (thus when an Order is being inserted the private method Check Order Referential Integrity is fired). These methods could also be explicitly invoked by the public methods maintaining the objectbase.
- the occurrence of a business condition occurring. A typical example could be that when the stock level is too low then place an order to purchase some more stock.
- the passage of time, such as end of month processing.

The last two types of events are unpredictable as to when they occur. They therefore require to be modelled as random events outside the standard object state model.

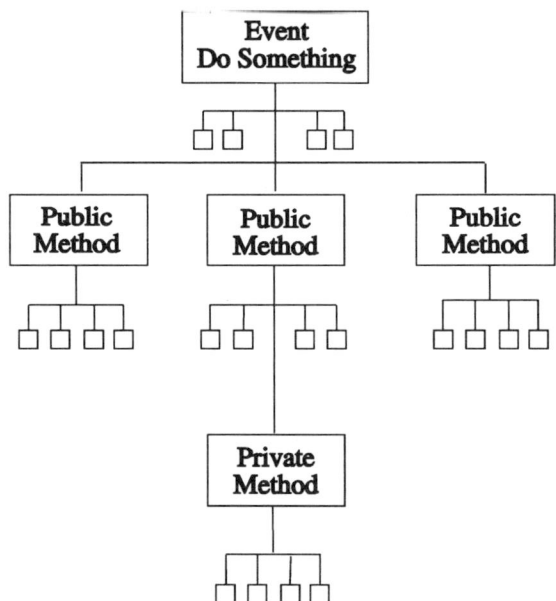

Figure 6.43 Modelling publicly triggered private methods

Enhancements to SSADM 387

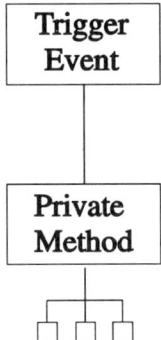

Figure 6.44 Modelling privately triggered private methods

When this situation occurs then the modelling of the private method can be as illustrated in figure 6.44

6.8.10 Genericity

This is parameter-driven software. This supports the facility of describing a class in terms of parameters, with the object class to be processed is specified as a parameter. In the context of SSADM genericity is very useful for supporting the facility of the functional level process. There is a general function for the management of motor vehicles, thus covering the different types of vehicles of car, trucks and buses. There is a common process which does not know which class is to be supported when it is triggered. This is decided by the parameter that is passed to the method Manage Vehicle, say a parameter of the value "C" for car and so on. A model of a generic function is illustrated in figure 6.45.

6.8.11 The deliverables integration

The logical design deliverables integrate in the following manner:

- for each class object in the class object model that contains user data there requires there to be an object state model and a document detailing the data properties and the methods (the logic properties);
- for each class object in the class object model that contains the normalised logic of an event/business requirement there requires there to be an event process model;

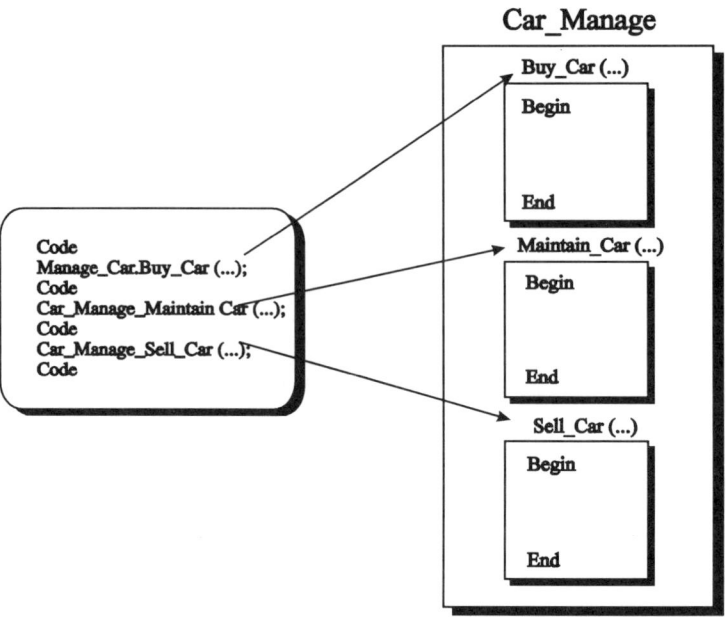

Figure 6.45 Genericity (parameter driven software)

- for each class object in the class object model that contains a method with logic pertinent to the business application there requires there to be an event process model. If the class object is bought in or is part of the class libraries attached to an object oriented programming language there is no need for an event process model.
- for each method in the event process model there is a method to be defined for a class object containing user data or a method only class object;
- common operations in the methods of the event process models should result in a Common Procedure class object in the class object model;
- for each method that affects an object there needs to be pre- and post-condition defined either through the use of the state indicators in the object state models or through the validation of the required data properties;
- for each menu and dialogue screen there requires to be a method defined as a class object in the class object model.

6.8.12 Objectbase design

In line with the concept of logical design = physical design objectbase design has already been done in the class object model. Each class object in the

Enhancements to SSADM

model becomes a class object in the objectbase. Abstract class objects are only definitions in the objectbase schema whereas class objects with user keys and data properties are also object instances in the objectbase.

The rules of objectbase design have not changed from those long applied for database design. Such design aspects as clustering the detail object in the same page/block the most frequently accessed master object, storing the object instances in a key sequence if that is the most frequent access for a given class object and its instances and larger page/blocks for a class object and its instances if the bulk of access is serial rather than direct are as true for objectbase design as for database design. But an objectbase is more than the traditional database. It contains additional facilities, including:

- definitions of the methods, either as properties of the class objects containing user data or as method-only class objects;
- the schema requires to be part of the locking mechanism of the objectbase. Given the need for property inheritance, if there is a change in the schema definition the schema needs to be locked while the changes are being made, otherwise the old version of the schema may be used for property inheritance as the changes are being made.
- the storage of the methods as part of the objectbase.

6.8.13 Object oriented program design

The basic mechanism of program identification and batch program sequencing is unchanged for object oriented design. In a practical world module identification in an object oriented design is simple. The processing that is at the event level is within the application program event object.[1] Typically this logic event is concerned with initialisation of program variables, the synchronisation of the message responses, any processing dependent on the synchronised responses and any transaction screen formatting. The "modules" are the normalised logic in the public and private methods within the other system and business objects the event object message. Methods can in turn issue further messages to other methods in the same or other objects, thus creating the classic program structure chart as described in section 6.8.7. The format of the messages and their returned information can be defined as in conventional programming. Program structure charts are perfectly valid in object oriented programming—just base them directly on the event process models.

---

[1] As modelled in the Event Process models in section 6.5.8.4.

INDEX

ADABAS, 181, 200
Aggregation, 317, 360

Boyce/Codd, 144, 146
Business System Option, 14, 38

CCTA, 3, 8, 9, 34, 251
Chaining
  backward, 261
  forward, 261, 378
Chen, Peter, 233
Class, 314, 360
Client–server, 160, 179, 230
CODASYL, 20, 26, 118
Codasyl file handlers, 26
Codd, Dr Edgar, 55, 148, 307, 366
Composite Logical Data Design, 17, 48, 82
Concurrency Control, 188
Conflict Graph Analysis, 192, 193
Conflict Resolution, 270
Control Processes, 247
Critical Success Factors, 33, 75

Database Design, 100
DATACOM/DB, 56, 100, 149
Dataflow Diagrams, 18, 47, 53, 74, 85, 209, 237, 370
Date, Chris, 144
DB2, 177, 200, 203

DBMS-32, 185
Dee, Ed, 186
Dialogue Design, 16, 19, 61, 97, 169, 230, 247, 379
Dialogue Structure Diagrams, 44, 69, 98, 108, 161, 172
Digital Equipment Corporation, 177
DEC, 103
Distributed Database, 178
Distributed Dictionary, 203
Distributed Processing, 182, 230
Distributed Query Optimisation, 200
Distributed Transaction Rollback, 194
Distributed Transaction Rollforward, 198
Distributed Transaction Utilisation Monitor, 202
DL/1, 30, 78

Effect Correspondence Diagrams, 15, 19, 32, 43, 45, 47, 53, 58, 59, 72, 74, 77, 81, 93, 102, 106, 108, 130, 161, 164, 371
Elementary Process Descriptions, 15, 75, 86
Encapsulation, 332
Enquiry Access Paths, 15, 18, 43, 45, 47, 53, 59, 72, 74, 77, 81, 91, 102, 108, 130, 161, 163, 371
Entity/Event Modelling, 18

Entity Life Histories, 15, 18, 39, 57, 92, 106, 148, 217, 372
Event Process Models, 375
Event Recognition, 245
Expert Systems, 5, 249
Expert systems, 5

Fagin, Dr, 84
Feasibility Module, 14
Function Component Implementation Map, 20, 68, 69, 137, 172, 173
Function Definition, 18, 54, 89, 102

Gateways, 185
GEMINI, 8, 287
GENERIS, 250, 253, 265, 277
GOLDWORKS, 250, 265, 280

Hewlett Packard, 103
HOOD, 251

I/O Descriptions, 15
I/O Structures, 15, 16, 46, 47, 54, 59, 89, 108, 161, 162, 166, 247, 370
IBM, 103, 177
ICL, 103
IDMS, 26
IMS, 23
Information Engineering, 42, 56, 57, 74, 75
INGRES, 67, 98, 200, 203

Jackson, Michael, 9, 17, 108, 171
JSD, 232

KADS, 251
Knowledge
  access, 300
  data, 291
  logic, 294
Knowledge map, 297
Kobler unit, 9, 36

LGDEs, 44, 98
Location Transparency, 204
Logic Normalisation, 366
Logical Data Structuring, 17
Logical Database Process Design, 19
Logical Design, 13
Logical System Specification, 13
Logical Systems Specification, 13, 15, 33

MASCOT, 232
Menu Structure Diagrams, 44
Messages, 336
Methods, 338, 340, 381
  private, 338
  public, 338
Module
  Logical Systems Specification, 15
  Physical Design, 16
  Requirements Analysis, 14
  Requirements Specification, 15

NATURAL EXPERT, 302

Object Object, 346
Object State Models, 372
ORACLE, 10, 79, 92, 178, 200

Physical Design, 13, 16, 19, 46, 62, 230
Physical Design Control, 20
Polymorphism, 341, 378
PRINCE, 16, 40
Process Data Interface, 43, 67, 68, 69
Process Model, 15
Process Models, 10, 17, 20, 32, 43, 47, 58, 60, 65, 94, 108, 161, 162, 172, 247, 376
Process Outlines, 10, 17, 77, 92
Product Breakdown Structure, 11
Product Descriptions, 11
Program-to-program communication, 183
Property Inheritance, 269, 320

Rapid Application Development (RAD), 99
Realtime, 5, 232
Relational Data Analysis, 15, 17, 48, 55, 82, 84, 142, 217, 247, 364
Remote Database Access, 182
Requirements Analysis, 13, 37
Requirements Specification, 13, 15
Retraction, 271
Rockart, 33
Rule, 269
  action, 260
  daemon, 260
  inference, 259
  I/O, 269
  storage and access, 266
  truth maintenance, 260

Ruleset
    access mechanisms, 263
    access strategies, 260

Semantic Nets, 269
Set Level Processing, 187
Software AG, 181
SMALLTALK, 322
STAGES, 251, 287
Structure Clashes, 171
Summary Access Path Maps, 118, 121, 222
SYBASE, 98

Technical Systems Options, 13, 14, 15, 38, 80
Timestamping, 188

TOTAL, 23, 24
Transaction Access Path Analysis, 150, 218
Two-phase commit, 190, 195, 197, 198

Uncertainty, 272, 273, 274, 275, 276
    Bayesian probability, 275
    certainty factors, 274
    classical probability, 273
    fuzzy logic, 275
    uncertain data, 276

VSAM, 128

Wide area network, 26

Yourdon, 4, 5, 7, 45, 232, 235

TITLES IN THIS SERIES

Fletcher J. Buckley • Implementing Software Engineering Practices
John J. Marciniak and **Donald J. Reifer**
   • Software Acquisition Management
John S. Hares • SSADM for the Advanced Practitioner
Martyn A. Ould • Strategies for Software Engineering
   The Management of Risk and Quality
David P. Youll • Making Software Development Visible
   Effective Project Control
Charles P. Hollocker • Software Review and Audits Handbook
Robert L. Baber • Error-free Software
   Know-how and Know-why of Program Correctness
Charles R. Symons • Software Sizing and Estimating
   Mk II FPA (Function Point Analysis)
Robert Berlack • Software Configuration Management
David Whitgift • Methods and Tools for Software
   Configuration Management
John S. Hares • Information Engineering for the Advanced Practitioner
Lowell Jay Arthur • Rapid Evolutionary Development Requirements,
   Prototyping and Software Creation
K.C. Shumate and **M.M. Keller** • Software Specification and Design
   A Disciplined Approach for Real-Time Systems
Michael Dyer • The Cleanroom Approach to Quality
   Software Development
Jean Paul Calvez • Embedded Real-time Systems
   A Specification and Design Methodology
Lowell Jay Arthur • Improving Software Quality
   An Insider's Guide to TQM

John S. Hares • SSADM Version 4
　　　　　　　　The Advanced Practitioner's Guide

Keith Edwards • Real-Time Structured Methods
　　　　　　　Systems Analysis

John S. Hares and **John D. Smart** • Object Orientation
　　　　　　　　　　　　　　　Technology, Techniques, Management and
　　　　　　　　　　　　　　　Migration